INTRODUCTION TO
ELECTRODYNAMICS
AND RADIATION

This is Volume 34 in
series: PURE AND APPLIED PHYSICS

A Series of Monographs and Textbooks

Consulting Editors: H. S. W. MASSEY AND KEITH A. BRUECKNER

A complete list of titles in this series appears at the end of this volume.

INTRODUCTION TO

ELECTRODYNAMICS AND RADIATION

WALTER T. GRANDY, JR.

Department of Physics
University of Wyoming
Laramie, Wyoming

1970

ACADEMIC PRESS New York and London

ACADEMIC PRESS, INC.
111 Fifth Avenue, New York, New York 10003

United Kingdom Edition published by
ACADEMIC PRESS, INC. (LONDON) LTD.
Berkeley Square House, London W1X 6BA

LIBRARY OF CONGRESS CATALOG CARD NUMBER: 78-117077

PRINTED IN THE UNITED STATES OF AMERICA

For Chris, Neal, Susan, and Jeanne
in the hope that they shall one day understand it

CONTENTS

PREFACE

The continuing rapid advance of both theoretical and experimental developments in physics induces a parallel need for continually appraising the presentation of these subjects. As it becomes necessary for the professional physicist to absorb an ever-increasing amount of material, it is also important to reevaluate the traditional order and content of the so-called "core" graduate courses. This book is an attempt in that direction, in that it embraces the subject of electrodynamics as a very large and vital part of modern physics whose significance should be grasped early in the game.

Nonrelativistic quantum electrodynamics is presented here as a logical outgrowth of the classical theory, both relativistic and nonrelativistic. Although the detailed exposition does not extend into the area of covariant quantum electrodynamics (QED), the resulting integrated picture forms a sound foundation for the step up to relativistic quantum field theory—which, of course, must finally be met on its own terms. Nevertheless, the advanced mathematical and diagrammatic techniques of the latter theory are here introduced in a simple and easily understood manner, and one is able to grasp the essential aspects of QED much more quickly than in the usual graduate sequence. The final chapter consists of a relatively comprehensive survey of the covariant theory, serving as a stimulus to further study, as well as to complete the spectrum of ideas.

The first seven chapters concern classical relativistic electrodynamics, with particular emphasis placed on presentation and choice of topics. Experts will note that these chapters are by no means exhaustive within the classical context. Rather, they focus on those aspects deemed essential to an understanding of the classical theory of charged particles. For example, Chapter II on tensor analysis and Riemannian spaces is included for two reasons: the subject, along with its physical implications, is now an essential ingredient of theoretical physics and, at least in the author's experience, a proper understanding of special relativity relies heavily on a deep appreciation of the geometrical structure of space–time. Moreover, it would seem that it is this type of course and on this level at which such topics should be mastered. In general, the first part of the book is intended to provide an elementary foundation for the second.

The second part of the book is entirely quantum mechanical in outlook, beginning with the familiar quantization of the Hamiltonian formulation of classical electrodynamics. Chapter X is a rather self-contained treatment of the many-body formalism leading to Fock-space techniques. This development then leads to the ensuing diagrammatic analysis of nonrelativistic QED, the consequences of which comprise the remainder of the book.

The text has been developed with an eye to the currently prevailing undergradute background of most graduate students: courses in elementary mechanics (classical and quantum), electricity and magnetism, and modern physics. In addition, one semester each of graduate quantum mechanics and electromagnetic theory (including extensive applications of Maxwell's equations) is assumed, along with the concomitant mathematical maturity. Exercises in elementary group-theoretical arguments are included in order to introduce the tool, although no expertise in the subject is assumed. At the University of Wyoming, the course based on this book is usually taken in the second or third graduate semester, in place of the traditional second semester of electricity and magnetism.

ACKNOWLEDGMENTS

It is quite impossible to recall here the great number of papers, books, and people which have influenced this work. Nevertheless, the indirect but large influences of Dr. David M. Kerns, Professor Franz Mohling, Professor C. Ray Smith, and many perceptive students must be acknowledged, although all are accorded the customary release from responsibility for any shortcomings of the book. More directly, I am grateful to Professor Sergio Mascarenhas of the University of São Paulo (Departamento de Física, Escola de Engenharia de São Carlos) whose gracious hospitality helped nourish the original ideas for the book, and to Professor Derek Prowse for his encouragement during the more tedious aspects of the undertaking.

I THE SPECIAL THEORY OF RELATIVITY

A. NEWTONIAN PHYSICS

The object of this book is to study on a fundamental level the microscopic theory of electricity and magnetism. That is, we shall be concerned with the ultimate sources of the electromagnetic field—the charged particles, their kinematics and dynamics, and to some extent their structure. Classically, the only elementary particle is the charged particle, so that in a sense the first part of this work deals with the classical theory of elementary particles.

Lorentz and Abraham developed the first comprehensive theory of charged particles prior to the turn of the century. However, the attempt to include the associated electromagnetic fields in the framework of Newtonian mechanics led to tremendous difficulties and contradictions, which were not resolved until Einstein suggested a revision of our basic notions of space and time. Thus, in order to develop an adequate understanding of the physics of charged particles, it is first necessary to understand the modifications in Newtonian mechanics brought about by the special theory of relativity.

To begin, recall Newton's first law of motion: a body remains at rest, or in motion with constant velocity, if and only if it is not subjected to the influence of other bodies or forces. This state of the body can be verified classically by measuring its acceleration, or change in velocity. Newton postulated this law relative to *absolute* space and time, so that classical mechanics is already a theory of relativity. That is, the laws of motion are also valid in any frame of reference which is itself moving at constant velocity with respect to absolute space. Therefore, if the reference frame S is at rest with respect to absolute space and $\bar{S}$ is a coordinate system in uniform motion with velocity $\mathbf{v}$ relative to S, then the radius vec-

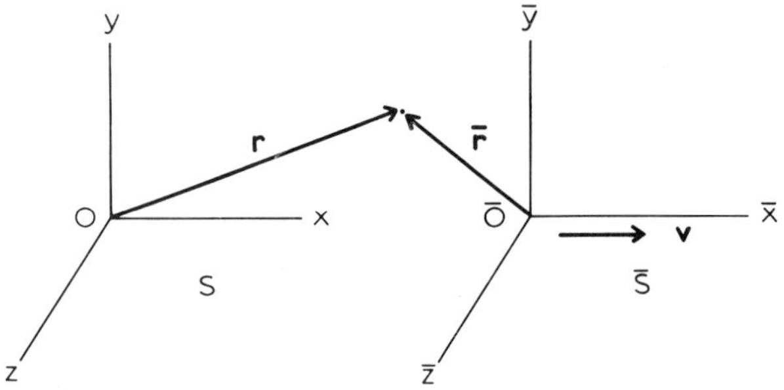

Fig. 1. Two Galilean reference systems in relative motion along their common x axes with velocity **v**.

tors to some absolute point are related by (see Fig. 1)

$$\bar{\mathbf{r}} = \mathbf{r} - \mathbf{v}t\,. \tag{1-1}$$

Newton's laws, at least for velocity-independent forces, are form-invariant under the transformation (1-1), because he incorporated this principle of relativity (first stated by Galileo) into them. Equation (1-1) is called a *Galilean transformation*, and the (mathematical) group of all such transformations is called the *Galilean group.*

Exercise. Show that Newtonian mechanics is invariant under this group.

Newton's first law is sometimes referred to as Galileo's law of inertia, and the next problem to be faced is that of testing this law. In order to do this, one must consider a particle in force-free surroundings, in which case it has no acceleration relative to absolute space, by Newton's second law. If the law of inertia is found to be true, then the coordinate system describing the particle is said to be an *inertial system.* All other inertial systems are obtained from this one by making Galilean transformations.* A good approximation to such a system is one anchored in the "fixed" stars, which are sufficiently far removed to exert essentially no forces on a body. We are led to conclude, then, that Newton's laws of motion are valid in *Galilean reference frames.*

* This definition of an inertial frame is actually *not* operational, a result that can be inferred from the principle of equivalence. One *cannot* make a Galilean transformation to a freely falling observer in a gravitational field. The concept of inertial frame is really a very local one, and the definition must be made more carefully than we care to do here. Therefore, we shall follow Einstein and refer to these systems as *Galilean reference frames*.

B. INCLUSION OF THE ELECTROMAGNETIC FIELD

While it was clear that all of Newtonian mechanics followed the Galilean principle of relativity, it was observed that the wave equation derived from Maxwell's equations was *not* invariant under the transformation (1-1). Thus, as with other wave phenomena, electromagnetic waves were assumed to propagate through a medium called the "ether" with respect to which the velocity of propagation was c—in other inertial frames the velocity is determined by the Galilean principle of relativity.

Michelson and Morley[1] set out to establish the existence of the ether in 1887 by observing fringe shifts in an interferometer due to motion through the ether. They obtained a null result, as have many other observers since that time,[2] and the ether hypothesis seemed untenable. Supporting this experiment was one performed by Trouton and Noble[3] in 1903, which essentially involves two equal but opposite charges at opposing ends of a rigid rod. If the apparatus is moving through the ether, there should be a magnetic interaction on the two charges, producing a couple, and therefore a torque on the system. Again, a null result was obtained.

In an attempt to save the ether theory, Lorentz[4] and Fitzgerald[5] postulated independently that material bodies might contract in the direction of their motion, and that this could account for the null result of the Michelson–Morley experiment. They suggested, therefore, that the experiment should be done using an interferometer with arms of unequal length, in which case a fringe shift would be detected if the ether were indeed physical. Kennedy and Thorndike[6] constructed such an instrument, and again found a null result. Interestingly enough, we shall see below that the contraction postulated by Lorentz and Fitzgerald does occur in the special theory of relativity, but that it alone is not sufficient to form the basis of the theory.

A final suggestion to save the ether theory concerns the possibility that the ether is dragged along with material bodies, so that nothing other than a null result could be obtained in the above experiments. This argument fails, however, when one observes the aberration of starlight. If the ether were dragged along by the earth, we would expect to see no apparent motion of the distant stars. It must finally be concluded that no physically measurable ether exists; that is, any physical quantities associated with matter-free space have no coordinates or velocities.

The downfall of the ether concept did not, however, solve the problem of incorporating the electromagnetic field into the realm of Newtonian mechanics. A further step in this direction was taken by Ritz[7] and others[8] by proposing the so-called emission theories of light. It was postulated that the velocity of light relative to the *source* is always c, and remains associated with the source. This assumption clearly agrees with the result of the Michelson–Morley experiment. De Sitter,[9] however, demonstrated that a careful analysis of eclipsing binary stars refutes the basic postulate of Ritz's theory. For, were it correct, the orbits of the twin stars would have large eccentricities, and this effect is not observed. Thus, the proposal that the velocity of light depends on the source appears untenable.*

In view of these and other difficulties encountered in trying to synthesize all the known physical laws into the Newtonian picture of nature, Einstein was led to reexamine the fundamental principles on which mechanics was based, and we shall now see how these principles needed to be modified.

C. RESOLUTION OF THE PROBLEM

In deriving the contraction hypothesis mentioned above, Lorentz[10] had suggested that it may be necessary to consider a local time in a moving reference system, along with some absolute time. (Voight[11] had also considered this earlier.) In fact, he showed that Maxwell's equations were indeed invariant under the transformation given by Eqs. (1-8) below, but he failed to notice the generality of these equations. Poincaré[12] filled in the gap in Lorentz's work by postulating the general validity of the *relativity principle*:

(*A*) ***The Laws of Physics Are Independent of the Uniform Translational Motion of the System in Which They Operate.***

Lorentz and Poincaré had both taken Maxwell's equations as a basis for their arguments. But Einstein[13] recognized that any theorem as fundamental as (**A**) must transcend a particular physical system and lead to a theory derivable from the simplest possible assumptions. He demonstrated this by making a postulate directly opposite to that of Ritz's emission theory:

(*B*) ***The Velocity of Light Is Independent of the Motion of the Source.***

The experimental evidence for the validity of these two postulates is abundant. The Michelson–Morley experiment[1] and its many repetitions[2] have clearly demonstrated the absence of any physical effects due to an ether. More recently, Jaseja *et al.*,[14] have looked for an ether drift by comparing the frequencies in two He–Ne masers and have demonstrated that no effect larger than $(v/c)^2 \times 10^{-3}$ exists, where v is the earth's orbital speed. Thus, Postulate (**A**) seems well verified.

As mentioned above, de Sitter[9] (among others[15]) fairly clearly de-

* See, however, the discussion following Postulate (**B**).

molished the emission theories, such as that of Ritz, thereby confirming Postulate (**B**). Nevertheless, Fox[16] has recently raised the question as to whether these previous experiments really represented a valid test of the constancy of the velocity of light, basing the question on the extinction theorem of Ewald and Oseen.[17] Thus, if direct light from a source is extinguished by intervening stationary material, then the measured velocity of light may depend on the characteristics of the material. Several new experiments have been made in recent years which seek to avoid the extinction problem, the most precise of which has been done by Alväger *et al.*[18] They measured the speed of γ-rays with energies >6 GeV from the decays of fast π^0 mesons with speed v. The conclusion reached is that, if the speed of the photons is $c + kv$, then $k = 0$ within an accuracy of 0.0001. This probably represents the best confirmation of Postulate (**B**) to date.

Postulates (**A**) and (**B**) can be incorporated *separately* into Newtonian physics, since the data obtained from the experiments discussed above would tend to support *either* one. Einstein, however, showed that they can *both* be adopted if Galilean reference frames are related, not by Eq. (1-1), but by the transformations introduced by Lorentz (and called "Lorentz transformations" by Poincaré). We shall now derive these new transformation laws.

Consider two coordinate systems S and $\bar{S}$ which coincide at $t = 0$, and let $\bar{S}$ move in the direction of the positive x axis with speed v. Suppose that a point electromagnetic pulse flashes on and off at the common origin at the precise instant the two systems coincide, such that an observer in each frame can detect the spherical wavefront when it arrives at his location. Postulate (**B**) implies that the equation of the wave front in the respective frames will be

$$
\begin{aligned}
S: &\qquad x^2 + y^2 + z^2 - c^2t^2 = 0\,, \\
\bar{S}: &\qquad \bar{x}^2 + \bar{y}^2 + \bar{z}^2 - c^2\bar{t}^2 = 0\,.
\end{aligned}
\tag{1-2}
$$

It is clear that the two observers see the same spherical wave, for otherwise Postulate (**A**) would be violated. Now, the existence of the wave is independent of any coordinate system, so that there must exist a transformation between the two descriptions (1-2). Moreover, and this is a fundamental assumption of the theory, space and time are considered to be homogeneous and isotropic; therefore, the transformation between the two equations should be linear. In order to allow for a possible change of scale between the two coordinate systems, we shall introduce a velocity-dependent scale factor $\alpha(v)$ such that $\alpha(0) = 1$, and write

$$
\bar{x}^2 + \bar{y}^2 + \bar{z}^2 - c^2\bar{t}^2 = \alpha^2(x^2 + y^2 + z^2 - c^2t^2)\,. \tag{1-3}
$$

Since the motion is in the x direction, there is no change in the transverse coordinates. This can be seen by considering measuring rods lying along the two y axes and observing that an asymmetry could arise in the conclusions of the two observers, thereby violating Postulate (**A**). Thus (see Fig. 1),

$$\bar{y} = \alpha y , \qquad \bar{z} = \alpha z \tag{1-4}$$

independent of time.

The most general homogeneous, linear transformation that we can write for the other two coordinates is

$$\bar{x} = \alpha(a_1 x + a_2 t) , \tag{1-5a}$$

$$\bar{t} = \alpha(b_1 t + b_2 x) . \tag{1-5b}$$

Note that in the limit $v \to 0$, $(a_1, b_1) \to 1$, $(a_2, b_2) \to 0$. Also, by considering the coordinates of the origin of the $\bar{S}$ system in the S frame, we can see that

$$a_2 = -va_1 . \tag{1-6}$$

Combination of Eqs. (1-3)–(1-6) then yields

$$\begin{aligned} a_1^2 v^2 - c^2 b_1^2 &= -c^2 , \\ a_1^2 - c^2 b_2^2 &= 1 , \\ -a_1^2 v - c^2 b_1 b_2 &= 0 . \end{aligned}$$

Solving these equations in a straightforward manner, one finds

$$a_1 = b_1 = \left[1 - \frac{v^2}{c^2}\right]^{-1/2} \equiv \gamma , \tag{1-7a}$$

$$b_2 = -\frac{v}{c^2} a_1 . \tag{1-7b}$$

Now consider the reverse transformation from $\bar{S}$ to S, in which the latter has a relative velocity $-\mathbf{v}$ with respect to $\bar{S}$. One concludes that

$$\alpha(v) \cdot \alpha(-v) = 1 .$$

Since α cannot depend on the sign of $\mathbf{v}$, from Eq. (1-4), it follows that $\alpha(v) = 1$. The transformation equations between S and $\bar{S}$ are therefore

$$\begin{aligned} \bar{x} &= \gamma(x - vt) , & \bar{y} &= y , \\ \bar{t} &= \gamma\left(t - \frac{v}{c^2} x\right) , & \bar{z} &= z , \end{aligned} \tag{1-8}$$

called a *Lorentz transformation.* Note that Eqs. (1-8) reduce to a Galilean transformation in the limit $(v/c) \to 0$, and that they also leave Maxwell's equations invariant (see Problem 1-2 at the end of the chapter). It should be obvious that the inverse transformation back to the S frame is obtained by interchanging the barred and unbarred coordinates and making the replacement $v \to -v$ in Eq. (1-8).

Exercise. Show that the homogeneous Lorentz transformation corresponding to arbitrary direction of translation with velocity $\mathbf{v}$ is

$$\bar{\mathbf{r}} = \mathbf{r} + (\gamma - 1)\mathbf{v}\frac{\mathbf{r}\cdot\mathbf{v}}{\mathbf{v}^2} - \mathbf{v}t\gamma\,, \qquad \bar{t} = \gamma\left[t - \frac{\mathbf{v}\cdot\mathbf{r}}{c^2}\right]. \tag{1-9}$$

Equation (1-9) still does not represent the most general Lorentz transformation, in which one can also consider the axes of $\bar{S}$ to be rotated with respect to those of S. Furthermore, we have only discussed the homogeneous transformation $\mathsf{L}x$, and can obviously generalize this to the inhomogeneous form $\mathsf{L}x + a$. There are many special Lorentz transformations which have application to various physical situations, and we shall discuss some of them in Chapter III.

Finally, the relativity principle, Postulate (**A**), leads to a generalization of the statement made at the beginning of this chapter regarding the Galilean group; namely, the laws of physics are invariant under the group of Lorentz transformations. There are, of course, several such groups corresponding to the different types of Lorentz transformations mentioned above, but the proof that the collinear Lorentz transformations which we have been considering possess the group property is fairly straightforward. If we define a parameter

$$\boldsymbol{\beta} = \mathbf{v}/c \tag{1-10}$$

characterizing each Lorentz transformation, then two successive collinear transformations with parameters $\boldsymbol{\beta}_1$ and $\boldsymbol{\beta}_2$ are equivalent to a single transformation with parameter

$$\beta = \frac{\beta_1 + \beta_2}{1 + \beta_1\beta_2}\,. \tag{1-11}$$

This is known as the *law of addition of velocities*, and its verification will be left as a problem.

D. SOME CONSEQUENCES OF SPECIAL RELATIVITY

As mentioned previously, the contraction hypothesis of Lorentz and Fitzgerald is actually a real effect contained within the special theory of relativity. Referring again to Fig. 1, we consider a rigid* rod of length l_0 in the $\bar{S}$ frame and oriented parallel to the direction of relative motion along the positive x axis. In this frame the length of the rod is given by the coordinates $l_0 = \bar{x}_2 - \bar{x}_1$, while in the S system the length is $l = x_2 - x_1$ when the members of each pair of coordinates are observed at the *same* time t. Then, from Eq. (1-8) there immediately follows the relation

$$l_0 = \gamma l \, , \tag{1-12}$$

and the observer in the S frame sees a shorter rod than is observed in the $\bar{S}$ system.

The above contraction effect when taken at face value has led to some interesting conjectures regarding the appearance of material bodies moving at high velocities. Unfortunately, these conjectures quite often neglect other, equally important, aspects of relativity theory which invalidate their conclusions. It seems in order, therefore, to examine a bit more closely the consequences of Eq. (1-12).

Since contraction affects only the dimension in the direction of the velocity, it is clear from Eq. (1-12) that a solid body such as a sphere moving at high velocity should appear as an ellipsoid to a stationary observer. Likewise, a high-velocity cube will appear to the stationary observer as a rectangular block. These interpretations of the Lorentz–Fitzgerald contraction were generally accepted until Terrell[19] pointed out in 1959 that such views did not properly account for the method by which we would observe the moving body—that is, the interception of photons by our eye or a photographic film. As we shall see below [see Eq. (4-30)], the wave-vector **k** describing light coming from the moving object is *not* an invariant under Lorentz transformations. When the distortion effects associated with this observation are coupled with the contraction effect of Eq. (1-12) it is found that *there is no distortion* of the object, but there is a rotation. Thus, a sphere will always appear as a sphere, while a cube will appear as a cube, but rotated so that we see the side opposite to that facing in the direction of the motion. For further study of this problem the reader is referred to the lucid article by Weisskopf.[20]

Special relativity also forces a change in the Newtonian concept of a rigid body. Indeed, one finds it virtually impossible to define rigidity in the usual sense, as can be seen by the following argument due to von Laue[21]: since no action can be propagated with a velocity greater than c, an impulse imparted to a body simultaneously at n different places will *at the beginning* produce a motion which must be described by at least n degrees of freedom. Thus, we must clearly reject the Newtonian statement that the motion of a rigid body is completely determined when the motion of one of its points

* See, however, the discussion of rigidity below.

is specified. Rindler[22] has constructed a very nice paradox which can arise if one neglects this change in the concept of rigidity induced by relativity.

A final consequence of the special theory of relativity which we wish to discuss at this point concerns the performance of clocks in different Galilean reference frames or, what is probably now a better description, Lorentz frames. As an important example consider a meson, the lifetime of which represents a very precise clock, and suppose the meson to be created at the origin of coordinates when the frames S and $\bar{S}$ coincide at $t = \bar{t} = 0$. We shall consider the meson to be at rest in the $\bar{S}$ frame with lifetime τ_0. From Eq. (1-8), then, an observer in the S frame observes that the particle has position $x = vt$ and that its lifetime is

$$\bar{t} = \tau_0 = \frac{t - vx/c^2}{(1 - v^2/c^2)^{1/2}},$$

or

$$\tau = \gamma\tau_0 . \tag{1-13}$$

Thus, a *time dilation* is predicted by the theory of relativity, so that clocks will appear to run at different rates in different Lorentz frames. The above time dilation has actually been observed in meson decay in cosmic rays.[23] This point constitutes an excellent check on one of the key predictions of relativity theory.

Other experimental checks on the special theory of relativity are numerous, including the explanation of the aberration of starlight,* the transverse Doppler shift[24] (discussed below), and the Thomas precession[25] describing some of the fine-structure splitting in atomic spectra. One is therefore led to conclude that the theory is will founded in fact, within its limitations of restriction to Galilean reference frames.

PROBLEMS

1-1. Verify Eq. (1-11) of the text.

1-2. Show directly that Maxwell's equations are form invariant under the transformation (1-9). (See, e.g., Lorentz.[26])

1-3. Derive the transformation equations for velocities between two Lorentz frames with relative velocity $\mathbf{v}$.

1-4. In a particular Lorentz frame K a particle has acceleration $\dot{\mathbf{u}}$. Find the acceleration in a second frame $\bar{K}$ which is moving with an arbitrary velocity $\mathbf{v}$ relative to K. If the frame $\bar{K}$ represents the instantaneous rest frame of the particle, show that the square of the acceleration in $\bar{K}$ is

$$\gamma^6[\dot{\mathbf{u}}^2 - (\dot{\mathbf{u}} \times \boldsymbol{\beta})^2] .$$

* Actually, sufficiently refined measurements of the aberration of starlight have yet to be made in a manner which would completely eliminate an explanation within the scope of Newtonian mechanics.

1-5. Suppose an observer on earth sees a rocket going north at a speed $v_1 = \frac{3}{4}c$ and another going south at $v_2 = \frac{3}{4}c$. Analyze the possible conclusions of the earthly observer as to the relative velocity of the two rockets.

1-6. Consider a long pole lying in the $\bar{x}\bar{y}$ plane of Fig. 1 such that it makes an angle $\bar{\phi}$ with the $\bar{x}$ axis. What is the angle with the x axis as measured by an observer in the S frame?

REFERENCES

1. A. A. Michelson and E. W. Morley, *Am. J. Sci.* **34,** 333 (1887); *Phil. Mag.* **24,** 449 (1887).
2. R. Tomaschek, *Ann. Physik* **73,** 105 (1924); K. K. Illingworth, *Phys. Rev.* **30,** 692 (1927); R. J. Shankland, S. W. McKuskey, F. C. Leone, and G. Kuerti, *Rev. Mod. Phys.* **27,** 167 (1955).
3. F. T. Trouton and H. R. Noble, *Phil. Trans. Roy. Soc. London Ser. A* **202,** 165 (1903); *Proc. Roy. Soc. (London)* **72,** 132 (1903).
4. H. A. Lorentz, *Verslag Gewone Vergader. Akad. Amsterdam* **1,** 74 (1892).
5. G. F. Fitzgerald, quoted in O. Lodge, *London Trans.* **A184,** 727 (1893).
6. R. J. Kennedy and E. M. Thorndike, *Phys. Rev.* **42,** 400 (1932).
7. W. Ritz, *Ann. Chim. Phys.* **13,** 145 (1908).
8. R. C. Tolman, *Phys. Rev.* **30,** 291 (1910); J. J. Thomson, *Phil. Mag.* **19,** 301 (1910); O. M. Stewart. *Phys. Rev.* **32,** 418 (1911).
9. W. de Sitter, *Koninkl. Ned. Akad. Wetenschap. Proc.* **15,** 1297; **16,** 395 (1913).
10. H. A. Lorentz, *Arch. Néerl. Sci.* **25,** 363 (1892).
11. W. Voight, *Nachr. Kgl. Ges. Wiss. Göttingen,* 41 (1887).
12. H. Poincaré, *Compt. Rend.* **140,** 1504 (1905).
13. A. Einstein, *Ann. Physik* **17,** 891 (1905).
14. T. S. Jaseja, A. Javan, J. Murray, and C. H. Townes, *Phys. Rev.* **133A,** 1221 (1964).
15. Q. Majorana, *Phil. Mag.* **35,** 163 (1918); **37,** 145 (1919); D. Comstock, *Phys. Rev.* **30,** 267 (1910).
16. J. G. Fox, *Am. J. Phys.* **30,** 297 (1962).
17. P. P. Ewald, *Ann. Physik* **49,** 1 (1916); C. W. Oseen, *ibid.* **48,** 1 (1915).
18. T. Alväger, A. Nilsson, and J. Kjellman, *Nature* **197,** 1191 (1963).
19. J. Terrell, *Phys. Rev.* **116,** 1041 (1959).
20. V. Weisskopf, *Phys. Today* **13,** No. 9, 24 (1960).
21. M. von Laue, *Physik. Z.* **12,** 85 (1911).
22. W. Rindler, *Am. J. Phys.* **29,** 365 (1961).
23. B. Rossi and D. B. Hall, *Phys. Rev.* **59,** 223 (1941).
24. H. E. Ives and C. R. Stilwell, *J. Opt. Soc. Am.* **28,** 215 (1938); **31,** 369 (1941).
25. L. H. Thomas, *Phil. Mag.* **3,** 1 (1927).
26. H. A. Lorentz, *Proc. Acad. Sci. Amsterdam* **6,** 809 (1904).

GENERAL REFERENCES

Excellent discussions of the experimental and historical foundations of special relativity can be found in:

W. G. V. Rosser, "An Introduction to the Special Theory of Relativity." Butterworths, London, 1964.

W. K. H. Panofsky and M. Phillips, "Classical Electricity and Magnetism." Addison-Wesley, Reading, Massachusetts, 1962.

II TENSOR ANALYSIS AND N-DIMENSIONAL GEOMETRY

If one accepts the relativistic view of nature developed in the last chapter, then the next logical step is to reformulate the laws of physics so that they reflect the principle of relativity directly. To do this most efficiently, it is necessary to develop a language particularly suited for such a description, and since mathematics is the foremost tool of theoretical physics, it is appropriate to look for a mathematical language adaptable to treating time and space coordinates on an equal footing. Thus, in this chapter and in the next we shall discuss tensor analysis and its application to relativistic physics. The present treatment will actually be more general than necessary for the purposes of the following exposition, but it is such an easy matter to give a fairly complete treatment that it would seem remiss to be too brief.

A. *N*-DIMENSIONAL SPACES AND THEIR TRANSFORMATIONS

We consider an ordered set of N independent, real variables (x^1, $x^2, \ldots, x^N$) which are said to be the coordinates of a point in an N-dimensional space. The superscripts are used advisedly and are not to be confused with exponents. A curve in the space is defined as a collection of points satisfying the equations

$$x^i = x^i(s)\,, \qquad i = 1, 2, \ldots, N\,, \tag{2-1}$$

where $x^i(s)$ is a real function of the parameter s, satisfying certain fairly general continuity conditions. A surface in this space is defined as a function of $M = N - 1$ parameters $(u^1, u^2, \ldots, u^M)$. By introducing single-valued, continuously differentiable functions of the coordinates, ϕ^i, we can define a transformation of coordinates by the set of equations

$$\bar{x}^i = \phi^i(x^1, \ldots, x^N), \qquad i = 1, 2, \ldots, N. \tag{2-2}$$

Moreover, we shall be primarily interested in transformations possessing an inverse, which requires the functions ϕ^i to be independent. A necessary and sufficient condition is given by the requirement that the Jacobian of the transformation be nonvanishing. Then we can invert Eq. (2-2) to get

$$x^i = \Phi^i(\bar{x}^1, \ldots, \bar{x}^N), \qquad i = 1, 2, \ldots, N. \tag{2-3}$$

For convenience we shall adopt the notation that *all* indices run from 1 to N, and that repeated indices appearing in an equation shall always be summed over. Thus, differentiation of Eq. (2-2) yields

$$d\bar{x}^i = \sum_{j=1}^{N} \frac{\partial \phi^i}{\partial x^j}\, dx^j = \frac{\partial \bar{x}^i}{\partial x^j}\, dx^j . \tag{2-4}$$

The superscript j in this case is called a dummy index. Equation (2-4) gives the transformation equations for the coordinate differentials.

In a particular coordinate system it is clear that $\partial x^i/\partial x^k$ vanishes unless $i = k$; otherwise, the coordinates would not be independent. This suggests introducing a useful quantity called the *Kronecker delta*, defined as

$$\delta_k{}^i = \begin{cases} 1, & i = k \\ 0, & i \neq k. \end{cases} \tag{2-5}$$

The above condition on the coordinates can then be written

$$\frac{\partial x^i}{\partial x^k} = \delta_k{}^i . \tag{2-6}$$

Exercise. Show that

$$\frac{\partial x^k}{\partial \bar{x}^i}\frac{\partial \bar{x}^i}{\partial x^j} = \delta_j{}^k \tag{2-7}$$

and that

$$\frac{\partial^2 \bar{x}^k}{\partial x^i\, \partial x^j} + \frac{\partial^2 x^m}{\partial \bar{x}^n\, \partial \bar{x}^p}\frac{\partial \bar{x}^n}{\partial x^i}\frac{\partial \bar{x}^p}{\partial x^j}\frac{\partial \bar{x}^k}{\partial x^m} = 0 . \tag{2-8}$$

Let us now consider a set of N functions A^i of the N coordinates, x^i.

These are said to form the components of a *contravariant vector** if they transform as

$$\bar{A}^i = \frac{\partial \bar{x}^i}{\partial x^j} A^j \tag{2-9}$$

that is, in the same manner as the coordinate differentials [see Eq. (2-4)]. In fact, dx^i/ds is the contravariant tangent vector to the curve of Eq. (2-1). If we multiply Eq. (2-9) by $\partial x^k/\partial \bar{x}^i$, the inverse transformation is obtained:

$$A^k = \frac{\partial x^k}{\partial \bar{x}^i} \bar{A}^i . \tag{2-10}$$

Exercise. Show that the transformations of contravariant vectors form a group.

Contravariant vectors will always be denoted by using single superscript indices for the components, which serve to distinguish them from covariant vectors to be introduced below. In general, the coordinates x^i themselves do *not* form the components of a contravariant vector. A simple example in three-dimensional rectangular Cartesian coordinates is the position vector which has components (x, y, z). The components in spherical coordinates are *not* (r, θ, ϕ). If, however, we restrict ourselves to linear transformations with constant coefficients of the form $\bar{x}^i = a_j{}^i x^j$, where the $a_j{}^i$ are a set of N^2 constants, then we can indeed write

$$\bar{x}^i = \frac{\partial \bar{x}^i}{\partial x^j} x^j .$$

Only in this case do the x^i form the components of a contravariant vector.†

From ordinary vector analysis the reader is aware that there are various ways to construct a vector quantity. One of the most common is to form

* This is actually a misnomer, since the concept of a vector should be taken to have absolute meaning. A more precise wording is to say that the A^i are the *contravariant components* of a vector, which are to be distinguished from the covariant components of the same vector to be introduced below. For a vector space in which no definition of length is specified the two types of vector are indeed distinct, but below we will always define length, and, therefore, when the metric is introduced into the space the vector itself has a unique meaning. However, usage seems to sanction the words "contravariant vector" and "covariant vector" and so we will accede to tradition, remembering all the time what is really meant.

† We emphasize that the contravariant components of a vector are neither more nor less than defined by Eq. (2-9). For the relation to the usual vector components of mechanics, see the discussion of physical components below Eq. (2-72).

the gradient of a scalar function f of the coordinates $\partial f/\partial x^i$. Upon a change of coordinates we find that

$$\frac{\partial f}{\partial \bar{x}^i} = \frac{\partial x^j}{\partial \bar{x}^i}\frac{\partial f}{\partial x^j} . \tag{2-11}$$

This is quite different from the transformation law (2-9), so that we are led to define a set of N functions A_i as the components of a covariant vector (although, by the previous footnote we mean the covariant components of a vector) if they transform as

$$\bar{A}_i = \frac{\partial x^j}{\partial \bar{x}^i} A_j . \tag{2-12}$$

A single subscript index will always denote a covariant vector, so that the index i in $\partial f/\partial x^i$ will be regarded as a subscript for the quantity as a whole. Note that we do not introduce the notion of a covariant coordinate, so that we shall never have use for the notation x_i.

Exercise. Show that the transformations of covariant vectors form a group.

Equations (2-9) and (2-11) very clearly exhibit the distinction between the components of contravariant and covariant vectors. It is not clear, however, in looking back at our parenthetical remarks regarding terminology, that the gradient vector should also have contravariant components. Therefore, it may be in order to briefly clarify further the need for the two types of vector or, rather, the two types of component.

In Fig. 2 we have drawn a two-dimensional oblique coordinate system with axes inclined at an angle θ with one another. The vector $\mathbf{A}$ lies wholly in the x^1x^2 plane and has *parallel* components A^1 and A^2 along these respective axes. There is some ambiguity in deciding whether to describe the vector $\mathbf{A}$ by parallel or perpendicular components along the axes, a problem which would not occur were θ a right angle and the system orthogonal. The vector can be represented as

$$\mathbf{A} = A^1\mathbf{x}^1 + A^2\mathbf{x}^2 , \tag{2-13}$$

in terms of the *basis vectors* $\mathbf{x}^1$ and $\mathbf{x}^2$, and its length* is clearly

$$|\mathbf{A}|^2 = (A^1)^2 + (A^2)^2 + 2A^1A^2\cos\theta . \tag{2-14}$$

Now, one would like to have a formula for the scalar product of a vector with itself (and therefore of one vector with another) which is in-

* Although we have not yet introduced the concept of length in our N-dimensional space, the usual definition of Euclidean geometry is certainly applicable here.

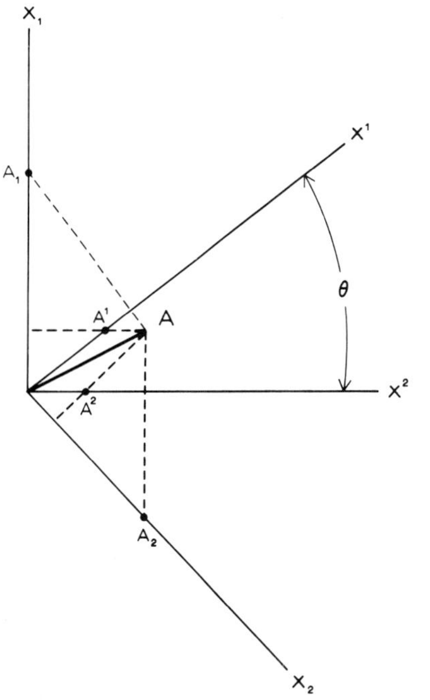

Fig. 2. Representation of the two-dimensional vector **A** in both an oblique coordinate system and its reciprocal. The parallel components of the vector are indicated in both systems.

dependent of the angle between the axes and merely a sum of products of components. To accomplish this, one introduces the *reciprocal coordinate system* indicated in Fig. 2, in which **A** can be represented as

$$\mathbf{A} = A_1\mathbf{x}_1 + A_2\mathbf{x}_2 \,. \tag{2-15}$$

Clearly,

$$\mathbf{x}^1 \cdot \mathbf{x}_2 = \mathbf{x}^2 \cdot \mathbf{x}_1 = 0 \,. \tag{2-16}$$

Exercise. Show that

$$(A^1\mathbf{x}^1 + A^2\mathbf{x}^2) \cdot (A_1\mathbf{x}_1 + A_2\mathbf{x}_2) = (A^1)^2 + (A^2)^2 + 2A^1A^2 \cos\theta \,. \tag{2-17}$$

The basis vectors in neither of these two coordinate systems are necessarily of unit length. If, however, we scale them as follows:

$$|\mathbf{x}^i| = |\mathbf{x}_i| = \sin^{-1/2}\theta \,, \tag{2-18}$$

then obviously

$$\mathbf{x}^1 \cdot \mathbf{x}_1 = \mathbf{x}^2 \cdot \mathbf{x}_2 = 1 \,. \tag{2-19}$$

From the work in the previous exercise one immediately see that this scaling then gives

$$|\mathbf{A}|^2 = A^1A_1 + A^2A_2 = A^iA_i \,. \tag{2-20}$$

From the generalization of this idea it follows that the scalar product between two vectors should be written

$$\mathbf{A} \cdot \mathbf{B} = A^i B_i = A_i B^i \,. \tag{2-21}$$

Note that, if $\theta = 90°$, the two coordinate systems coincide, there is no difference between the two types of vector component, and the basis vectors are unit vectors. One designates the A^i as contravariant components and the A_i as covariant components of the vector **A**.

The foregoing example demonstrates quite clearly the need for defining the two types of vector (or vector components) in general curvilinear coordinates, which will occupy our interest in what follows. Further motivation can be found in the formalism of quantum mechanics, where the need for an adjoint vector space arises in the operator formalism. Thus, the reciprocal components of a state vector are related by complex conjugation, and scalar product is defined only *between* the two spaces, as in Eq. (2-21). (For further study of this matter the reader is referred to the books by Halmos[1] or Merzbacher.[2])

Finally, the above discussion leads one to ask exactly *when* there is no difference between covariant and contravariant vectors. The answer is that they are identical if the transformations of coordinates are restricted to linear transformations with constant coefficients

$$\bar{x}^i = a_k{}^i x^k + b^i \,, \tag{2-22}$$

such that

$$a_m{}^i a_n{}^i = \delta_n{}^m \,. \tag{2-23}$$

In the case of three-dimensional rectangular coordinates these reduce to the usual orthogonal transformations.

Exercise. Show that for the transformations (2-22)

$$\frac{\partial \bar{x}^i}{\partial x^j} = \frac{\partial x^j}{\partial \bar{x}^i} = a_j{}^i \,. \tag{2-24}$$

B. SCALARS, VECTORS, AND TENSORS

In the preceding section covariant and contravariant vectors were defined in an N-dimensional space and the definitions were seen to be straightforward generalizations from the familiar Euclidean space of three dimensions to general N-dimensional spaces of curvilinear coordinates. The transformation laws (2-9) and (2-12) are extremely general. Suppose now that there exists a function T of the N coordinates x^i such that after the transformation (2-2) it is found that

$$\bar{T} = T, \tag{2-25}$$

that is, T is *invariant* under an arbitrary coordinate transformation. Functions or quantities possessing the property (2-25) are called *scalars* or *invariants*, and physical examples are mass and charge.*

A very important invariant can be formed from the components A^i of a contravariant vector and the components B_i of a covariant vector. Consider the product $A^i B_i$ and the same product of the two transformed vectors in the $\bar{x}^i$ system. We find

$$\bar{A}^i \bar{B}_i = \frac{\partial \bar{x}^i}{\partial x^k} A^k \frac{\partial x^j}{\partial \bar{x}^i} B_j = \delta_k{}^j A^k B_j = A^j B_j , \tag{2-26}$$

meaning that the product $A^i B_i$ has the same value in any coordinate system. This is why the product (2-26) is called a *scalar product*, although we shall shortly generalize this operation and refer to it as an *inner product*.

Consider now the set of N^2 quantities $\mathsf{T}^{ij} = B^i C^j$ formed from the *direct* product of the components of the contravariant vectors B^i and C^j. From the transformation law (2-9) this quantity transforms as

$$\bar{\mathsf{T}}^{ij} = \frac{\partial \bar{x}^i}{\partial x^m} \frac{\partial \bar{x}^j}{\partial x^n} \mathsf{T}^{mn} . \tag{2-27}$$

Regardless of the way they are formed, we shall say that any set of N^2 functions A^{ij} which transform according to Eq. (2-27) are the components of a *contravariant tensor* of *second rank*. In like manner, a set of N^2 functions T_{ij} form the components of a *covariant tensor* of second rank if they transform as

$$\bar{\mathsf{T}}_{ij} = \frac{\partial x^m}{\partial \bar{x}^i} \frac{\partial x^n}{\partial \bar{x}^j} \mathsf{T}_{mn} , \tag{2-28}$$

and a set of N^2 functions $\mathsf{T}_j{}^i$ form the components of a *mixed tensor* of second rank if they transform as

$$\bar{\mathsf{T}}_j{}^i = \frac{\partial \bar{x}^i}{\partial x^m} \frac{\partial x^n}{\partial \bar{x}^j} \mathsf{T}_n{}^m . \tag{2-29}$$

* Strictly speaking, one should distinguish between the words "scalar" and "invariant," such that a function of the coordinates is scalar if its value at the same physical point is the same in any coordinate system, and is an invariant if it also has the same functional dependence on the coordinates in any system. In this sense, Eq. (2-25) defines an invariant. We shall use the words interchangeably in the present discussion.

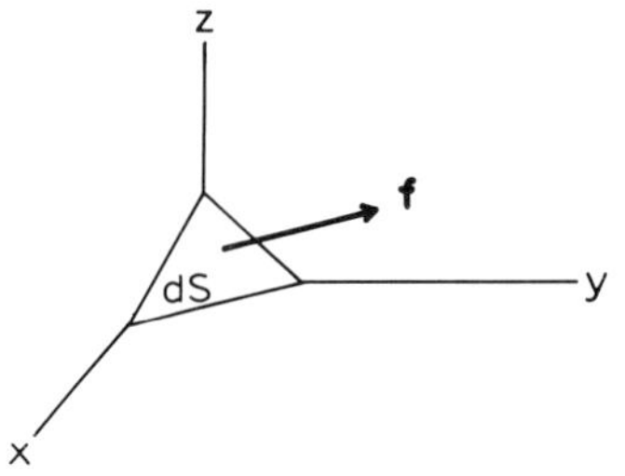

Fig. 3. An arbitrary force **f** acting on a surface element $d\mathbf{S}$.

Exercise. Show that the Kronecker delta $\delta_j{}^i$ is a mixed tensor of second rank. Is δ_{ij} a tensor?

One might here inquire as to the point of introducing tensor quantities. Do these objects have any physical significance? Part of the answer lies in the deduction made from Eqs. (2-27)–(2-29) that if the components of a tensor are zero at a point in one system, or identically zero everywhere, then they are zero at that point, or identically zero, in every coordinate system. Referring to Postulate (**A**) in Section C of Chapter I, we conclude that the equations of physics will have the same *form* in all Lorentz frames if they are written as tensor equations. Another motivation for introducing tensor quantities stems from the consideration of the stress on a material body due to a force acting in some direction. The resulting strain will not generally be the same at each point, nor will it in general be the same in all directions at any given point. From Fig. 3 we see that the pressure on a given component of the surface element $d\mathbf{S}$ due to the x component of the force can be written p^{xi}. In turn, the force on the surface due to the p^{xi} is

$$f^x = \mathsf{p}^{xx}\, dS_x + \mathsf{p}^{xy}\, dS_y + \mathsf{p}^{xz}\, dS_z\, .$$

Hence, any component of the force can be written

$$f^i = \mathsf{p}^{ij}\, dS_j\, , \tag{2-30}$$

and one finds it convenient to define the second-rank pressure tensor by its components p^{ij}. In like manner, many problems in electromagnetic theory are conveniently formulated in terms of the Maxwell stress tensor, which we shall study later [see Eq. (4-38)].

According to the definition above, the rank of a tensor is apparently associated with the number of indices ascribed to its components. In this sense one may refer to a vector as a tensor of rank one, T^i or T_i, and a scalar as a tensor of rank zero, T. Although second-rank tensors and lower will be eminently the most important to us in this book, it is of value to completely generalize the tensor concept to higher ranks by considering a set of N^{m+n} functions of the coordinates. Thus, quantities $\mathsf{T}^{k_1\cdots k_m}_{l_1\cdots l_n}$ are said to be the components of a mixed tensor of rank $(m + n)$ if they transform as

$$\bar{T}^{r_1 \cdots r_m}_{s_1 \cdots s_n} = \frac{\partial \bar{x}^{r_1}}{\partial x^{k_1}} \cdots \frac{\partial \bar{x}^{r_m}}{\partial x^{k_m}} \frac{\partial x^{l_1}}{\partial \bar{x}^{s_1}} \cdots \frac{\partial x^{l_n}}{\partial \bar{x}^{s_n}} T^{k_1 \cdots k_m}_{l_1 \cdots l_n}. \tag{2-31}$$

Exercise. How many components must be specified to define a third-rank contravariant tensor in a five-dimensional space?

Second-rank tensors in an N-dimensional space have a particularly useful representation in terms of matrices. For, it is clear that we can always arrange the N^2 components of T^{ij} in an $N \times N$ square array, which we will designate by

$$[T^{ij}] = \begin{bmatrix} T^{11} & T^{12} & \cdots & T^{1N} \\ T^{21} & T^{22} & \cdots & T^{2N} \\ \vdots & & & \\ T^{N1} & T^{N2} & \cdots & T^{NN} \end{bmatrix}. \tag{2-32a}$$

The determinant of the matrix will be designated by $|T^{ij}|$, and, if this quantity is nonzero, we know from the theory of matrices that the inverse matrix J_{ij} exists and that

$$T^{ij} J_{ik} = \delta_k{}^j. \tag{2-32b}$$

The covariant tensor J_{ij} is said to be the *conjugate* of T^{ij}. The coordinate transformations themselves can, of course, also be represented as $N \times N$ matrices, but the matrix elements, or transformation coefficients, *do not* necessarily form the components of a second-rank tensor.

Exercise. Demonstrate that the quantities J_{ik} form the components of a tensor.

Development of a tensor algebra follows in a straightforward manner. Addition and subtraction of two tensors is defined only for tensors having the same covariant and contravariant character, such as

$$\alpha A^i_{jk} \pm \beta B^i_{jk}, \tag{2-33}$$

where α and β are invariants.

One must pay due attention to the order of indices in a tensor because, for example, A^{ij} is not necessarily equal to A^{ji} (in matrix notation, the transpose of A^{ij}). If, however, a tensor is unaltered by interchanging a pair of contravariant *or* a pair of covariant indices, then it is said to be *symmetric* in those two indices. If the interchange merely changes the sign of the tensor without changing the magnitude, then the tensor is *antisymmetric* in the two indices. A whole spectrum of results can now be proved quite easily. Show that the following statements are true:

Exercise. The above symmetry properties of tensors are preserved under coordinate transformations.

Exercise. A second-rank symmetric tensor has at most $\frac{1}{2}N(N+1)$ different components.

Exercise. Every contravariant (or covariant) second-rank tensor can be written as the sum of a symmetric tensor and an antisymmetric tensor.

Exercise. Symmetry properties *between* contravariant and covariant indices are not in general definable. However, the Kronecker delta is a mixed tensor symmetric in its two indices.

Division of one tensor by another is not defined in the usual sense, but there are two types of products. The *outer*, or *direct product* of two tensors,

$$\mathsf{T}^{ijk}_{mnrs} = \mathsf{A}^{ij}_{m}\mathsf{B}^{k}_{nrs}\,, \tag{2-34}$$

is a tensor of the type indicated. The second type of product is based on a process called *contraction*. If, in the tensor T^{ij}_{kmn}, we change the index j to n, then we can perform the indicated sum to obtain

$$\mathsf{T}^{in}_{kmn} = \mathsf{T}^{i}_{km}\,. \tag{2-35}$$

Exercise. Use the transformation laws to demonstrate that the right-hand side of Eq. (2-35) is a tensor.

The *inner product* of two tensors is now defined as multiplication followed by contraction over one or more indices, such as

$$\mathsf{A}^{ij}_{km}\mathsf{B}^{n}_{il} = \mathsf{C}^{jn}_{kml}\,, \qquad \mathsf{T}^{ij}A_i = T^j\,. \tag{2-36}$$

Clearly, a single contraction always reduces the rank of a tensor by two.

Finally, it is important to have some way of testing a set of functions to see if they form the components of a tensor. The obvious test, of course, is to ascertain whether or not they satisfy the transformation law (2-31). This can be quite tedious in practice and it would be more convenient to have reference to a simpler test. Such a procedure is provided by the so-called *quotient law*: N^n functions form the components of an nth-rank tensor if the inner product of these functions with an arbitrary tensor is itself a tensor. The arbitrary nature of the test tensor is to be emphasized. It will suffice to prove the theorem for the following case. Consider the set of N^2 quantities T^{ij} and let A_i be an arbitrary covariant vector. Then, if

$$B^j = \mathsf{T}^{ij} A_i \tag{2-37}$$

is a contravariant vector, the set T^{ij} is a second-rank contravariant tensor. But this means that

$$\bar{B}^j = \bar{\mathsf{T}}^{ij} \bar{A}_i \,, \tag{2-38}$$

and from Eq. (2-12)

$$\bar{A}_i = \frac{\partial x^m}{\partial \bar{x}^i} A_m \,.$$

Substitution into Eq. (2-38) yields

$$\bar{B}^j = \bar{\mathsf{T}}^{ij} \frac{\partial x^m}{\partial \bar{x}^i} A_m = \frac{\partial \bar{x}^j}{\partial x^n} B^n \,,$$

or

$$\bar{\mathsf{T}}^{ij} \frac{\partial x^m}{\partial \bar{x}^i} A_m = \frac{\partial \bar{x}^j}{\partial x^n} \mathsf{T}^{kn} A_k \,. \tag{2-39}$$

Changing the dummy index m to k, we have

$$\left(\bar{\mathsf{T}}^{ij} \frac{\partial x^k}{\partial \bar{x}^i} - \mathsf{T}^{kn} \frac{\partial \bar{x}^j}{\partial x^n} \right) A_k = 0 \,,$$

and, since A_k is completely arbitrary,

$$\bar{\mathsf{T}}^{ij} \frac{\partial x^k}{\partial \bar{x}^i} = \mathsf{T}^{kn} \frac{\partial \bar{x}^j}{\partial x^n} \,. \tag{2-40}$$

Now take the scalar product of this last equation with $\partial \bar{x}^r / \partial x^k$ and apply Eq. (2-7) to obtain

$$\bar{\mathsf{T}}^{rj} = \frac{\partial \bar{x}^r}{\partial x^k} \frac{\partial \bar{x}^j}{\partial x^n} \mathsf{T}^{kn} \,. \tag{2-41}$$

Thus, T^{ij} is a second-rank contravariant tensor according to Eq. (2-27).

C. RIEMANNIAN GEOMETRY

It should by now be conspicuously evident that we have avoided discussing the concept of distance in the N-dimensional space introduced at the beginning of the chapter. In fact, distance or length is *not* an intrinsic property of a space, and one can develop a logically consistent (mathematical) theory in which the concept never appears. However, we are here ultimately interested in the physical properties of space, and the location of physical events therein, so that it is necessary to introduce, *ourselves*, a definition of distance. This choice can be guided by the expression for a line element in a three-dimensional Euclidean space in rectangular Cartesian coordinates:

$$ds^2 = (dx^1)^2 + (dx^2)^2 + (dx^3)^2 = g_{ij}\, dx^i\, dx^j\,, \tag{2-42}$$

where $g_{ij} = \delta_{ij}$.

From the suggestion given by Eq. (2-42) we will take as the most general space to be considered a *Riemannian space*, which is an N-dimensional space in which is defined a symmetric covariant *metric tensor* of second rank, g_{ij}. Distance in this space is then given by the *metric form*

$$ds^2 = g_{ij}\, dx^i\, dx^j\,. \tag{2-43}$$

Note that there is no requirement for the metric form to be positive definite unless g_{ij} is so specified, so that in a Riemannian space with an indefinite metric two points may be at zero distance from one another without being coincident. We emphasize that Riemannian geometry is built on the concept of distance between neighboring points rather than on the concept of finite distance.

In order to complete the definition of a Riemannian space, we must ensure that the conditions laid down above can be met. To do this, one must postulate that the distance between two points is independent of the coordinate system—that is, ds (or ds^2) is an invariant. It now follows immediately from the quotient law that g_{ij} is a covariant tensor of second rank.

Exercise. Show that, if A^{ij} is antisymmetric and B_{ij} is symmetric, then

$$A^{ij}B_{ij} = 0\,, \qquad \text{identically.}$$

Thus, since the product $dx^i\, dx^j$ is clearly symmetric, any antisymmetric portion of g_{ij} will give a zero contribution to the metric form, and g_{ij} can always be taken to be symmetric.

Equation (2-43) is a quite general quadratic form; it is not necessarily a sum of squares, nor need the coordinate differentials have constant coefficients. Moreover, it can happen that the metric is indefinite, contrary to the positive-definite form (2-42) in Euclidean 3-space. Note that in this latter space with spherical polar coordinates (r, θ, ϕ) the metric tensor has components

$$\begin{gathered} g_{11} = 1\,, \qquad g_{22} = r^2\,, \qquad g_{33} = r^2 \sin^2\theta\,, \\ g_{12} = g_{13} = g_{23} = 0\,. \end{gathered} \tag{2-44}$$

Referring to Eq. (2-31b), we introduce the tensor conjugate to g_{ij} by writing

$$\mathsf{g}_{ij}\mathsf{g}^{ik} = \delta_j^{\ k} . \tag{2-45}$$

This equation defines the contravariant metric tensor, or, more accurately, the contravariant components of the metric tensor. Clearly, the respective matrices of these two quantities are inverses. Moreover, Eq. (2-43) can also be written as*

$$ds^2 = \mathsf{g}^{ij}\, dx_i\, dx_j .$$

The definition (2-45) suggests a general procedure: given a contravariant vector A^j, we define its *associated vector* as

$$A_i = \mathsf{g}_{ij}A^j . \tag{2-46}$$

Thus, as mentioned in an earlier footnote, a vector quantity has existence independently of its coordinate representation, and in a particular coordinate system it can have both covariant and contravariant components. This observation could not really be made, however, until the notion of distance was introduced via the metric tensor. In like manner, associated with a covariant vector B_j is a contravariant vector

$$B^i = \mathsf{g}^{ij}B_j . \tag{2-47}$$

The operations exhibited by Eqs. (2-46) and (2-47) are called lowering and raising of indices, respectively. These two operations can be generalized to abritrary tensors by writing, for example,

$$\mathsf{T}_j^{\ iklm} = \mathsf{g}_{rj}\mathsf{g}^{sl}\mathsf{g}^{tm}\mathsf{T}^{irk}{}_{st} . \tag{2-48}$$

One must note that the subscripts in this last equation were *not* written directly underneath any of the superscripts, which was necessary in order to avoid confusion. Whenever it is possible for such an ambiguity to arise, we shall adopt this convention as exhibited in Eq. (2-48). Finally, although g_{ij} and g^{ij} are conjugate tensors, it does not follow that T^{ij} and T_{ij} need be conjugates, although they are associated.

Exercise. If the determinant of the metric tensor is designated by g, show that†

$$\mathsf{g}^{ik}\frac{\partial \mathsf{g}_{ik}}{\partial x^j} = \frac{\partial}{\partial x^j}\ln|g| . \tag{2-49}$$

* We have agreed previously to always write the coordinates as x^i. However, dx^i is a tensor of rank one, so that if we remember that the index applies to the differential as a whole, and not to the coordinate, then there is no inconsistency in writing dx_i.

† The absolute value of g is used to account for a possible minus sign should the metric form be indefinite.

As a final matter for this section, let us consider the problem of defining the magnitude of a vector. One cannot merely square the components, add them together, and take the square root because the result will depend on the coordinate system chosen. In a nonmetric space it is not even possible to define the magnitude, but in a Riemannian space one can quite easily set up the desired expression via the metric tensor. Referring to the metric form (2-43), we shall define the magnitude of a vector A in terms of either its covariant or contravariant components as

$$(A)^2 = \mathsf{g}^{ij} A_i A_j = \mathsf{g}_{ij} A^i A^j \,. \tag{2-50}$$

This is a reasonable definition in that it reduces to the usual form in a three-dimensional Euclidean space, and, moreover, the magnitude is an invariant. Clearly, Eq. (2-50) is equivalent to

$$(A)^2 = A_i A^i = A^i A_i \,. \tag{2-51}$$

One should note carefully that the magnitude of a vector may be zero or imaginary if the metric is indefinite, and in the former case we refer to a *null vector*. These various possibilities have important physical consequences, as will be seen later.

Exercise. Show that the angle between two vectors in a Riemannian space can be defined as

$$\cos\theta = \frac{\mathsf{g}_{ij} A^i B^j}{(\mathsf{g}_{ij} A^i A^j \mathsf{g}_{mn} B^m B^n)^{1/2}}\,, \tag{2-52}$$

and therefore, the two vectors are *orthogonal* if

$$\mathsf{g}_{ij} A^i B^j = 0 \,. \tag{2-53}$$

D. TENSOR CALCULUS

If a tensor is defined at every point in a region of an N-dimensional space, then it constitutes a *tensor field* in that region, say, $\mathsf{T}^{kj}(x^i)$. The objective of this section will be the development of a calculus for tensor fields, in anticipation of their representation of physical fields.

The first two questions which come to mind, and which are closely related, are: (i) how does one calculate the derivative of a tensor? and (ii) is the derivative itself a tensor? For tensors* of rank zero the answers are

* We shall generally not use the cumbersome expression "tensor field," for by the word "tensor" it is always clear in context whether it is a field or a tensor defined only at a point which is being considered.

immediate, for we have seen from Eq. (2-11) that $\partial f/\partial x^i$ forms the components of a covariant vector. In this case, differentiation raises the rank of the tensor by one, and the result itself is a tensor. Now consider the covariant vector $\bar{A}_i$ and differentiate its transformation law (2-12) with respect to $\bar{x}^k$:

$$\frac{\partial \bar{A}_i}{\partial \bar{x}^k} = \frac{\partial A_j}{\partial \bar{x}^k}\frac{\partial x^j}{\partial \bar{x}^i} + A_j \frac{\partial^2 x^j}{\partial \bar{x}^k\,\partial \bar{x}^i}$$

$$= \frac{\partial A_j}{\partial x^m}\frac{\partial x^m}{\partial \bar{x}^k}\frac{\partial x^j}{\partial \bar{x}^i} + A_j \frac{\partial^2 x^j}{\partial \bar{x}^i\,\partial \bar{x}^k}\,. \tag{2-54}$$

Because of the second term on the right-hand side of this equation, it is clear that the derivative does not transform as a tensor, and this is also true for derivatives of tensors of higher rank.

The prime motivation for digressing into a discussion of tensor calculus in this chapter was to be able to formulate the laws of physics in a manner such that they take the same form in all Lorentz reference frames. But the result of Eq. (2-54) clearly frustrates this attempt, since derivatives enter into almost all equations of physics and so must be required to have tensor character. Thus, the answer to question (ii) above is in general "no," unless question (i) is answered differently than indicated by Eq. (2-54).

To find the correct answer, one must first realize that in the very general type of metric space which we are considering the usual method of forming a derivative may not be invariant under a coordinate transformation. In fact, one can convince oneself that the method *is not* invariant by considering a vector having the same components at two different points in the space—reference to Eq. (2-50) shows that the vector will *not* have the same length at the two points, unless the metric tensor has constant components over the space, such as in a three-dimensional Euclidean space referred to Cartesian coordinates. In this latter situation, in fact, the second term on the right-hand side of Eq. (2-54) vanishes, and the derivatives indeed transform among these coordinate systems as tensors.

The notion of derivative in a Riemannian space, then, must be examined with some care, for, in general curvilinear coordinates the usual difference in a vector or tensor evaluated at two different points breaks down. The curved nature of the space induces an overestimate (or underestimate) in the calculation, and so must be corrected in order to ensure the proper interpretation of derivative as the actual change in something with respect to the coordinates. So as to discuss these correction terms in a convenient way, we shall now introduce some very useful functions of the metric tensor and its derivatives.

Let us first introduce some better notation by agreeing to write partial derivatives from now on as follows:

$$\frac{\partial A^i}{\partial x^j} = \partial_j A^i \,; \qquad \frac{\partial \mathsf{T}^{ij}}{\partial \bar{x}^k} = \bar{\partial}_k \mathsf{T}^{ij} \,;$$

$$\frac{\partial B_i}{\partial t} = \partial_t B_i \,; \qquad \frac{\partial^2 \bar{x}^i}{\partial x^j \, \partial x^k} = \partial^2_{jk} \bar{x}^i \,, \tag{2-55}$$

and so on. Then we define the *Christoffel symbol of the first kind* as

$$[ij, k] = \tfrac{1}{2}(\partial_j \mathsf{g}_{ik} + \partial_i \mathsf{g}_{jk} - \partial_k \mathsf{g}_{ij}) = [ji, k] \,, \tag{2-56}$$

and the *Christoffel symbol of the second kind* as

$$\{m, ij\} = \mathsf{g}^{mk}[ij, k] = \{m, ji\} \,. \tag{2-57}$$

One can readily obtain a number of useful relations for Christoffel symbols.

Exercise. Show that*

$$\partial_j \mathsf{g}_{ik} = [ij, k] + [kj, i] \,, \tag{2-58a}$$

$$\partial_n \mathsf{g}^{mk} = - \mathsf{g}^{mi}\{k, in\} - \mathsf{g}^{ki}\{m, in\} \,, \tag{2-58b}$$

$$\{i, ij\} = \partial_j \ln \sqrt{|g|} = \{i, ji\} \,. \tag{2-58c}$$

The transformation laws for the Christoffel symbols can now be derived by first considering the transformation equation for the covariant metric tensor,

$$\bar{\mathsf{g}}_{mn} = \bar{\partial}_m x^i \, \bar{\partial}_n x^j \, \mathsf{g}_{ij} \,, \tag{2-59}$$

and differentiating with respect to $\bar{x}^p$ to get

$$\begin{aligned} \bar{\partial}_p \bar{\mathsf{g}}_{mn} = {} & \bar{\partial}_m x^i \cdot \bar{\partial}_n x^j \cdot \partial_k \mathsf{g}_{ij} \cdot \bar{\partial}_p x^k + \bar{\partial}^2_{pm} x^i \cdot \bar{\partial}_n x^j \cdot \mathsf{g}_{ij} \\ & + \bar{\partial}_m x^i \cdot \bar{\partial}^2_{pn} x^j \cdot \mathsf{g}_{ij} \,. \end{aligned} \tag{2-60}$$

Now form the two equations obtained from Eq. (2-60) by cyclically interchanging the indices m, n, and p, add these two together, and subtract Eq. (2-60) from the sum. By changing the dummy indices appropriately, one finds

$$\overline{[mn, p]} = [ij, k] \, \bar{\partial}_m x^i \cdot \bar{\partial}_n x^j \cdot \bar{\partial}_p x^k + \mathsf{g}_{ij} \bar{\partial}_p x^i \cdot \bar{\partial}^2_{mn} x^j \,, \tag{2-61}$$

the desired transformation law. The quantity on the left-hand side of this equation is the Christoffel symbol of the first kind evaluated in the coordinate system $\bar{x}^i$ using the metric tensor $\bar{\mathsf{g}}_{ij}$.

* By the square root, we shall always infer that the positive root is to be taken, unless explicitly noted otherwise.

Starting with the transformation law of the contravariant metric tensor, one obtains in a similar manner

$$\overline{\{p, mn\}} = \{s, ij\}\, \partial_s \bar{x}^p \cdot \bar{\partial}_m x^i \cdot \bar{\partial}_n x^j + \partial_j \bar{x}^p \cdot \bar{\partial}^2_{mn} x^j \,, \tag{2-62}$$

the transformation law for the Christoffel symbol of the second kind. It is clear from these last two equations that the Christoffel symbols transform as tensors *only* in the special case of linear transformations with constant coefficients.*

It is now possible to obtain a relation which is fundamental to achieving our goal of finding a suitable definition of derivative in a Riemannian space. Take the inner product of $\bar{\partial}_p x^r$ with Eq. (2-62), which yields

$$\bar{\partial}^2_{mn} x^r = \overline{\{p, mn\}}\, \bar{\partial}_p x^r - \{r, ij\}\, \bar{\partial}_m x^i \cdot \bar{\partial}_n x^j \,, \tag{2-63}$$

the needed result. Now return to Eq. (2-54) and eliminate the second derivative by means of Eq. (2-63), obtaining

$$\bar{\partial}_k \bar{A}_i - \overline{\{p, ik\}}\, \bar{A}_p = [\partial_n A_m - A_j \{j, mn\}]\, \bar{\partial}_i x^m \cdot \bar{\partial}_k x^n \,. \tag{2-64}$$

Finally, introduce the notation

$$\nabla_k A_j = \partial_k A_j - \{i, jk\} A_i \,, \tag{2-65}$$

which we call the *covariant derivative* of A_j with respect to x^k. From Eq. (2-64) one sees that the covariant derivative transforms as a covariant second-rank tensor:

$$\bar{\nabla}_k \bar{A}_i = \bar{\partial}_i x^m\, \bar{\partial}_k x^n \cdot \nabla_n A_m \,. \tag{2-66}$$

Following the same procedure, one can readily verify that the covariant derivative of a contravariant vector is given by

$$\nabla_j A^k = \partial_j A^k + \{k, ij\} A^i \,, \tag{2-67}$$

which is a mixed tensor of second rank.

With Eqs. (2-65) and (2-67) we have succeeded in finding a definition of derivative in a Riemannian space which has the tensor characteristic of having the same form in all coordinate systems. Moreover, at least with tensors of rank zero and one, the covariant derivative increases the rank of the tensor by one (if we *define* covariant differentiation of an invariant to be the ordinary partial derivative). One can think of the terms containing the Christoffel symbols in Eqs. (2-65) and (2-67) as being the necessary "correction terms" needed to give the derivative its usual meaning in a Riemannian space. The reader should, of course, ask if the definitions

* In this respect note that Eq. (2-58c) implies that g is *not* an invariant (see Section E).

of covariant derivative really satisfy the intuitive meaning of derivative in terms of changes in a quantity with respect to changes in the coordinates, the answer to which is certainly not obvious from the above treatment. In order to actually verify that Eqs. (2-65) and (2-67) satisfy our intuitive notions of variation, we must examine more deeply the concept of transplantation of a vector in a Riemannian space, and this leads to a discussion of parallel displacement. It does not appear desirable to spend the time here to discuss this problem, but in Appendix A we develop the concept of derivative from the basic geometry of the space and find that the resultant expression agrees with the above definition of covariant derivative. The interested reader can pursue the discussion further there.

As a check on our expressions for the derivative, Eqs. (2-65) and (2-67), one can ascertain that they reduce to the ordinary and expected partial derivatives in the rectangular Cartesian coordinates of three-dimensional Euclidean space. Furthermore, the ∇_i notation for covariant derivative is well-chosen because it reduces to the usual ∇ operator where expected. For example, the *divergence* of a contravariant vector A^j is defined by the contraction procedure

$$\begin{aligned} \nabla_j A^j &= \partial_j A^j + \{j, kj\} A^k \\ &= \partial_j A^j + A^k \,\partial_k \ln \sqrt{|g|}\,, \end{aligned} \tag{2-68}$$

using Eq. (2-58c), and, if the determinant of the metric tensor is independent of the coordinates, this reduces to the usual definition of divergence. The divergence of a covariant vector is defined as

$$\operatorname{div} A_i = \mathsf{g}^{jk}\, \nabla_k A_j\,. \tag{2-69}$$

Exercise. The concept of the *Laplacian operator* can be generalized to a Riemannian space. Show that its action on scalar and vector fields is as follows:

$$\nabla^2 f = \mathsf{g}^{jk}[\partial^2_{jk} f - \{p, jk\}\partial_p f]\,, \tag{2-70a}$$

$$\begin{aligned} \nabla^2 A^j = \mathsf{g}^{ik}[\partial^2_{ik} A^j &+ 2\{j, pi\}\,\partial_k A^p + \partial_k\{j, pi\}\cdot A^p - \{q, ik\}\,\partial_q A^j] \\ &+ \{j, qk\}\{q, pi\}(\mathsf{g}^{ik} A^p - \mathsf{g}^{ip} A^k)\,. \end{aligned} \tag{2-70b}$$

Exercise. Show that the right-hand sides of Eqs. (2-68) and (2-69) are equal.

It should be pointed out here that the covariant derivative is *not* an esoteric quantity related only to arbitrary coordinate systems in abstract N-dimensional spaces. In fact, it is a necessary and useful definition of derivative quite common to everyday classical mathematical physics—it is often merely disguised much better than is done here. For instance, let us determine the components of the covariant derivative in three-dimensional spherical polar coordinates, in which case the components of the metric tensor are given by Eq. (2-44). The divergence of A^j can be calculated from either the first or second line of Eq. (2-68). In the first case one observes that the only nonvanishing, independent Christoffel symbols are

$$\begin{aligned} \{1, 22\} &= -r\,, & \{1, 33\} &= -r\sin^2\theta\,, \\ \{2, 12\} &= r^{-1}\,, & \{2, 33\} &= -\sin\theta\cos\theta\,, \\ \{3, 13\} &= r^{-1}\,, & \{3, 23\} &= \cot\theta\,, \end{aligned} \tag{2-71a}$$

and in the second case one needs the value

$$|g| = r^4\sin^2\theta\,. \tag{2-71b}$$

In either case we find

$$\begin{aligned} \nabla_r A^r &= \partial_r A^r \\ \nabla_\theta A^\theta &= \partial_\theta A^\theta + \frac{1}{r}A^r \\ \nabla_\phi A^\phi &= \partial_\phi A^\phi + \frac{1}{r}A^r + \cot\theta\cdot A^\theta\,, \end{aligned} \tag{2-72}$$

which appears only vaguely familiar.

The divergence does not have the usual form[3] in this last expression because it is not expressed in terms of the *physical components* of a vector, which give the vector components correct physical significance and dimensions. In general, physical quantities (for example, those of mechanics) are defined in terms of rectangular Cartesian coordinates and only preserve their meaning under orthogonal transformations. If one makes a general transformation to curvilinear coordinates, the vector components along the new coordinate axes do not necessarily possess the same physical meaning.

In order to maintain this significance, we define new components of a vector by considering a set of curvilinear coordinates in three dimensions described by a metric tensor g_{ij}. Let A^k be the components of a vector in this system, and let n^k be a unit vector along the coordinate axes. The physical component of A^k in the direction n^k is then defined to be the invariant

$$\mathsf{g}_{ij}A^i n^j = A_j n^j = A^j n_j\,. \tag{2-73}$$

In a rectangular Cartesian system the vector has components B^j along coordinate axes with unit vectors u^j, and, since $A^j n_j$ is an invariant, we must have $A^j n_j = B^j u_j$. In the usual interpretation, B^j is the component in the direction u_j in the sense of orthogonal projection, so that $\mathsf{g}_{ij}A^i n^j$ must represent the physical component of A^i along n^i in the same sense.

In particular, if the curvilinear coordinate system is orthogonal, then

$$(ds)^2 = \mathsf{g}_{11}(dx^1)^2 + \mathsf{g}_{22}(dx^2)^2 + \mathsf{g}_{33}(dx^3)^2 .$$

According to Eq. (2-50) the magnitude of n^1 is

$$1 = \mathsf{g}_{11}(n^1)^2 ,$$

so that $n^1 = 1/\sqrt{\mathsf{g}_{11}}$. In like manner, $n^2 = 1/\sqrt{\mathsf{g}_{22}}$, $n^3 = 1/\sqrt{\mathsf{g}_{33}}$. Then, referring to Eq. (2-73), one sees that the physical components of a vector in an orthogonal curvilinear coordinate system are

$$A^1\sqrt{\mathsf{g}_{11}}\,, \qquad A^2\sqrt{\mathsf{g}_{22}}\,, \qquad A^3\sqrt{\mathsf{g}_{33}}\,, \tag{2-74a}$$

or

$$A_1/\sqrt{\mathsf{g}_{11}}\,, \qquad A_2/\sqrt{\mathsf{g}_{22}}\,, \qquad A_3/\sqrt{\mathsf{g}_{33}}\,. \tag{2-74b}$$

Thus, if we identify the physical components in Eq. (2-74) as $a^r = A^r$, $a^\theta = rA^\theta$, $a^\phi = (r \sin\theta)A^\phi$, we find immediately the familiar result[3]

$$\nabla_i A^i = \partial_r a^r + \frac{2}{r}\,a^r + \frac{1}{r}\,\partial_\theta a^\theta + \frac{\cot\theta}{r}\,a^\theta + \frac{1}{r\sin\theta}\,\partial_\phi a^\phi .$$

Following the above discussion one can now develop the laws of covariant differentiation for tensors of arbitrary rank. Consider the mixed tensor $\mathsf{T}_n{}^m$ and its transformation law (2-29), and take the inner product with $\bar{\partial}_i x^p$ to obtain

$$\bar{\mathsf{T}}_j{}^i\,\bar{\partial}_i x^p = \mathsf{T}_n{}^p\,\bar{\partial}_j x^n . \tag{2-75}$$

Now differentiate with respect to $\bar{x}^k$, use Eq. (2-63) to eliminate the second partial derivatives, and reintroduce Eq. (2-75) so that

$$[\bar{\partial}_k\bar{\mathsf{T}}_j{}^i + \bar{\mathsf{T}}_j{}^n\overline{\{i, nk\}} - \bar{\mathsf{T}}_m{}^i\overline{\{m, jk\}}]\,\bar{\partial}_i x^p$$
$$= [\partial_s\mathsf{T}_n{}^p + \mathsf{T}_n{}^r\{p, rs\} - \mathsf{T}_r{}^p\{r, ns\}]\,\bar{\partial}_k x^s\,\bar{\partial}_j x^n .$$

Taking another inner product with $\partial_p\bar{x}^t$, we see that the quantity

$$\nabla_s\mathsf{T}_n{}^p = \partial_s\mathsf{T}_n{}^p + \mathsf{T}_n{}^r\{p, rs\} - \mathsf{T}_r{}^p\{r, ns\} \tag{2-76}$$

transforms as a third-rank tensor, and we call this the covariant derivative of $\mathsf{T}_n{}^p$ with respect to x^s. By generalizing this process we can then define the covariant derivative with respect to x^n of a general mixed tensor of rank $(s + t)$ as

$$\nabla_n\mathsf{T}^{u_1\cdots u_s}_{v_1\cdots v_t} = \partial_n\mathsf{T}^{u_1\cdots u_s}_{v_1\cdots v_t}$$
$$+ \sum_{\alpha=1}^{s}\{u_\alpha, kn\}\,\mathsf{T}^{u_1\cdots u_{\alpha-1}ku_{\alpha+1}\cdots u_s}_{v_1\cdots v_t}$$
$$- \sum_{\alpha=1}^{t}\{k, v_\alpha n\}\,\mathsf{T}^{u_1\cdots u_s}_{v_1\cdots v_{\alpha-1}kv_{\alpha+1}\cdots v_t} . \tag{2-77}$$

The tedious (but straightforward) proof of this expression will be left for the problems. However, a much simpler proof that the right-hand side of Eq. (2-76) is a tensor is available based on the notion of parallel displacement and can be found in Appendix A.

One can deduce immediately from Eq. (2-77) the simple laws of covariant differentiation:

(i) The covariant derivative of the sum (or difference) of two tensors is the sum (or difference) of their covariant derivatives.

(ii) The covariant derivative of the inner or outer product of two tensors follows the product law for ordinary differentiation.

(iii) Unlike ordinary differentiation, successive covariant derivatives do not necessarily commute.

Exercise. Prove that*

$$\nabla_k \mathsf{g}_{ij} = \nabla_k \mathsf{g}^{ij} = \nabla_k \delta_j{}^i = 0\,. \tag{2-78}$$

E. PSEUDOTENSORS

At this point it would seem obvious that in order to complete the development of tensor calculus one should next say something about integrals of tensors. From the above discussion concerning derivatives, however, it might be suspected that some care must be taken to ensure that the integral of a tensor is indeed a tensor. In order to have such a guarantee, it is necessary to consider another quantity which is closely related to, but differs slightly in its transformation properties from, a tensor. Representative of the simplest type of such objects is the determinant of the covariant metric tensor g introduced in Eq. (2-49), which, according to the discussion following Eq. (2-62) and the related footnote, is not an invariant. Direct confirmation of this last statement is obtained by examining the transformation equations of g:

$$\begin{aligned} \bar{g} = |\bar{\mathsf{g}}_{ij}| &= |\mathsf{g}_{mn}\,\bar{\partial}_i x^m\,\bar{\partial}_j x^n| \\ &= |\mathsf{g}_{mn}|\,|\bar{\partial}_i x^p|\,|\bar{\partial}_j x^q|\,, \end{aligned} \tag{2-79}$$

by the product rule for determinants. But the last two factors on the right-hand side of Eq. (2-79) are just the inverses of the Jacobian of the transformation, and therefore

$$\bar{g} = J^{-2} g\,. \tag{2-80a}$$

Only if the Jacobian is unity does g transform as an invariant. In par-

* This very important result, that the metric tensor is an absolute constant with respect to covariant differentiation, is due to Ricci.

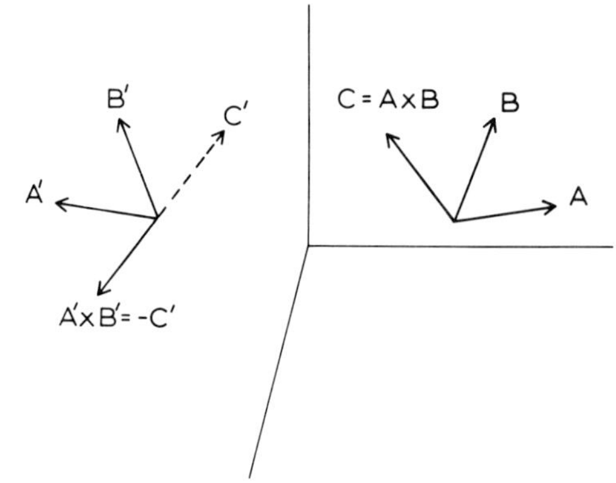

Fig. 4. Reflection of an axial vector $\mathbf{C} = \mathbf{A} \times \mathbf{B}$ through the xz coordinate plane.

ticular, taking absolute values, we can write

$$\sqrt{|\bar{g}|} = J^{-1}\sqrt{|g|}\,. \tag{2-80b}$$

Another familiar example of this type of behavior arises in elementary vector analysis when one considers the vector $\mathbf{C} = \mathbf{A} \times \mathbf{B}$, formed from the cross product of two vectors. Upon reflection through a coordinate plane, the vector $\mathbf{C}$ changes sign (see Fig. 4), as it also does if one considers a "left-handed" rather than a "right-handed" coordinate system. In this case $\mathbf{C}$ is called an *axial vector*, in contrast to the usual *polar vector*. The curl of a polar vector is also an axial vector. Thus, in a three-dimensional Euclidean space, if an orthogonal transformation contains a reflection, the Jacobian of the transformation will be (-1), and this suggests that J can be used to indicate the axial nature of the vector.

As is well known, in three dimensions one can avoid introducing special transformation properties for axial vectors by introducing equivalent antisymmetric tensors of rank two. This procedure, however, disguises the "true" nature of axial vectors and, moreover, impedes a clear discussion of such characteristics of tensors of higher rank. For this reason, we shall define a set of quantities $\mathfrak{T}^{ij\cdots}_{rs\cdots}$ to be the components of a *pseudotensor*, or *tensor density* of weight W, if they transform according to the equation

$$\bar{\mathfrak{T}}^{ij\cdots}_{rs\cdots} = J^W \mathfrak{T}^{mn\cdots}_{pq\cdots}\, \partial_m \bar{x}^i \cdots \bar{\partial}_r x^p \cdots, \tag{2-81}$$

where W is an integer (otherwise J^W would not be single valued). In this sense, $\sqrt{|g|}$ is a pseudoscalar of weight (-1), while g is a pseudoscalar of weight (-2).

The properties of pseudotensors are completely analogous to those of tensors, with the following important exceptions: one can only add or subtract pseudotensors of the same type, rank, *and weight*; and the product of a tensor and a pseudotensor is a pseudotensor, while the product of two pseudotensors of opposite weights is a tensor.

Exercise. Show that the determinant of the *contravariant* metric tensor is a pseudoscalar of weight $(+2)$.

It is not our purpose here to pursue too deeply the concept of pseudotensors, but it is of value to mention briefly a useful example, the *Levi-Civita pseudotensor*

$$\varepsilon_{r_1 \cdots r_N} = \begin{cases} 0, & \text{if any two indices are equal} \\ \pm 1, & \text{according to whether } (r_1 \cdots r_N) \text{ is an even or odd permutation of } (1, 2, \ldots, N). \end{cases} \tag{2-82}$$

This quantity is actually quite familiar from the problem of evaluating determinants. Using the Jacobian determinant as an example, one has

$$J = \varepsilon_{r_1 \cdots r_N} \bar{\partial}_1 x^{r_1} \cdots \bar{\partial}_N x^{r_N} . \tag{2-83}$$

Now recall that our basic assumption from the beginning has been that only those transformations are to be considered for which J is neither zero nor infinity. Then note that the right-hand side of Eq. (2-83) vanishes if two of the indices $(1, 2, \ldots, N)$ are made equal, and changes sign if any two are interchanged. Dividing both sides of the equation by J and replacing the set of indices $(1, 2, \ldots, N)$ by $(t_1, t_2, \ldots, t_N)$, we see that the left side must be equal to $\varepsilon_{t_1 \cdots t_N}$, or

$$\varepsilon_{t_1 \cdots t_n} = J^{-1} \varepsilon_{r_1 \cdots r_N} \bar{\partial}_{t_1} x^{r_1} \cdots \bar{\partial}_{t_N} x^{r_N} , \tag{2-84}$$

and $\varepsilon_{r_1 \cdots r_N}$ transforms as a covariant pseudotensor of weight (-1), as advertised.

Exercise. Prove that $\varepsilon^{r_1 \cdots r_N}$ transforms as a contravariant pseudotensor of weight $(+1)$, and that

$$\varepsilon^{r_1 \cdots r_N} \varepsilon_{r_1 \cdots r_N} = N! . \tag{2-85}$$

In order to return to the stated objective of this section, let us now consider the volume element in an N-dimensional Riemannian space

$$d^N x = dx^1 \, dx^2 \cdots dx^N . \tag{2-86}$$

It is known from elementary calculus that this quantity transforms as

$$d^N \bar{x} = J \, d^N x ; \tag{2-87}$$

that is, as a pseudoscalar of weight $(+1)$. Combining this result with that of Eq. (2-80b), one finds

$$\sqrt{|\bar{g}|}\, d^N\bar{x} = \sqrt{|g|}\, d^N x\,, \tag{2-88}$$

a true invariant called the *invariant volume element*. Thus, the integral

$$\int \sqrt{|g|}\, d^N x \tag{2-89}$$

gives the (invariant) volume of the space.

One can now write down a generalization of Gauss's integral theorem in a Riemannian space by considering an N-dimensional domain D bounded by a surface S of $(N - 1)$ dimensions. With the aid of Eq. (2-68) we have

$$\int_D \nabla_j A^j \sqrt{|g|}\, d^N x = \int_D \partial_j (A^j \sqrt{|g|})\, d^N x\,.$$

Integration by parts gives

$$\int_D \nabla_j A^j \sqrt{|g|}\, d^N x = \int_S \sqrt{|g|}\, A^j dS_j\,, \tag{2-90}$$

where dS_j is the $(N - 1)$-dimensional surface element normal to x^j and directed out of the volume D.

From the preceding discussion two important conclusions can be drawn: (i) we can always obtain a pseudotensor by multiplying a tensor by $\sqrt{|g|}$; and (ii) one must integrate pseudoscalars to obtain invariants. Note, however, that the integral of a tensor field has no well-defined transformation properties, because its value is not confined to a single point. Hence, we conclude that volume integrals of general tensor fields *do not* possess tensor character and must be treated accordingly when they arise. An important case in which we *can* ascribe tensor characteristics to an integral occurs if we integrate, for instance, a covariant tensor of rank m over an m-dimensional manifold as follows:

$$\int \mathsf{T}_{r_1 \cdots r_m} \sqrt{|g|}\, dx^{r_1} \cdots dx^{r_m}\,. \tag{2-91}$$

The integrand is an invariant and, therefore, so is the integral.

PROBLEMS

2-1. By generalizing the method of the text, verify explicitly Eq. (2-77).

2-2. If T_i is a covariant vector,

(a) show that $\nabla_j T_i - \nabla_i T_j = \partial_j T_i - \partial_i T_j$;

(b) show that this combination of derivatives is an antisymmetric covariant tensor of second rank; and

(c) show that for this tensor

$$\nabla_k \mathsf{T}_{ij} + \nabla_i \mathsf{T}_{jk} + \nabla_j \mathsf{T}_{ki} = 0 .$$

2-3. If I and J are invariants, show that

$$\operatorname{div}[J(\partial_i I)] = J\,\nabla^2 I + \mathsf{g}^{ij}(\partial_i I)(\partial_j J) ,$$

where ∇^2 is given by Eq. (2-70a).

2-4. Consider an arbitrary covariant vector T_i and form its second-order covariant derivatives.

(a) Show that one can express these derivatives in the form

$$\nabla_n\, \nabla_m T_i - \nabla_m\, \nabla_n T_i = \mathsf{R}^j_{imn} T_j .$$

(b) Prove that R^j_{imn} is a tensor of the type indicated.

(c) What are the components of the tensor R^j_{imn} in a Euclidean three-dimensional space referred to a Cartesian coordinate system? What are the components in the same space referred to spherical polar coordinates? (Hint: It will be useful to use the associated tensor $\mathsf{R}_{jimn} = \mathsf{g}_{jp}\mathsf{R}^p_{imn}$ and examine its symmetry properties so as to reduce the number of independent components to be calculated.)

REFERENCES

1. P. R. Halmos, "Finite-Dimensional Vector Spaces." Van Nostrand, Princeton, New Jersey, 1958; see, in particular, Section 13.
2. E. Merzbacher, "Quantum Mechanics," Chapter 14. Wiley, New York, 1961.
3. W. K. H. Panofsky and M. Phillips, "Classical Electricity and Magnetism," 2nd ed., p. 475. Addison-Wesley, Reading, Massachusetts, 1962.

GENERAL REFERENCES

Almost all texts and treatises on the general theory of relativity have lengthy discussions of tensor calculus. Probably the best treatment is that of following authors:

R. Adler, M. Bazin, and M. Schiffer, "Introduction to General Relativity." McGraw-Hill, New York 1965.

Two very good books dedicated completely to the tensor calculus and its applications to physics are:

B. Spain, "Tensor Calculus." Oliver & Boyd, Edinburgh and London, 1965.

J. L. Synge and A. Schild, "Tensor Calculus." Univ. of Toronto Press, Toronto, 1952.

III SPACE–TIME OF SPECIAL RELATIVITY

In the preceding chapter the powerful machinery of tensor analysis was developed with the primary purpose in mind of formulating mathematical expressions for the laws of physics so as to exhibit the principle of relativity explicitly. It may appear that the treatment was overly detailed, since the most general aspects of Riemannian spaces are not actually needed to discuss the special theory of relativity. However, in order to fully appreciate the significance of the four-dimensional geometry of space–time, it is desirable to have a good feeling for the fundamental aspects of such spaces, and so we have attempted to describe the general picture in some detail. Moreover, there is always a certain satisfaction in seeing a theory as a special case of some higher level of thought. In this respect, the reader is at this stage quite well prepared (after being introduced to a few more mathematical and physical notions) to undertake a study of the invariance of physical laws in general reference frames, known as the general theory of relativity. The point being made here is that if real physical space can be described by a set of coordinates, both spatial and temporal, then the properties of that space must be determined by the physical objects themselves. We have seen that all the properties of a Riemannian space are determined by the metric tensor, so that it seems reasonable to say that physics determines the correct form of the metric tensor and, therefore, the type of space which mathematically describes the physical world. Our first task, then, is to infer from Chapter I the metric form for special relativity, and to then deduce therefrom the physical implications.

A. SPACE–TIME

From the development in Section I-C we can infer that the measure of distance which remains invariant in all Galilean reference systems is the quantity

$$x^2 + y^2 + z^2 - c^2t^2\,,$$

so that we will take as a definition of length the *invariant interval**

$$ds^2 = -dx^2 - dy^2 - dz^2 + c^2\,dt^2 = g_{ij}\,dx^i\,dx^j\,. \tag{3-1}$$

This equation defines the (indefinite) metric form in the four-dimensional space of special relativity. In accordance with the formulation discussed in the preceding chapter, the coordinates of an *event* will be taken to be

$$x^1,\, x^2,\, x^3,\, x^4 = ct\,. \tag{3-2}$$

At this point it is well to digress a bit concerning notation. Many authors choose the time coordinate to be imaginary, $x^4 = ict$, primarily to simplify the ensuing mathematical description. This choice has the advantage of rendering the metric form positive definite [if the arbitrary association of signs in Eq. (3-1) is reversed], and also obviates the need for distinguishing between covariant and contravariant components, since they are equivalent in this case. One also does not have to worry about signs associated with the different components of vectors. Nevertheless, we shall here adopt the view that the real world is indeed described by an indefinite metric, and that deeper physical insight is obtained by not disguising the fact—although space and time are here placed on more or less *the same footing*, they are still different physical quantities, a difference indicated by the indefinite metric form of space–time. Moreover, the imaginary coordinate is not a very useful concept in general relativity, and it seems quite desirable to have the metric in the general case reduce to that of the present case under the appropriate physical conditions.

Poincaré[1] was the first to use the coordinate $x^4 = ict$, and the notation was later adopted by Minkowski[2] and used throughout his development of the four-dimensional formalism, primarily because this choice gives the Lorentz transformations the appearance of rotations in space–time. When this choice is made one usually refers to the 4-space as Minkowski space. In order to differentiate between this system and that of Eq. (3-2), we shall merely refer to the latter as space–time, although the term "Lorentz space" has also been suggested.[3]

There is also some arbitrariness in choosing the time coordinate as x^4, the choice being x^0 in many places. This particular point seems to be only a matter of taste, however, and so we shall use x^4 in order to remain consistent with the notation of the last chapter. Furthermore, we shall adopt the convenient and conventional notation that Greek letters when used as

* The choice of associating the minus signs with either the space coordinates or the time coordinate is arbitrary, but must be consistently maintained once the choice is made.

indices will always run over the values 1–4, while Latin indices run only from 1 through 3. Also, we shall replace the letter g by η in denoting the metric tensor in special relativity, so that from Eq. (3-1) we deduce for the matrix form of the *Lorentz metric*

$$\eta_{\mu\nu} = \begin{bmatrix} -1 & 0 & 0 & 0 \\ 0 & -1 & 0 & 0 \\ 0 & 0 & -1 & 0 \\ 0 & 0 & 0 & 1 \end{bmatrix} \tag{3-3}$$

The corresponding contravariant form is found from*

$$\eta^{\mu\nu}\eta_{\nu\alpha} = \delta_\alpha{}^\mu \tag{3-4}$$

A vector A^μ in space–time is called a *4-vector*, and the scalar product $A^\mu B_\mu$ between two 4-vectors is left invariant by Lorentz transformations [see Eq. (2-26)]. In fact, the homogeneous Lorentz transformations can be defined as the class of linear transformations under which the metric tensor (3-3) is invariant. Lorentz transformations in general will be denoted by the letter L, so that in space–time Eq. (2-9) becomes

$$\bar{A}^\mu = \mathsf{L}_\nu{}^\mu A^\nu \,. \tag{3-5}$$

In this notation the components of the transformation (1-8) in matrix form are

$$\mathsf{L}_\nu{}^\mu = \begin{bmatrix} \gamma & 0 & 0 & -\beta\gamma \\ 0 & 1 & 0 & 0 \\ 0 & 0 & 1 & 0 \\ -\beta\gamma & 0 & 0 & \gamma \end{bmatrix} \tag{3-6}$$

and the inverse transformation is obtained by replacing β by $(-\beta)$. There will be many occasions in the sequel for referring to Eq. (3-6) as a particularly simple example of a Lorentz transformation to be used in working examples. We shall at times also adopt the notation, found convenient when working in the framework of special relativity, of writing 4-vectors so as to explicitly indicate their spatial and temporal components. Thus (see Fig. 5),

$$A^\mu = (\mathbf{A}, A^4) \tag{3-7}$$

will be taken as interchangeable notation. Note that the corresponding covariant vector A_μ is obtained in this metric by merely changing the sign of the spatial component.

* Thus, $\eta^{\mu\nu} = \eta_{\mu\nu}$, in the space–time of special relativity.

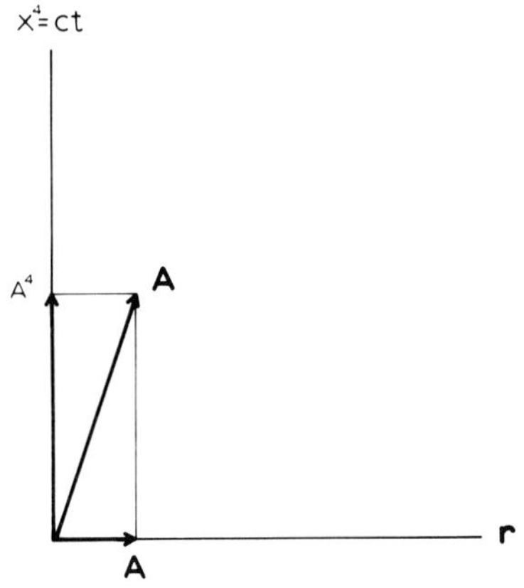

Fig. 5. Space–time representation of the 4-vector **A**.

The Lorentz transformation denoted by Eq. (3-5) is an example of a *homogeneous Lorentz transformation,* though L need not take the simple form of Eq. (3-6). As noted above, these transformations can be generated from the matrix equation

$$\mathsf{L}^{\dagger}\eta\mathsf{L} = \eta \,, \tag{3-8}$$

and from the rules of matrix multiplication we must have

$$\det \mathsf{L} = \pm 1 \,. \tag{3-9}$$

Clearly, the quantity in Eq. (3-6) has determinant $(+1)$, and we say that it belongs to the class of *proper homogeneous Lorentz transformations.* Equation (3-5) can be generalized to include a translation in space–time:

$$\bar{A}^{\mu} = \mathsf{L}_{\nu}{}^{\mu} A^{\nu} + B^{\mu} \,, \tag{3-10}$$

and these are classified as *inhomogeneous Lorentz transformations.*

One can also classify the transformation according to the sign of $\mathsf{L}_4{}^4$. Thus, if $\mathsf{L}_4{}^4 \geq 1$, the transformation is *orthochronous,* and if $L_4{}^4 \leq -1$, it is *antichronous.* These inequalities may seem strange, but the reason for their form will become clear below.

Exercise. The set of all inhomogeneous Lorentz transformations forms a group. Prove this, and examine the group properties of the other types of Lorentz transformations, such as: which transformations are subgroups of other groups.

It follows from the definition of invariant length, Eq. (3-1), that sepa-

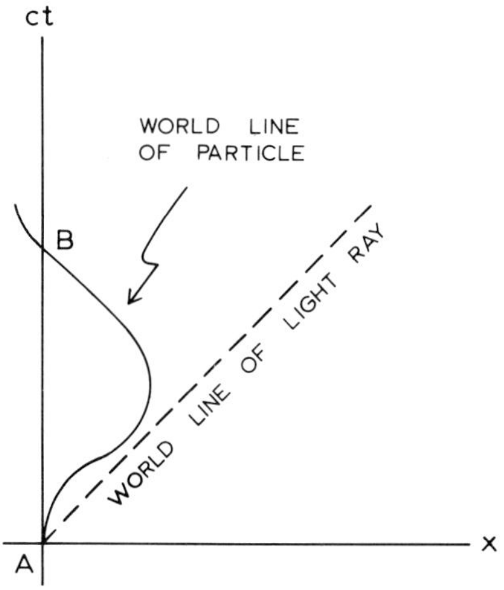

Fig. 6. Minkowski diagram illustrating the world lines of a particle and a light ray.

rations in space–time can be classified as being of three separate types:

$$\begin{aligned} ds^2 &> 0\,, &&\text{timelike separation} \\ ds^2 &= 0\,, &&\text{lightlike (or null) separation} \\ ds^2 &< 0\,, &&\text{spacelike separation.} \end{aligned} \tag{3-11}$$

The law for addition of velocities, Eq. (1-11), indicates that a material particle can never move with a speed exceeding that of light, c, and this observation has never been seen to be violated.* Thus, it follows that any 4-vector describing the motion of a particle in space–time must be timelike, and that spacelike events cannot be reached from each other by signals with velocities equal to or less than c. This latter observation is known as the *causality requirement*.†

The spatially linear path of a particle in space–time can be represented on a two-dimensional *Minkowski diagram* (Fig. 6), and this path is known as the *world line* of the particle. Hence, world lines of particles are time-like, while the path of a light ray must be described by a null vector. In fact, it is interesting to note that the indefinite form of the metric (3–1) implies that two events in space–time can have zero separation.

Consider a reference frame in which a material particle is at rest. Then, from Eq. (3–1), the invariant interval describing the motion of the particle through space–time is given by

$$ds^2 = c^2\,d\tau^2\,, \tag{3–12}$$

* This observation also explains the form of the inequalities delineating orthochronous and antichronous transformations above.

† See, however, the article by Feinberg,[4] in which he discusses the possibility of faster-than-light particles.

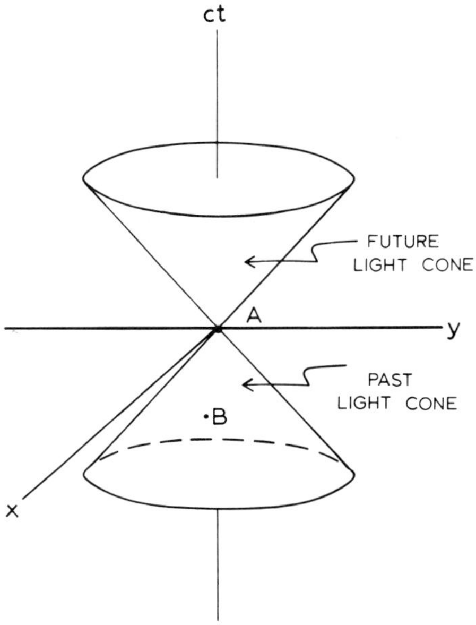

Fig. 7. Future and past light cones associated with an event *A*.

where we have used the symbol τ to represent the time variable in this special situation. This particular coordinate system is known as the *rest frame*, or *proper frame* of the particle, and τ is called its *proper time*. The relation of the proper time to the time in a moving frame can be obtained by transforming the contravariant differential dx^4 of space–time by means of Eq. (2-4). Referring to Fig. 1 as an example, and using Eq. (3-6), one finds that

$$d\bar{t} = \gamma \, d\tau \, . \tag{3-13}$$

A proper time interval in the moving frame is therefore given by

$$\Delta\bar{t} = \int_{\tau_1}^{\tau_2} \gamma(\tau) \, d\tau \, . \tag{3-14}$$

Figure 6 exhibits the space–time path of a particle between events *A* and *B*. One can observe here a unique feature of the space–time geometry, in that the proper time interval on the curved world line is actually shorter than on the direct straight line along the *t* axis. This is a manifestation of the indefinite metric of space–time.

Finally, one can generalize the space–time diagram to a three-dimensional representation using two spatial and one time coordinate, as in Fig. 7. The diagram is centered on an event *A* and the two conical surfaces represent all events that can be connected with *A* by a light ray. All events

B that can have an effect on A lie in or on the *past light cone*, while all those that can be affected by A lie in or on the *future light cone*. The circle intercepted by a plane passed through the future light cone perpendicular to the t axis represents the locus of an electromagnetic pulse emitted at A—at a later time this circle will have expanded to a larger radius, as expected. If one were able to draw a full four-dimensional diagram, this locus would be a sphere. A more detailed discussion of the geometry of space–time can be found in the excellent book by Taylor and Wheeler.[5]

B. RELATIVISTIC CLASSICAL MECHANICS

Prior to taking up a study of electrodynamics, which is really the purpose of this book, it is worthwhile to examine the equations of mechanics as a prelude. We have set out to reformulate the equations of physics so as to incorporate into them the principle of relativity. Thus, we wish to write these equations so that they have the same form in all inertial coordinate systems, and, when so written, an equation will be said to be written in *covariant form*. That is, the coordinates vary in such a coordinated manner as to leave the form of the equation unchanged.*

Consider first the classical definition of the linear momentum of a particle with mass m:

$$\mathbf{p} = m\mathbf{u} \, .$$

This, however, is not a covariant expression, because, as is evident, $\mathbf{u} = d\mathbf{r}/dt$ does not represent the first three components of a 4-vector. Rather,

$$u^\nu = \frac{dx^\nu}{d\tau} \tag{3-15}$$

is a contravariant 4-vector, where $d\tau$ is given by Eq. (3-13). The contravariant 4-velocity then has components

$$u^\nu = (\gamma\mathbf{u}, c\gamma) \, . \tag{3-16}$$

Let us now introduce a quantity m_0, called the *rest mass* of the particle, which is the mass in its rest frame. The 4-momentum is then defined as

$$p^\nu = (\gamma m_0\mathbf{u}, \gamma m_0 c) \, , \tag{3-17}$$

and the spatial part can be written as usual

$$\mathbf{p} = m\mathbf{u} \tag{3-18}$$

* It is to be emphasized that the equations of physics can be written in other ways and still be relativistically invariant, in the sense that they describe the same physical laws in different coordinate systems. When they are written so as to have the same *form* in all coordinate systems they are said to be *manifestly covariant*.

if we define

$$m = \gamma m_0 \,. \tag{3-19}$$

The quantity m is called the *relativistic mass* of the particle, and this velocity dependence of the mass is a direct consequence of the covariant formulation.

It is still necessary to interpret the fourth component of p^ν, or $p^4 = mc$. In a given reference frame its time derivative can be written

$$\frac{dp^4}{dt} = \frac{1}{c}\mathbf{u}\cdot\frac{d\mathbf{p}}{dt} \tag{3-20}$$

Exercise. Demonstrate the validity of Eq. (3-20).

The force in this frame is $\mathbf{F} = d\mathbf{p}/dt$, so that the right-hand side of Eq. (3-20) must be proportional to the rate at which work is done in a particular frame. Hence, with E the total energy,

$$\frac{dp^4}{dt} = \frac{1}{c}\frac{dE}{dt}\,, \tag{3-21}$$

or

$$p^4 = (E/c) + \text{const.}$$

Since we only measure energy differences, $p^4 = E/c$, or

$$E = mc^2\,, \tag{3-22}$$

a well-known result. We can now rewrite Eq. (3-17), and refer to p^ν as the (contravariant) *energy–momentum 4-vector*,

$$p^\nu = (\mathbf{p}, E/c)\,. \tag{3-23}$$

Using the two different expressions for p^4, one can form the very useful Lorentz invariant

$$p^\nu p_\nu = -\mathbf{p}^2 + (E^2/c^2) = m_0{}^2c^2\,, \tag{3-24}$$

which yields the relativistic energy–momentum relation

$$E = (c^2\mathbf{p}^2 + m_0{}^2c^4)^{1/2} \xrightarrow[u \ll c]{} m_0c^2 + \tfrac{1}{2}m_0u^2 + \cdots\,. \tag{3-25}$$

The quantity m_0c^2 is the intrinsic, or *rest energy* of the particle, which it possesses over and above any kinetic energy due to motion. From Eq. (3-24) it is clear that rest energy is an invariant, and that it represents the magnitude of the 4-vector p^ν. By way of contrast, energy is related to the time component of p^ν, so that while energy and mass can essentially be transformed into one another, there is a clear distinction between the two.

It is well to make a comment here about the conservation laws in relativistic mechanics. The rest mass m_0 is an invariant, and is precisely the quantity measuring inertia in every Lorentz frame. On the other hand, the total energy and total linear momentum are not invariant quantities—it is only the 4-vector p^ν combining energy and momentum that retains its form in different Lorentz frames, in accordance with the covariance principle. This covariance, however, is to be distinguished from conservation of these quantities in, say, a collision between particles. Energy or momentum, or both, may or may not be conserved in a collision—this is a dynamical consideration to be considered independently of reference frames.* However, the covariant formulation *does* tell us, via Eq. (3-24), that if *either* energy or momentum is conserved in a process, the other is necessarily conserved. One should note, though, that the rest energy must be included in order to ensure conservation.

As an example of these principles, consider the decay of a π meson to a μ meson and a neutrino,

$$\pi \rightarrow \mu + \nu \,. \tag{3-26}$$

For a general process of this type, energy and momentum conservation are expressed by

$$P^\nu = p_1{}^\nu + p_2{}^\nu \,, \tag{3-27}$$

with respective invariant magnitudes†

$$P^\nu P_\nu = M^2c^2 \,, \qquad p_1{}^\nu(p_1)_\nu = m_1{}^2c^2 \,, \qquad p_2{}^\nu(p_2)_\nu = m_2{}^2c^2 \,. \tag{3-28}$$

Then,

$$\begin{aligned} p_1{}^\nu(p_1)_\nu &= (P - p_2)^\nu(P - p_2)_\nu = m_1{}^2c^2 \\ &= M^2c^2 + m_2{}^2c^2 - 2p_2{}^\nu P_\nu \,. \end{aligned} \tag{3-29}$$

In the rest frame of M the space part of $p_2{}^\nu P_\nu$ vanishes, so that

$$p_2{}^\nu P_\nu = ME_2 \,. \tag{3-30}$$

Combining this with Eq. (3-29), we find for one of the total energies

$$E_2 = (M^2c^2 + m_2{}^2c^2 - m_1{}^2c^2)/2M \,. \tag{3-31a}$$

In like manner,

$$E_1 = (M^2c^2 + m_1{}^2c^2 - m_2{}^2c^2)/2M \,. \tag{3-31b}$$

We can define the relativistic kinetic energy of a particle as

$$T = E - m_0c^2 = m_0c^2(\gamma - 1) \,. \tag{3-32}$$

* This question has recently been studied in some detail by Currie and Jordan.[6]

† The quantities M, m_1, and m_2 are rest masses.

One then writes from Eq. (3-31)

$$T_1 = c^2(\Delta M)\left(1 - \frac{m_1}{M} - \frac{\Delta M}{2M}\right), \tag{3-33a}$$

$$T_2 = c^2(\Delta M)\left(1 - \frac{m_2}{M} - \frac{\Delta M}{2M}\right), \tag{3-33b}$$

where $\Delta M = M - m_1 - m_2$, and $\Delta M/2M$ is a purely relativistic correction.

For the process of Eq. (3-26), the rest energy of the pion is 139.6 MeV, that of the muon is 105.7 MeV, and the neutrino has zero rest mass. Hence, the respective kinetic energies are $T_\mu = 4.1$ MeV and $T_\nu = 29.8$ MeV.

Finally, it is clear that $\mathbf{F} = d\mathbf{p}/dt$ is not the space component of a 4-vector but, rather, one must define the force as the time rate of change of of the 4-momentum with respect to the proper time of the particle:

$$\begin{aligned} F^\nu &= dp^\nu/d\tau \\ &= \left(\gamma\mathbf{F}, (\gamma/c)\mathbf{u}\cdot\mathbf{F}\right). \end{aligned} \tag{3-34}$$

This is called the *Minkowski force* and represents the relativistic generalization of Newton's second law. To make the generalization even more transparent, we can define a 4-acceleration in terms of the 4-velocity as

$$a^\mu = \frac{dv^\mu}{d\tau} = \gamma^2\left(\mathbf{a} + \gamma^2\mathbf{v}\frac{\mathbf{v}\cdot\mathbf{a}}{c^2}, \gamma^2\frac{\mathbf{v}\cdot\mathbf{a}}{c}\right), \tag{3-35}$$

where $\mathbf{a} = d\mathbf{v}/dt$ is the 3-acceleration in the instantaneous rest frame of the particle. A short calculation then demonstrates that we can write

$$F^\nu = m_0 a^\nu, \tag{3-36}$$

completely equivalent to Eq. (3-34). Note that it is the *rest mass* of the particle which enters into Eq. (3-36).

PROBLEMS

3-1. The Compton effect[7] predicts a shift in wavelength $\Delta\lambda$ for a photon scattered from an electron. Calculate $\Delta\lambda$ in terms of the scattering angle in the rest frame of the electron.

3-2. For the homogeneous Lorentz transformation of Eq. (3-5), use Eq. (3-8) to prove the relations

$$(\mathsf{L}_4{}^4)^2 - \sum_{i=1}^{3}(\mathsf{L}_4{}^i)^2 = 1$$

$$(\mathsf{L}_k{}^4)^2 - \sum_{m=1}^{3}(\mathsf{L}_k{}^m)^2 = -1, \qquad k = 1, 2, 3$$

$$\mathsf{L}_\mu{}^\zeta \eta_{\zeta\sigma} \mathsf{L}_\nu{}^\sigma = 0, \qquad \mu \neq \nu.$$

3-3. Show that the components of a Lorentz transformation connecting two frames moving with respect to each other in an arbitrary direction with velocity **v** can be written in the matrix form

$$\mathsf{L}_{\mu}{}^{\nu} = \begin{bmatrix} 1+\alpha\beta_1\beta_1 & \alpha\beta_1\beta_2 & \alpha\beta_1\beta_3 & -\beta_1\gamma \\ \alpha\beta_2\beta_1 & 1+\alpha\beta_2\beta_2 & \alpha\beta_2\beta_3 & -\beta_2\gamma \\ \alpha\beta_3\beta_1 & \alpha\beta_3\beta_2 & 1+\alpha\beta_3\beta_3 & -\beta_3\gamma \\ -\beta_1\gamma & -\beta_2\gamma & -\beta_3\gamma & \gamma \end{bmatrix},$$

where $\alpha = (\gamma - 1)/\beta^2$, and $c\beta_i = v_i$ is the component of **v** along the x^i axis.

3-4. Assume that two photons with different energies annihilate in a vacuum, creating an electron–positron pair. For what ranges of initial photon energies and angles between their directions of propagation can this reaction take place? (The physical basis for this process is discussed in Chapter XVI.)

REFERENCES

1. H. Poincaré, *Rend. Circ. Mat. Palermo* **21,** 129 (1906).
2. H. Minkowski, *Math. Ann.* **68,** 472 (1910).
3. A. O. Barut, "Electrodynamics and Classical Theory of Fields and Particles." Macmillan, New York, 1964.
4. G. Feinberg, *Phys. Rev.* **159,** 1089 (1967).
5. E. F. Taylor and J. A. Wheeler, "Spacetime Physics." Freeman, San Francisco, California, 1966.
6. D. G. Currie and T. F. Jordan, *Phys. Rev.* **167,** 1178 (1968).
7. A. H. Compton, *Phys. Rev.* **22,** 411 (1923).

IV THE COVARIANCE OF CLASSICAL ELECTRODYNAMICS

A. FOUR-VECTORS OF ELECTRODYNAMICS

In order to gain some perspective for formulating electrodynamics in a manifestly covariant manner, let us recall Maxwell's equations for the electromagnetic field *in vacuo:*

$$\begin{aligned} \nabla \cdot \mathbf{E} &= 4\pi\rho\,, & \nabla \times \mathbf{E} &= -(1/c)\,\partial_t \mathbf{B}\,, \\ \nabla \cdot \mathbf{B} &= 0\,, & \nabla \times \mathbf{B} &= (4\pi/c)\mathbf{J} + (1/c)\,\partial_t \mathbf{E}\,. \end{aligned} \tag{4-1}$$

Although the fields are considered to be the primary measurable physical quantities in classical electromagnetic theory, it is found convenient to introduce potential functions as an aid in solving Eq. (4-1). Thus, the fields can be specified by a vector potential **A** and a scalar potential ϕ:

$$\mathbf{B} = \nabla \times \mathbf{A}\,, \qquad \mathbf{E} = -(1/c)\,\partial_t \mathbf{A} - \nabla\phi\,. \tag{4-2}$$

This specification is not unique, however, because Eqs. (4-2) are invariant under the *gauge transformation*

$$\mathbf{A} \rightarrow \bar{\mathbf{A}} = \mathbf{A} + \nabla\Lambda\,, \tag{4-3a}$$

$$\phi \rightarrow \bar{\phi} = \phi - (1/c)\,\partial_t \Lambda\,, \tag{4-3b}$$

where $\Lambda(\mathbf{r}, t)$ is the gauge function. If this arbitrariness in the choice of gauge is used to require that the potentials satisfy the *Lorentz condition*

$$\nabla \cdot \mathbf{A} + (1/c)\,\partial_t \phi = 0\,, \tag{4-4}$$

then the potentials are related to the sources by the inhomogeneous wave equations

$$\nabla^2\phi - (1/c^2)\,\partial_t^2\phi = -4\pi\rho\,, \tag{4-5a}$$

$$\nabla^2\mathbf{A} - (1/c^2)\,\partial_t^2\mathbf{A} = -(4\pi/c)\,\mathbf{J}\,. \tag{4-5b}$$

Finally, conservation of charge is embodied in the *continuity equation*

$$\nabla\cdot\mathbf{J} + \partial_t\rho = 0\,. \tag{4-6}$$

We shall approach the covariant formulation of these equations by first examining the variation of physical quantities in space–time, which is a special case of a general Riemannian space. Upon referring to Section II-D, one sees that the covariant derivative is particularly simple when the Lorentz metric, Eq. (3-3), is appropriate. In this case, the Christoffel symbols all vanish and the divergence, Eqs. (2-68) and (2-69), becomes

$$\operatorname{div} T^\mu = \partial_\mu T^\mu = \partial_1 T^1 + \partial_2 T^2 + \partial_3 T^3 + (1/c)\,\partial_t T^4\,, \tag{4-7a}$$

$$\operatorname{div} T_\mu = \eta^{\mu\nu}\,\partial_\nu T_\mu = -\partial_1 T_1 - \partial_2 T_2 - \partial_3 T_3 + (1/c)\,\partial_t T_4\,. \tag{4-7b}$$

In like manner, from Eq. (2-70) the four-dimensional Laplacian of a contravariant vector is

$$\Box T^\mu = \eta^{\alpha\beta}\,\partial_\alpha\partial_\beta T^\mu\,, \tag{4-8}$$

where we have used the standard notation referred to as the D'Alembertian operator.

Equations (4-4)–(4-8) suggest that we define as contravariant vectors the *4-current density*

$$j^\alpha = (\mathbf{J},\, c\rho)\,, \tag{4-9}$$

and the 4-*potential*

$$A^\alpha = (\mathbf{A},\, \phi)\,. \tag{4-10}$$

One must, of course, ask if these quantities actually are 4-vectors, and the answer must be supplied by physics. By writing the current density as $\mathbf{J} = \rho\mathbf{u}$, we see that

$$j^\alpha j_\alpha = \rho^2 c^2/\gamma^2\,. \tag{4-11}$$

If this quantity is to be an invariant, then it must have the same value in the proper frame of the charge distribution, so that

$$\rho = \gamma\rho_0\,, \tag{4-12}$$

giving the transformation equation for charge density. Now, the physical criterion to be invoked is the conservation of charge—a neutral atom must remain so in every reference frame. From the Lorentz contraction phenomenon of Chapter I we know that the spatial volume element transforms as*

$$dV_0 = \gamma\, dV, \tag{4-13}$$

so that the amount of charge contained in a given volume remains constant:

$$\rho\, dV = \rho_0\, dV_0. \tag{4-14}$$

Thus, charge conservation forces j^α to be a 4-vector, and this conclusion is reinforced by the covariance of the continuity equation, Eq. (4-18).

In a similar manner, the 4-vector nature of A^α follows from the physical requirement that the Lorentz condition, Eq. (4-4), be identically satisfied in every Lorentz frame. This covariant choice of gauge is called the *Lorentz gauge* because of its importance to relativistic electrodynamics.

We have belabored the question of the 4-vector character of the definitions (4-9) and (4-10) because these definitions are so crucial to the ensuing covariant formulation. Equations (4-7)–(4-10) now allow us to write immediately

$$\bar{A}^\mu = A^\mu - \eta^{\mu\nu}\,\partial_\nu \Lambda\,, \tag{4-15}$$

$$\partial_\mu A^\mu = 0\,, \tag{4-16}$$

$$\Box A^\mu = (4\pi/c) j^\mu\,, \tag{4-17}$$

$$\partial_\mu j^\mu = 0 \tag{4-18}$$

corresponding, respectively, to Eqs. (4-3)–(4-6).

Turning now to the fields and their equations, it is to be noted from Eqs. (4-1) that **E** and **B** cannot *both* be chosen as polar vectors. Because of the curl relations between the fields, one must be a vector field and the other a pseudovector field. The requirement of charge conservation also includes the sign of the charge, so that the electric force, and therefore **E**, should be taken as a vector. Hence, **B** must be described by a pseudovector, in the manner discussed in Section II-E. However, it was also noted there that a pseudovector could equally well be described by an antisymmetric tensor of second rank, and, from the point of view of covariance, it is preferable to accept that alternative here.

The covariant field equations can be found by referring to Eq. (4-2) and defining an *electromagnetic field-strength tensor* in terms of the covariant 4-potential

$$\mathsf{F}_{\mu\nu} = \partial_\mu A_\nu - \partial_\nu A_\mu = -\mathsf{F}_{\nu\mu}\,. \tag{4-19}$$

* Observe that this is *not* the 4-volume element, which must transform according to Eq. (2-87); in the particularly simple case under discussion here $|g| = 1$.

The contravariant tensor is

$$\mathsf{F}^{\mu\nu} = \eta^{\mu\alpha}\eta^{\nu\tau}\mathsf{F}_{\alpha\tau} \,. \tag{4-20}$$

Explicitly, the matrix associated with $\mathsf{F}_{\mu\nu}$ is

$$\mathsf{F}_{\mu\nu} = \begin{bmatrix} 0 & -B_3 & B_2 & -E_1 \\ B_3 & 0 & -B_1 & -E_2 \\ -B_2 & B_1 & 0 & -E_3 \\ E_1 & E_2 & E_3 & 0 \end{bmatrix} . \tag{4-21}$$

Exercise. Verify the tensor character of $\mathsf{F}_{\mu\nu}$, and write out explicitly the components of $\mathsf{F}^{\mu\nu}$.

In terms of the electromagnetic field tensor the field strengths are given explicitly by

$$E_k = \mathsf{F}_{4k} \,, \tag{4-22}$$

$$B_k = -\tfrac{1}{2}\mathsf{F}^{ij}\varepsilon_{ijk} \,, \tag{4-23}$$

where ε_{ijk} is given by Eq. (2-82), and Maxwell's equations become

$$\partial_\mu \mathsf{F}^{\mu\nu} = (4\pi/c)j^\nu \,, \tag{4-24}$$

$$\partial_\mu \mathsf{F}_{\nu\lambda} + \partial_\nu \mathsf{F}_{\lambda\mu} + \partial_\lambda \mathsf{F}_{\mu\nu} = 0 \,, \tag{4-25}$$

as is readily verified.* Note how Eq. (4-23) exhibits the pseudovector character of **B**.

Exercise. Show that only four of the 64 equations represented by Eq. (4-25) are nontrivial and distinct.

Equations (4-24) and (4-25) explicitly exhibit the Lorentz covariant nature of Maxwell's equations, and at the same time exhibit the consequent noninvariance of the fields themselves. One can, however, deduce the transformation properties of the field strengths from the field tensor, which must transform as

$$\bar{\mathsf{F}}^{\mu\nu} = \mathsf{L}_\alpha{}^\mu \mathsf{L}_\beta{}^\nu \mathsf{F}^{\alpha\beta} \,. \tag{4-26}$$

* Equation (4-25) follows immediately from the result of Problem 2-2, thereby demonstrating that two of Maxwell's equations are actually a consequence only of the intimate relationship between space and time.

Using the simplified transformation (3-6), one finds that

$$\begin{aligned} \bar{E}_x &= E_x\,, & \bar{B}_x &= B_x \\ \bar{E}_y &= \gamma(E_y - \beta B_z)\,, & \bar{B}_y &= \gamma(B_y + \beta E_z) \\ \bar{E}_z &= \gamma(E_z + \beta B_y)\,, & \bar{B}_z &= \gamma(B_z - \beta E_y)\,. \end{aligned} \tag{4-27}$$

The transformation for arbitrary **v** can now be guessed from Eq. (4-27), or worked out explicitly from the result of Problem 3-3. With an obvious notation indicating directions parallel and perpendicular to the direction of translation, one finds that in general

$$\begin{aligned} \bar{\mathbf{E}}_{\parallel} &= \mathbf{E}_{\parallel}\,, & \bar{\mathbf{B}}_{\parallel} &= \mathbf{B}_{\parallel} \\ \bar{\mathbf{E}}_{\perp} &= \gamma[\mathbf{E}_{\perp} + (1/c)\mathbf{v}\times\mathbf{B}]\,, & \bar{\mathbf{B}}_{\perp} &= \gamma[\mathbf{B}_{\perp} - (1/c)\mathbf{v}\times\mathbf{E}]\,. \end{aligned} \tag{4-28}$$

The major implication of these equations is that the fields **E** and **B** do not really have an independent existence, but are interrelated in the single physical quantity called the electromagnetic field. Thus, referring to Eqs. (4-9) and (4-10), we see that the special theory of relativity emphasizes the intimate relationships between charge and current, vector and scalar potentials, and electric and magnetic fields, demonstrating that what classically were thought to be different physical quantities are actually only different aspects of the same thing. Recognition of these facts has allowed us to exploit this "sameness" by developing the very beautiful covariant formulation of the physical laws. Some of the physical effects following from Eqs. (4-28) will be explored in the problems.

One further important consequence of the 4-vector formulation can be obtained from the covariant form of the wave equation (4-18). The form of the solutions must also be covariant so that a plane wave in one Lorentz frame will be a plane wave in any other Lorentz frame. However, this can only happen if the phases are equal at all points of space–time:

$$\bar{\mathbf{k}}\cdot\bar{\mathbf{r}} - \bar{\omega}\bar{t} = \mathbf{k}\cdot\mathbf{r} - \omega t\,. \tag{4-29}$$

It follows that the wave vector **k** and the frequency ω must form the space and time parts, respectively, of a lightlike 4-vector with contravariant components

$$k^{\nu} = (\mathbf{k}, \omega/c)\,, \tag{4-30}$$

which we shall call the *propagation 4-vector*.

From electromagnetic theory we know that the invariant associated with this 4-vector vanishes; that is, $k_{\mu}k^{\mu} = 0$. Comparing Eq. (4-30) with Eqs. (3-23) and (3-25), we see that **k** and ω can be interpreted as proportional to the momentum and energy of a particle with zero rest mass, with proportionality factor Planck's constant divided by 2π. For the electromagnetic field the realization of such a particle is the photon, and one sees that such a concept is both relativistic and quantum mechanical, and has no Newtonian analogy.

An important experimental test of special relativity is based on the 4-vector k^ν, which we can discuss by deriving the relativistic Doppler formulas. With reference to Fig. 1, let us consider a plane wave with frequency ω and wave vector $\mathbf{k}$ making an angle θ with the x axis. Then,

$$\bar{k}^\nu = \mathsf{L}_\mu{}^\nu k^\mu \,, \tag{4-31}$$

and, using the transformation (3-6), we find that the frequency in the $\bar{S}$ frame is

$$\bar{\omega} = \omega\gamma(1 - \beta\cos\theta)\,, \tag{4-32}$$

and the change in propagation direction is given by

$$\cos\theta = \frac{\beta + \cos\bar{\theta}}{1 + \beta\cos\bar{\theta}} \tag{4-33}$$

where $\bar{\theta}$ is the angle between $\bar{\mathbf{k}}$ and the x axis. Equation (4-32) differs from the nonrelativistic Doppler formula by the factor γ, which shows that there is also an effect in the transverse direction ($\theta = \pi/2$). This *transverse Doppler shift* was first observed by Ives and Stillwell[1] in 1938, and has since made a prominent appearance in experiments involving resonant absorption of nuclear γ-rays.[2]

B. CHARGED PARTICLES AND CONSERVATION LAWS

In classical, nonrelativistic electromagnetic theory one learns that the force per unit volume from the action of the electromagnetic field on a charge described by a charge density ρ and current density $\mathbf{J}$ is given by the Lorentz force density

$$\mathbf{f} = \rho\mathbf{E} + (1/c)\mathbf{J} \times \mathbf{B}\,. \tag{4-34}$$

Examination of Eqs. (4-9) and (4-21) reveals that Eq. (4-34) can be written as

$$f^i = (1/c)\,\mathsf{F}^{i\nu} j_\nu\,,$$

remembering that Latin indices take only the values 1, 2, 3. Thus, the quantity

$$f^\mu = (1/c)\,\mathsf{F}^{\mu\nu} j_\nu = (\mathbf{f}, f_0/c) \tag{4-35}$$

is a 4-vector, and we call it the *Lorentz 4-force density*. The fourth component is interpreted by writing explicitly

$$f_0 = cf^4 = \mathbf{E} \cdot \mathbf{J}\,, \tag{4-36}$$

which is just the work done per unit volume by the electric field on the charge.

Let us now define a symmetric *electromagnetic stress–energy–momentum tensor*

$$\begin{aligned} \mathsf{W}^{\mu\nu} &= (1/4\pi)(\mathsf{F}^{\mu\gamma}\mathsf{F}_\gamma{}^\nu + \tfrac{1}{4}\eta^{\mu\nu}\mathsf{F}_{\gamma\delta}\mathsf{F}^{\gamma\delta}) \\ &= \begin{bmatrix} -\mathsf{T}_{11} & -\mathsf{T}_{12} & -\mathsf{T}_{13} & cg_1 \\ -\mathsf{T}_{21} & -\mathsf{T}_{22} & -\mathsf{T}_{23} & cg_2 \\ -\mathsf{T}_{31} & -\mathsf{T}_{32} & -\mathsf{T}_{33} & cg_3 \\ cg_1 & cg_2 & cg_3 & u \end{bmatrix}, \end{aligned} \tag{4-37}$$

where

$$\mathsf{T}_{ij} = (1/4\pi)[E_iE_j + B_iB_j - \tfrac{1}{2}\delta_j{}^i(E^2 + B^2)] = \mathsf{T}^{ij} \tag{4-38}$$

is the three-dimensional Maxwell stress tensor of second rank,

$$\mathbf{g} = (1/4\pi c)\mathbf{E} \times \mathbf{B} = \mathbf{S}/c^2 \tag{4-39}$$

is the electromagnetic field momentum density in a particular Lorentz frame in terms of the Poynting vector **S**, and

$$u = (1/8\pi)(E^2 + B^2) \tag{4-40}$$

is the field energy density in a particular frame.

Exercise. Demonstrate explicitly that the components of $\mathsf{W}^{\mu\nu}$ are indeed those advertised, and that its trace vanishes: $\mathsf{W}_\mu{}^\mu = 0$.

The second-rank tensor $\mathsf{W}^{\mu\nu}$ has been named advisedly, as we will now show by deriving the conservation laws for the fields. Let us first calculate the divergence of $\mathsf{W}^{\mu\nu}$ as follows:

$$\partial_\mu \mathsf{W}^{\mu\nu} = (1/c)j_\gamma \mathsf{F}^{\gamma\nu} + (1/4\pi)(\mathsf{F}^{\mu\gamma}\partial_\mu\mathsf{F}_\gamma{}^\nu + \tfrac{1}{2}\mathsf{F}^{\gamma\delta}\eta^{\mu\nu}\partial_\mu\mathsf{F}_{\gamma\delta})\,, \tag{4-41}$$

where we have used Eq. (4-24) and the properties of the metric tensor $\eta^{\mu\nu}$. We now make use of the homogeneous field equations (4-25) to write for the expression in parentheses in Eq. (4-41)

$$\begin{aligned} \mathsf{F}^{\mu\gamma}\partial_\mu\mathsf{F}_\gamma{}^\nu + \tfrac{1}{2}\mathsf{F}^{\gamma\delta}\eta^{\mu\nu}\partial_\mu\mathsf{F}_{\gamma\delta} &= \mathsf{F}^{\gamma\delta}[-\partial_\delta\eta^{\nu\alpha}\mathsf{F}_{\gamma\alpha} + \tfrac{1}{2}\eta^{\mu\nu}\partial_\mu\mathsf{F}_{\gamma\delta}] \\ &= \tfrac{1}{2}\eta^{\mu\nu}\mathsf{F}^{\gamma\delta}[\partial_\gamma\mathsf{F}_{\mu\delta} + \partial_\delta\mathsf{F}_{\mu\gamma}]\,. \end{aligned} \tag{4-42}$$

For a given value of μ the quantity in brackets is symmetric, while $\mathsf{F}^{\gamma\delta}$ is antisymmetric. Therefore, from the exercise below Eq. (2-43) we conclude that Eq. (4-41) can be written

$$\partial_\mu \mathsf{W}^{\mu\nu} = -(1/c)\mathsf{F}^{\nu\mu} j_\mu = -f^\nu , \tag{4-43}$$

using the definition (4-35).

The important deduction to be made from this last result is that $\mathsf{W}^{\mu\nu}$ has zero divergence if there are no charges present. Hence, from Eq. (4-37) we can immediately write the 3-vector equations

$$\partial_i \mathsf{T}^{ij} - (1/c)\partial_t S^j = 0 , \tag{4-44a}$$

$$\partial_t u + \nabla \cdot \mathbf{S} = 0 , \tag{4-44b}$$

which are just differential conservation laws of energy and momentum for the electromagnetic field. If sources are present, then Eq. (4-43) merely tells us that the field energy decreases by the work it does on the charges, and the field momentum is diminished at a rate equal to the force it exerts on the charges.

One can also develop integral forms of the conservation laws by defining a quantity

$$P^\mu = \frac{1}{c}\int \mathsf{W}^{\mu\nu}\, d\sigma_\nu , \tag{4-45}$$

where the integral is over a three-dimensional spacelike plane, and identifying P^μ as the field 4-momentum. This identification is not trivial, however, as one will readily admit after recalling the observations made at the end of Chapter II to the effect that the integral of a tensor is not necessarily itself a tensor. For the free-field case it can be shown that the quantity P^μ defined by Eq. (4-45) is indeed a 4-vector, and the reader is referred to the excellent treatment of this topic given by Rohrlich[3] for further details. When charges are present ($j^\mu \neq 0$) the integral in Eq. (4-45) very definitely depends on the particular Lorentz frame being considered, and the conservation laws follow only when all the mutual effects among charges and fields are included. Equations (4-24), (4-25), and (4-35) together constitute the Maxwell–Lorentz theory of electrodynamics.

It is not our purpose here to exhaust all aspects of the covariant theory of classical electrodynamics, as this has been done quite nicely by both Rohrlich[3] and Barut.[4] One can, for example, consider the Lagrangian and Hamiltonian formulations and the relations between conservation laws and symmetry principles to which they lead. Rather, let us just mention qualitatively some of these developments.

The Maxwell–Lorentz equations can be derived from the Lagrangian density

$$\mathscr{L} = -(1/16\pi)\mathsf{F}_{\mu\nu}\mathsf{F}^{\mu\nu} + (1/c) j_\mu A^\mu \tag{4-46}$$

by means of the action principle

$$\delta \int \mathscr{L}\, d^4x = 0 . \tag{4-47}$$

The conservation laws then follows from Noether's theorem[5] and the requirement of invariance under inhomogeneous Lorentz transformations.

The problem of determining the motion of the charges, given the fields, can be discussed starting with the Lagrangian

$$L = (1/\gamma)[-mc^2 - (e/mc)P_\nu A^\nu] , \tag{4-48}$$

which yields the Lorentz 4-force when substituted into the Euler–Lagrange equations obtained from the action principle.

Classically, the Lagrangian formulation when interactions among charged particles are included encounters grave difficulties. The crux of the problem is that L is to be considered a function of the instantaneous velocities and coordinates of the particles, and, due to the finite value of c, the values of $\mathbf{A}$ and ϕ at the position of one particle due to the other particles depend on what happened at *retarded* times (which will be discussed further in the next section). This problem of developing a completely covariant Hamiltonian formalism is one that continues to plague theoretical physics today.[6]

It is possible, however, to find a Lagrangian in the case when retardation effects are small, which includes the lowest-order relativistic corrections. For two charged particles the interaction Lagrangian can be written

$$L_{\text{int}} = \frac{q_1 q_2}{r}\left\{-1 + \frac{1}{2c^2}\left[\mathbf{v}_1 \cdot \mathbf{v}_2 + \frac{(\mathbf{v}_1 \cdot \mathbf{r})(\mathbf{v}_2 \cdot \mathbf{r})}{r}\right]\right\} , \tag{4-49}$$

and was first obtained by Darwin in 1920.[7] The quantum-mechanical analog is due to Breit.[8] Derivation of the Hamiltonian function related to L_{int} will be left to the problems.

C. POTENTIALS AND FIELDS OF CHARGED PARTICLES

In the next chapter we wish to study the radiation from charged particles in motion, and to do this most conveniently it is desirable to have explicit expressions for the potentials and fields of a charged particle. As is well known, the inhomogeneous wave equation (4-17) can be solved for the 4-potential by the Green's function method.[9] Once this function is obtained, one can then write for the potential

$$A^\mu(x) = (4\pi/c) \int \mathsf{G}(x - x') j^\mu(x')\, d^4x' , \tag{4-50}$$

where x and x' are 4-vectors of position in space–time.

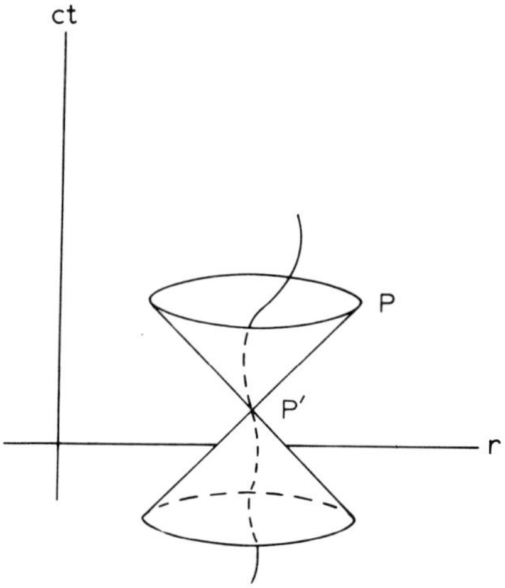

Fig. 8. Space–time representation of the causality condition relating events P and P'.

In finding G, however, one must here be very careful to observe the causality condition in space–time. With reference to Fig. 8, the electromagnetic fields at event P are due to the motion of the charge at event P', and are therefore called *retarded* fields because they depend on the motion of the charge at an earlier time. We mention this point because it is necessary to specify this condition explicitly in order to obtain the Green function leading to a retarded potential. That is, we demand that

$$\mathsf{G}(x_{P'} - x_P) = 0, \qquad t_P < t_{P'} . \tag{4-51}$$

Were the opposite condition specified, $t_{P'} < t_P$, then P would have to lie on the past light cone with vertex at the point P', and we would refer to *advanced* solutions. Although the retarded fields will be the physically important quantities of interest in the next chapter, it will be useful to have reference to the advanced solutions later, and so we shall retain them here. The two forms of G are well known from ordinary electromagnetic theory[10] and, designating advanced and retarded solutions by plus and minus signs, respectively, we find

$$A_{\pm}{}^{\mu}(x, t) = \frac{1}{c}\int d^3x' \int \frac{j^{\mu}(\mathbf{x}', t')}{R}\, \delta\!\left(t' \pm t + \frac{R}{c}\right) dt' , \tag{4-52}$$

where $\mathbf{R} = \mathbf{x} - \mathbf{x}'$.

In what follows we shall be interested primarily in the current density due to a particle with charge e in motion, which, in 4-vector notation, can be written

$$j^{\mu}(\mathbf{x}, t) = ec\int_{-\infty}^{\infty} \delta[x - r(\tau)]v^{\mu}(\tau)\, d\tau , \tag{4-53}$$

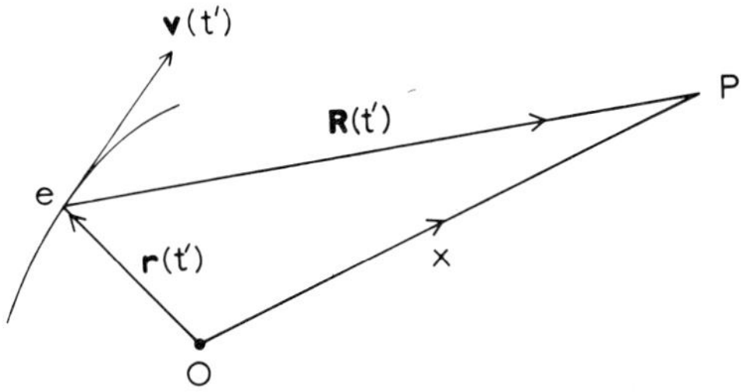

Fig. 9. Relationship of the various position vectors in Eq. (4-54) to the instantaneous velocity in 3-space.

where the δ function is four dimensional, and v^μ is the 4-velocity. Performing the temporal integration in Eq. (4-53), substituting the result into Eq. (4-52), and evaluating the volume integral, one finds

$$A_\pm{}^\mu(\mathbf{x}, t) = \frac{e}{c} \int \frac{\delta[t' \pm t + (R/c)]}{R} v^\mu(t')\, dt' , \tag{4-54}$$

where now $\mathbf{R}(t') = \mathbf{x} - \mathbf{r}(t')$. Figure 9 exhibits the relationships among the position vectors and the instantaneous velocity in 3-space.

Exercise. If $f(x)$ has a finite number of zeros at $x = x_i$, show that[11]

$$\int g(x)\, \delta[f(x)]\, dx = \sum_i \frac{g(x_i)}{|df/dx|_{x=x_i}} . \tag{4-55}$$

Applying the formula to the integral in Eq. (4-54), we find for the needed derivative of the argument of the δ function

$$1 - \frac{1}{c} \frac{\mathbf{R} \cdot \dot{\mathbf{r}}}{R} .$$

Hence, in the rest frame of the observer at P the potentials in noncovariant form are

$$\phi_\pm(\mathbf{x}, t) = [e/(R - \mathbf{R} \cdot \boldsymbol{\beta})]_\pm , \tag{4-56a}$$

$$\mathbf{A}_\pm(\mathbf{x}, t) = [e\boldsymbol{\beta}/(R - \mathbf{R} \cdot \boldsymbol{\beta})]_\pm , \tag{4-56b}$$

where

$$\boldsymbol{\beta} = \mathbf{v}/c , \tag{4-57}$$

and where the quantities in brackets are to be evaluated at the advanced or retarded times $(t \pm R/c)$. These potentials are known as the *Liénard–Wiechert potentials*,[12] and Eqs. (4-56) express the values of ϕ and $\mathbf{A}$ at the location P of the observer in terms of the physical events when the charge was at the point P'. The reader will note at this point that the equations have lost their covariance as a result of the above derivation, a lapse which was quite intentional, since we shall find the form of Eqs. (4-56) much better suited to our purposes in the following chapter.

As a final matter for this discussion, we should like to obtain the fields from the potentials, by means of Eq. (4-2), and this is accomplished most easily by differentiating in Eq. (4-54) directly. Defining a unit vector $\mathbf{n} = \mathbf{R}/R$ (see Fig. 9), we find that a straightforward calculation yields

$$\mathbf{E}^{\pm}(\mathbf{x}, t) = e \int \Big[\frac{\mathbf{n}}{R^2}\, \delta\Big(t' \pm t + \frac{R}{c}\Big) + \frac{1}{cR}\Big(\boldsymbol{\beta} - \mathbf{n}\Big)\delta'\Big(t' \pm t + \frac{R}{c}\Big)\Big]\, dt' \tag{4-58a}$$

$$\mathbf{B}^{\pm}(\mathbf{x}, t) = e \int (\mathbf{n} \times \boldsymbol{\beta}) \Big[-\frac{\delta[t' \pm t + (R/c)]}{R^2} + \frac{1}{cR}\, \delta'\Big(t' \pm t + \frac{R}{c}\Big)\Big]\, dt' , \tag{4-58b}$$

where the primes on the δ functions indicate derivatives with respect to their arguments. The formula of Eq. (4-55) and the observation that

$$\int_c^d f(x)\, \delta'(x - a)\, dx = -f'(a)\,, \qquad c < a < d\,, \tag{4-59}$$

can now be employed. Defining

$$K \equiv 1 - \mathbf{n} \cdot \boldsymbol{\beta}\,, \tag{4-60}$$

we readily find

$$\mathbf{E}^{\pm}(\mathbf{x}, t) = e\Big[\frac{\mathbf{n}}{KR^2} + \frac{1}{cK}\frac{d}{dt'}\frac{\mathbf{n} - \boldsymbol{\beta}}{KR}\Big]_{\pm}\,, \tag{4-61a}$$

$$\mathbf{B}^{\pm}(\mathbf{x}, t) = e\Big[\frac{\boldsymbol{\beta} \times \mathbf{n}}{KR^2} + \frac{1}{cK}\frac{d}{dt'}\frac{\boldsymbol{\beta} \times \mathbf{n}}{KR}\Big]_{\pm}\,. \tag{4-61b}$$

Exercise. Verify the relations

$$\frac{1}{c}\frac{d\mathbf{n}}{dt'} = \frac{\mathbf{n} \times (\mathbf{n} \times \boldsymbol{\beta})}{R}\,, \tag{4-62a}$$

$$\frac{1}{c}\frac{d}{dt'}(KR) = \beta^2 - \mathbf{n} \cdot \boldsymbol{\beta} - \frac{R}{c}\,\mathbf{n} \cdot \dot{\boldsymbol{\beta}}\,, \tag{4-62b}$$

where $\dot{\boldsymbol{\beta}} = d\boldsymbol{\beta}/dt'$.

With the results of this exercise one finally obtains

$$\mathbf{E}^{\pm}(\mathbf{x}, t) = e\left[\frac{(\mathbf{n} - \boldsymbol{\beta})}{\gamma^2 K^3 R^2}\right]_{\pm} + \frac{e}{c}\left[\frac{\mathbf{n} \times (\mathbf{n} \times \dot{\boldsymbol{\beta}} - \boldsymbol{\beta} \times \dot{\boldsymbol{\beta}})}{K^3 R}\right]_{\pm}, \qquad (4\text{-}63)$$

$$\mathbf{B}^{\pm}(\mathbf{x}, t) = \mathbf{n} \times \mathbf{E}^{\pm}(\mathbf{x}, t)\,. \qquad (4\text{-}64)$$

The first terms in these equations are called *velocity fields* and possess the R^{-2} dependence of static fields, whereas the second terms are the *acceleration fields* and have the R^{-1} dependence of radiation fields. These latter fields are transverse to **R** and will be of major importance to us in the following chapter.

PROBLEMS

4-1. If **E** is perpendicular to **B**, but not of the same magnitude, there exists a Lorentz frame in which the field is either purely electric or purely magnetic. Verify this statement.

4-2. Find the Hamiltonian function corresponding to the Darwin Lagrangian, Eq. (4-49).

4-3. A man runs a red light and then attempts to convince the judge that the light appeared green to him. What would his speed have had to be to justify his argument?

4-4. Consider an unconstrained, excited nucleus of rest mass m_0 (in the unexcited state) moving with velocity **v** directly away from an observer when it emits a gamma ray in the direction $(-\mathbf{v})$, thereby reducing its energy by ΔE. Describe the characteristics of the photon reported by the observer.

4-5. A particle of rest mass m_0 and charge e has an initial velocity parallel to the uniform acceleration $\mathbf{a} = \mathbf{b}\gamma^{-3}$ produced by a constant force, where **b** is a constant vector.

(a) If **b** is chosen in the x direction and the initial conditions are taken as $x(0) = \Delta = c^2/|\mathbf{b}|$ and $\mathbf{v}(0) = 0$, show that the equations of the orbit are

$$x = (\Delta^2 + c^2t^2)^{1/2}\,, \qquad y = z = 0\,,$$

which describes a hyperbola on a Minkowski diagram.

(b) This hyperbolic motion of the charged particle leads to a 4-current which can be calculated from Eq. (4-53). Calculate the electromagnetic field produced by this current, and express your result in cylindrical coordinates.

4-6. Show that Eq. (4-56a) can be obtained by applying a Lorentz transformation to the Coulomb potential of a charged particle in its proper frame.

4-7. From Maxwell's equations in free space, deduce the corresponding equations for the potentials in the Coulomb gauge ($\nabla \cdot \mathbf{A} = 0$). In particular:

(a) Obtain and interpret a formal solution for the scalar potential.

(b) Divide the current into longitudinal and transverse parts,

$$\mathbf{j} = \mathbf{j}_l + \mathbf{j}_t \,,$$

and show that the wave equation for **A** can be written entirely in terms of the *transverse current*

$$\mathbf{j}_t = \frac{1}{4\pi} \nabla \times \nabla \times \int \frac{\mathbf{j}}{|\mathbf{x} - \mathbf{x}'|} \, d^3x' \,.$$

REFERENCES

1. H. E. Ives and C. R. Stillwell, *J. Opt. Soc. Am.* **28,** 215 (1938); **31,** 369 (1941).
2. See, e.g., R. V. Pound and G. A. Rebka, *Phys. Rev. Letters* **4,** 274 (1960).
3. F. Rohrlich, "Classical Charged Particles." Addison-Wesley, Reading, Massachusetts, 1965.
4. A. O. Barut, "Electrodynamics and Classical Theory of Fields and Particles." Macmillan, New York, 1964.
5. E. Noether, *Nachr. Kgl. Ges. Wiss. Göttingen* 235 (1918); see also, R. Courant and D. Hilbert, "Methods of Mathematical Physics," p. 262. Wiley (Interscience), New York, 1953.
6. See, e.g., L. L. Foldy, *Phys. Rev.* **122,** 275 (1961).
7. C. G. Darwin, *Phil. Mag.* **39,** 537 (1920).
8. G. Breit, *Phys. Rev.* **36,** 383 (1930).
9. See, e.g., P. M. Morse and H. Feshbach, "Methods of Theoretical Physics," Chapter 7. McGraw-Hill, New York, 1953.
10. See, e.g., J. D. Jackson, "Classical Electrodynamics," Section 6.6. Wiley, New York, 1962.
11. See, e.g., B. Friedman, "Principles and Techniques of Applied Mathematics," p. 136. Wiley, New York, 1956.
12. A. Liénard, *L'Eclairage Elec.* **16,** 5, 53, 106 (1898); E. Wiechert, *Arch. Néerl.* **5,** 549 (1900).

V RADIATION FROM CHARGED PARTICLES

In the preceding chapter we obtained expressions for the electric and magnetic fields generated by a charged particle in motion, Eqs. (4-63) and (4-64), and observed that a part of these fields possessed the characteristic spatial dependence of radiation fields. There are two different points of view which could be taken at this point, the first of which is that the fields can be considered as only of intermediate interest, merely transmitting the interaction between charged particles. While this line of thought could certainly be accepted as reasonable, and a theory of charged particles developed therefrom, it would overlook the fact that the fields indeed possess physical attributes in their own right and are a significant physical phenomenon.

Thus, we shall adopt the alternative view that at this point it is of importance to examine the properties of the fields themselves and to extract their source-independent physical qualities. In particular, it is the radiation fields which are of fundamental significance here because, unlike the velocity fields which are permanently attached to the charge and move along with it, the radiation fields separate themselves from their source and lead an existence independent of it. Classically, radiation behaves in very many ways like matter and can be described in somewhat the same manner. In fact, with reference to Eq. (4-45), the vector P^μ can be shown[1] to be a timelike 4-vector with Lorentz transformation properties identical to those for a particle having the same energy and momentum, when the radiation fields are employed in the integrand. This particlelike behavior of radiation will attain full significance when we come to study its quantum aspects in Chapter IX, at which point the photon emerges as a particle every bit as fundamental as the electron.

In view of these remarks, the objective of the present chapter is to study the properties of the radiation fields generated by a charged particle at large distances from the particle itself—that is, in the radiation zone.

Although the radiation has independent existence at large distances from the source, its physical characteristics will quite clearly depend on the motion of the source at the time the radiation was emitted, so that our aim will be to describe the radiation in terms of the dynamical variables of the charged particle. Consequently, unless specifically stated otherwise, all fields in this chapter will be taken as retarded and the superscripts deleted.

A. ACCELERATED CHARGES

It is of interest to first make a qualitative investigation of the radiation emitted from charged particles in motion, so that in this section we shall concentrate on obtaining semiquantitative results only. With this thought in mind, consider a particle with charge e accelerated in a Lorentz frame in which its velocity remains much less than c. In this situation Eq. (4-63) reduces to

$$\mathbf{E} = \frac{e}{c}\left[\frac{\mathbf{n} \times (\mathbf{n} \times \dot{\boldsymbol{\beta}})}{R}\right]_{-}, \tag{5-1}$$

demonstrating that the radiation is polarized in the plane of $\mathbf{n}$ and $\dot{\mathbf{v}}$. Since $\mathbf{E}$ is normal to $\mathbf{n}$, Eq. (4-64) combined with (5-1) yields for the (instantaneous) Poynting vector describing the radiated energy*

$$\mathbf{S} = (c/4\pi)\,|\,\mathbf{E}\,|^2\,\mathbf{n}\,, \tag{5-2}$$

which is the energy per unit area per unit time. We shall take as the appropriate quantity characterizing the radiation the instantaneous† power radiated per unit solid angle, which can now be written

$$\begin{aligned}\frac{dW}{d\Omega} &= \frac{c}{4\pi}\,|\,R\mathbf{E}\,|^2 = \frac{e^2}{4\pi c}\,|\,\mathbf{n} \times (\mathbf{n} \times \dot{\boldsymbol{\beta}})\,|^2 \\ &= \frac{e^2}{4\pi c^3}\,\dot{v}^2 \sin^2\theta\,, \end{aligned}\tag{5-3}$$

where θ is the angle between $\mathbf{n}$ and $\dot{\mathbf{v}}$. Figure 10 exhibits this angular distribution of radiation.

It is now an easy matter to obtain the total instantaneous power radiated by doing the indicated angular integrations in Eq. (5-3), and one obtains

$$W = \tfrac{2}{3}e^2\dot{v}^2/c^3\,, \tag{5-4}$$

which is the famous *Larmor formula.*[3]

* This form follows from the transverse nature of the waves indicated by Eq. (4-64).

† That is, the power radiated with respect to the proper time of the particle, or in the particle's instantaneous rest frame. Note that this does *not* mean that the particle will not radiate. In this respect, see the discussion by Fulton and Rohrlich.[2]

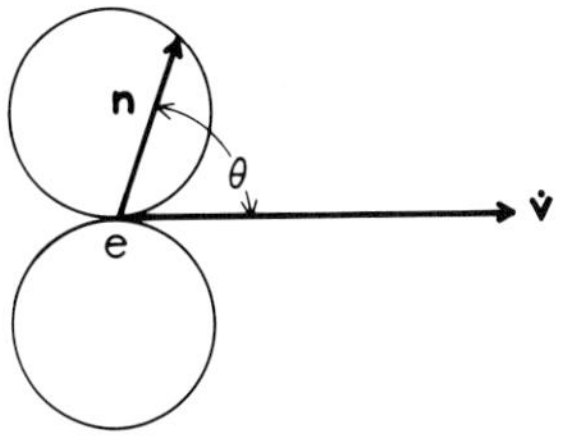

Fig. 10. Nonrelativistic angular distribution of radiation associated with the Larmor formula.

The nonrelativistic result of Eq. (5-4) can easily be generalized to the case of an arbitrary relativistic charged particle—one merely transforms the acceleration to a general Lorentz frame, using the results of Problem 1-4. The total instantaneous power radiated from an accelerated relativistic particle is then

$$W = \tfrac{2}{3}(e^2/c)\gamma^6[(\dot{\boldsymbol{\beta}})^2 - (\boldsymbol{\beta} \times \dot{\boldsymbol{\beta}})^2], \tag{5-5}$$

a result first obtained by Liénard.[4]

Exercise. Demonstrate that the quantity in square brackets in this last equation is positive definite.

There are two important observations to be made from Eq. (5-5), the first of which follows from the previous exercise. That is, we can deduce the criterion that a charge emits radiation relative to any Lorentz frame if and only if its acceleration is not identically zero. Moreover, and this is the second point to be made, the right-hand side of Eq. (5-5) is actually Lorentz invariant, although it is not exhibited in covariant form. It will be left for the problems to demonstrate that the positive-definite quantity of the previous exercise is actually the square of the magnitude of the particle's 4-acceleration. Thus, the invariant instantaneous radiation rate W is the same in all Lorentz frames. (This is not, of course, true under the transformations relating the more general frames of general relativity. In this respect, see the interesting article by Mould.[5])

The reader will have no doubt observed that the angular distribution of radiation in the nonrelativistic case has been obtained in Eq. (5-3), whereas for the relativistic charge only the total instantaneous power radiated is known from Eq. (5-5). Examination of the last three equations shows that the specific angular dependence of $dW/d\Omega$ in the relativistic case will be rather complicated due to the indicated relationship between the velocity and acceleration. However, it is possible to make an approximate calculation, from which we can draw some semiquantitative conclusions, by examining the quantity $[\mathbf{S} \cdot \mathbf{n}]_-$ evaluated using the full radiation field contained in Eq. (4-63). Although this is the energy flow which the

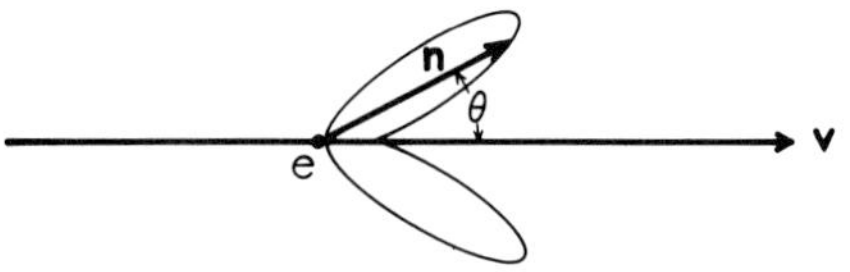

Fig. 11. Angular distribution of radiation as the speed of a charged particle, accelerated parallel to its velocity, approaches the speed of light.

observer would see in his frame, we can assume that the finite time interval over which the charged particle is accelerated in its frame is extremely small. Then, if the observer is at a very large distance from the charge during this period, the vectors **n** and **R** will remain sensibly constant, as will **v** and $\dot{\mathbf{v}}$. We can then write, in terms of the proper time of the particle,*

$$\frac{dW(t')}{d\Omega} \simeq R^2(\mathbf{S}\cdot\mathbf{n})\frac{dt}{dt'}$$

$$= \frac{e^2}{4\pi c}\frac{|\mathbf{n}\times(\mathbf{n}\times\dot{\boldsymbol{\beta}} - \boldsymbol{\beta}\times\dot{\boldsymbol{\beta}})|^2}{(1-\mathbf{n}\cdot\boldsymbol{\beta})^5}. \tag{5-6}$$

If the above approximations are invalid, then one must actually integrate $[\mathbf{S}\cdot\mathbf{n}]_-$ over the time interval of acceleration, and as will be seen below, this can be a complicated procedure. For the moment, however, we can use Eq. (5-6) to learn the essentially correct behavior of $dW/d\Omega$ in some particular situations of physical interest.

As a first example, consider the case when **v** and $\dot{\mathbf{v}}$ are parallel. This includes the situation discussed in Problem 4-5 but is really more general. If θ is the angle between $\boldsymbol{\beta}$ and **n**, then Eq. (5-6) becomes

$$\frac{dW(t')}{d\Omega} = \frac{e^2\dot{v}^2}{4\pi c^3}\frac{\sin^2\theta}{(1-\beta\cos\theta)^5}, \tag{5-7}$$

and, as $\beta \to 0$, this reduces to Eq. (5-3) and the associated angular distribution of Fig. 10. However, as β approaches unity, the distribution becomes tipped forward and elongated, yielding the characteristic radiation pattern for relativistic particles shown in Fig. 11.

Exercise. For β very close to unity show that Eq. (5-7) can be written approximately as

$$\frac{dW(t')}{d\Omega} \simeq \frac{8e^2\dot{v}^2}{\pi c^3}\gamma^8\frac{(\gamma\theta)^2}{(1+\gamma^2\theta^2)^5},$$

(Note that in this limit $\beta \simeq 1 - 1/2\gamma^2$.)

* Note that the quantity $(\mathbf{S}\cdot\mathbf{n})$ is *not* retarded, since we have changed the time variables.

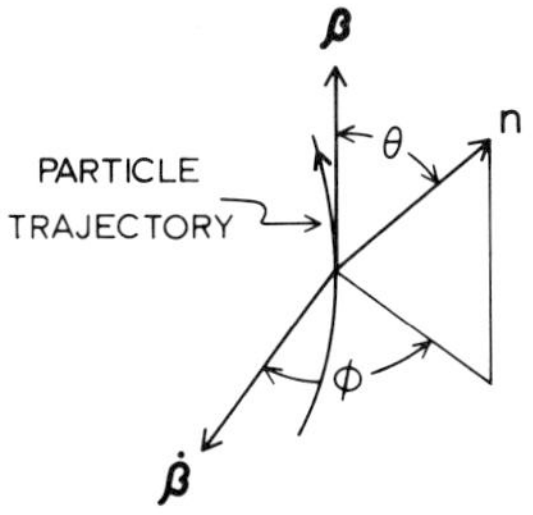

Fig. 12. A charged particle in instantaneous circular motion, with trajectory lying in the plane of the vectors $\boldsymbol{\beta}$ and $\dot{\boldsymbol{\beta}}$.

The total instantaneous power radiated by the relativistic charged particle can be obtained by integrating Eq. (5-7) over all angles, or directly from Eq. (5-5). In either case one obtains

$$W(t') = \tfrac{2}{3}(e^2\dot{v}^2/c^3)\gamma^6 . \tag{5-8}$$

Let us now take as a second example the situation contrary to the above, in which the acceleration is orthogonal to the velocity. Representative of such motion is the case of a charge in instantaneous circular motion, such as the trajectory indicated in Fig. 12. In order to evaluate Eq. (5-6) in this case, it is first necessary to expand the numerator in that equation using the appropriate vector identities.* In terms of the angular variables of Fig. 12, one readily obtains

$$\frac{dW(t')}{d\Omega} = \frac{e^2\dot{v}^2}{4\pi c^3}\frac{1}{(1-\beta\cos\theta)^3}\left[1 - \frac{\sin^2\theta\cos^2\phi}{\gamma^2(1-\beta\cos\theta)^2}\right]. \tag{5-9}$$

This angular distribution is not quite as simple as in the rectilinear case, but the total instantaneous radiated power can again be found by integrating over all angles, or directly from Eq. (5-5), and one obtains

$$W(t') = \tfrac{2}{3}(e^2\dot{v}^2/c^3)\gamma^4 . \tag{5-10}$$

It is important to compare this radiation rate with that of Eq. (5-8) for rectilinear motion. One observes that the radiation emitted with a transverse acceleration appears to be of order γ^2 *less* than that for linear acceleration. This, however, is deceiving, because the two expressions must be compared for the same magnitude of the applied force. Now, according to Eq. (3-36) and the result of Problem 5-2, we can write the invariant rate of instantaneous radiation in the alternative form

$$\begin{aligned} W(t') &= -\frac{2}{3}\frac{e^2}{m_0{}^2c^3}\frac{dp^\mu}{d\tau}\frac{dp_\mu}{d\tau} \\ &= \frac{2}{3}\frac{e^2}{m_0{}^2c^3}\left[\left(\frac{d\mathbf{p}}{d\tau}\right)^2 - \beta^2\left(\frac{dp}{d\tau}\right)^2\right]. \end{aligned} \tag{5-11}$$

* Recall that $(\mathbf{A}\times\mathbf{B})\cdot(\mathbf{C}\times\mathbf{D}) = (\mathbf{A}\cdot\mathbf{C})(\mathbf{B}\cdot\mathbf{D}) - (\mathbf{A}\cdot\mathbf{D})(\mathbf{B}\cdot\mathbf{C})$.

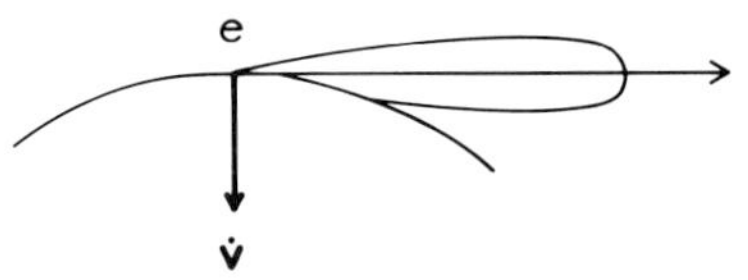

Fig. 13. "Searchlight-beam" distribution of radiation from an extreme relativistic charged particle due to the component of motion instantaneously along the arc of a circular path.

For rectilinear motion, only the magnitude of **p** changes, whereas, for uniform circular motion, only the direction of **p** changes. In the instantaneous rest frame of the particle we have for the two cases

$$W_r(t') = \frac{2}{3}\frac{e^2}{m_0^2c^3}\left(\frac{dp}{dt'}\right)^2, \tag{5-12a}$$

$$W_c(t') = \frac{2}{3}\frac{e^2}{m_0^2c^3}\gamma^2\left(\frac{dp}{dt'}\right)^2. \tag{5-12b}$$

Thus, for the same magnitude of the applied force we see that the radiation rate for transverse acceleration is actually a factor of γ^2 *larger* than that for parallel acceleration. One concludes, therefore, that for an extreme relativistic particle with an arbitrary direction of acceleration the dominant radiation will be due to the component of motion instantaneously along the arc of a circular path. There is then a narrow cone, or "searchlight beam," of radiation along the instantaneous velocity vector, and the observer will see a significant pulse of radiation only corresponding to the time when the velocity of the particle was directed toward him, as suggested by Fig. 13. Moreover, this pulse will be very short in terms of the particle's proper time, which, from general considerations of Fourier analysis, means that the radiation will be over a broad spectrum of frequencies. Hence, to accurately describe the radiation, it is really necessary to make a careful frequency analysis of the radiation process.

B. FREQUENCY AND ANGULAR DISTRIBUTIONS OF RADIATION*

Although the discussion of the preceding section has led to a reasonable physical understanding of the relativistic radiation processes, the results obtained were fairly qualitative and essentially useless from a practical point of view. There are two reasons for this criticism, the first of which is that the radiation rates have been calculated in the instantaneous rest frame of the particle and are not, therefore, what the observer would measure. Second, the radiation will not be monochromatic, so that it is really necessary to measure the frequency distribution in order to provide an adequate

* The analysis of this section, and, in fact, the entire detailed theory of classical radiation from accelerated charges which emerges, is originally due to Schwinger.[6]

test of the theory. Our aim in this section, then, will be to obtain equations suitable for making a comparison with experiment.

In terms of the observer's time, the quantity to be calculated is

$$dW(t)/d\Omega = [R^2\mathbf{S}\cdot\mathbf{n}]_- \equiv |\mathbf{G}(t)|^2 . \tag{5-13}$$

Under fairly general conditions[7] the function $\mathbf{G}(t)$ forms one of a Fourier transform pair

$$\mathbf{G}(t) = \frac{1}{(2\pi)^{1/2}}\int_{-\infty}^{\infty} \mathbf{F}(\omega)e^{-i\omega t}\,d\omega\,, \tag{5-14a}$$

$$\mathbf{F}(\omega) = \frac{1}{(2\pi)^{1/2}}\int_{-\infty}^{\infty} \mathbf{G}(t)\,e^{i\omega t}\,dt\,. \tag{5-14b}$$

Using the δ-function representation,

$$\delta(\omega-\omega') = (1/2\pi)\int_{-\infty}^{\infty} \exp[i(\omega'-\omega)t]\,dt\,,$$

along with the restriction that the observer be in the radiation zone, we can deduce from Eq. (5-13) the total energy radiated per unit solid angle*

$$dW/d\Omega = \int_{-\infty}^{\infty} |\mathbf{G}(t)|^2\,dt = \int_{-\infty}^{\infty} |\mathbf{F}(\omega)|^2\,d\omega\,. \tag{5-15}$$

In order to measure a frequency distribution, the significant physical quantity to be considered is the energy radiated per unit solid angle per unit frequency interval, $dI(\omega)/d\Omega$, where $I(\omega)$ is called the *intensity of radiation* at frequency ω. Since negative frequencies are not meaningful, this definition must be equivalent to

$$dW/d\Omega = \int_0^{\infty} [dI(\omega)/d\Omega]\,d\omega\,, \tag{5-16}$$

so that comparison with Eq. (5-15) yields the relation

$$dI(\omega)/d\Omega = 2\,|\mathbf{F}(\omega)|^2\,. \tag{5-17}$$

It is now necessary to calculate $\mathbf{F}(\omega)$ in order to evaluate Eq. (5-16), and this can be done by substituting from Eq. (4-63) into (5-13), which yields (in the radiation zone)

$$\mathbf{G}(t) = \left(\frac{e^2}{4\pi c}\right)^{1/2}\left[\frac{\mathbf{n}\times(\mathbf{n}\times\dot{\boldsymbol{\beta}}-\boldsymbol{\beta}\times\dot{\boldsymbol{\beta}})}{K^3}\right]_-\,. \tag{5-18}$$

* It is assumed that the acceleration takes place during a finite time interval, so that the total energy radiated is finite. In the case of uniform acceleration (hyperbolic motion) it can be shown that this is a necessary restriction.[2]

Now insert this result into the Fourier integral of Eq. (5-14b) and change the integration variable from $t = t' + [R(t')/c]$ to t', obtaining

$$\mathbf{F}(\omega) = \left(\frac{e^2}{8\pi^2 c}\right)^{1/2} \int_{-\infty}^{\infty} \left\{\exp\left[i\omega\left(t' + \frac{R}{c}\right)\right]\right\} \times \frac{\mathbf{n} \times [(\mathbf{n} - \boldsymbol{\beta}) \times \dot{\boldsymbol{\beta}}]}{K^2}\, dt' \,. \tag{5-19}$$

Recalling that the definition of radiation zone is essentially that $|\mathbf{r}| \ll |\mathbf{R}|$, $|\mathbf{r}| \ll |\mathbf{x}|$, we can refer to Fig. 9 and make the approximation

$$R = |\mathbf{R}(t')| = |\mathbf{x} - \mathbf{r}(t')| \simeq x - \mathbf{n} \cdot \mathbf{r}(t') \,, \tag{5-20}$$

where now $\mathbf{n} = \mathbf{x}/x \simeq \mathbf{R}/R$. Hence, to a very good approximation the unit vector $\mathbf{n}$ is directed from the origin toward the observer and is constant in time, as is $\mathbf{x}$. One now writes

$$\mathbf{F}(\omega) = \left(\frac{e^2}{8\pi^2 c}\right)^{1/2} e^{i\omega x} \int_{-\infty}^{\infty} \exp\left[i\omega\left(t' - \mathbf{n} \cdot \frac{\mathbf{r}}{c}\right)\right] \times \frac{\mathbf{n} \times [(\mathbf{n} - \boldsymbol{\beta}) \times \dot{\boldsymbol{\beta}}]}{K^2}\, dt' \,. \tag{5-21}$$

We can now integrate once by parts, in order to remove the explicit appearance of the acceleration, and substitute the resulting expression for $\mathbf{F}(\omega)$ into Eq. (5-17) to obtain the desired distribution*:

$$\frac{dI(\omega)}{d\Omega} = \frac{e^2\omega^2}{4\pi^2 c} \left| \int_{-\infty}^{\infty} \mathbf{n} \times (\mathbf{n} \times \boldsymbol{\beta}) \exp\left[i\omega\left(t - \mathbf{n} \cdot \frac{\mathbf{r}}{c}\right)\right] dt \right|^2, \tag{5-22}$$

where the primes on the time variables have been omitted for convenience. This is a very general expression for the intensity of radiation per unit solid angle per unit frequency interval, and can now be employed to study several physical situations of interest.

Exercise. Verify Eq. (5-22) explicitly.

Let us first apply Eq. (5-22) to a study of the radiation from a very-high-velocity charged particle undergoing arbitrary acceleration. As demonstrated in the preceding section, the dominant contribution to this radiation comes from the transverse component of acceleration. Therefore, the observer will see a short, intense pulse of radiation which is in all

* Although the acceleration no longer appears explicitly in this result, one cannot conclude that there is no radiation. The dependence of $dI/d\Omega$ on the particle acceleration is implicit in the factor of ω^2 on the right-hand side of Eq. (5-22).

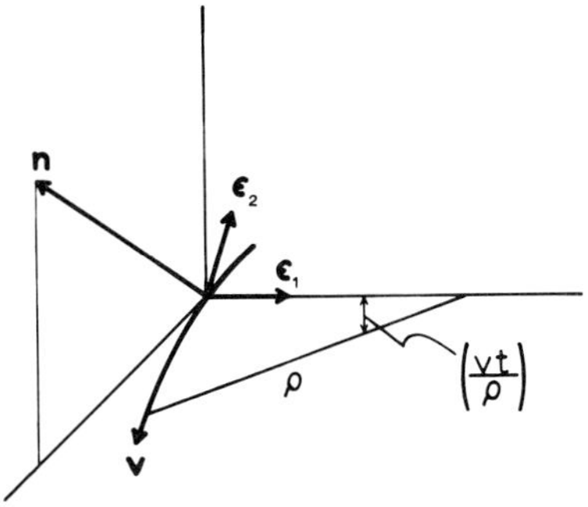

Fig. 14. Representation of the vector product in Eq. (5-22). Significant radiation occurs for very fast particles only when the angle between **n** and **v** is very small, and then the polarization vectors are almost perpendicular.

respects equivalent to that from a charge in instantaneous circular motion. [Remember that the time variable of integration in Eq. (5-22) is the proper time of the particle.]

Figure 14 exhibits the coordinate orientation of the small arc of particle trajectory to be considered, and, in keeping with the observations of the last section, significant amounts of radiation will be associated only with small angles when the particle is very relativistic. Let ρ be the instantaneous radius of curvature and take $t = 0$ when the particle is at the origin of the indicated coordinate system. Furthermore, the polarization of electromagnetic radiation will be introduced explicitly by choosing ϵ_1 to represent polarization in the plane of the orbit and $\epsilon_2 = \mathbf{n} \times \epsilon_1$ to represent polarization perpendicular to this plane (when θ is very small).

With these restrictions of very short radiation time and $\mathbf{r}(t)$ remaining close to the origin, we can approximate the vector product and phase in the integrand of Eq. (5-22). Expanding the sine and cosine (see Fig. 14), one finds

$$\begin{aligned} \omega\left(t - \frac{\mathbf{n}\cdot\mathbf{r}}{c}\right) &\simeq \omega\left[t - \frac{\rho}{c}\sin\left(\frac{vt}{\rho}\right)\cos\theta\right] \\ &\simeq \omega\left[t(1-\beta\cos\theta) + \frac{v^3t^3}{6\rho^2c}\cos\theta\right] \\ &\simeq \frac{\omega}{2}\left[(\gamma^{-2}+\theta^2)t + \frac{c^2}{3\rho^2}t^3\right], \end{aligned} \tag{5-23}$$

where terms of order $(t^3\theta^2)$ have been neglected.

Exercise. Verify the approximation

$$1 - \beta\cos\theta \simeq \tfrac{1}{2}(\gamma^{-2} + \theta^2)\,, \qquad \text{for} \quad \beta \sim 1\,, \quad \theta \ll 1\,. \tag{5-24}$$

We again refer to Fig. 14, and find that the vector product in Eq.

(5-22) is approximately

$$\mathbf{n} \times (\mathbf{n} \times \boldsymbol{\beta}) \simeq \beta \left| -\epsilon_1 \sin\left(\frac{vt}{\rho}\right) + \epsilon_2 \cos\left(\frac{vt}{\rho}\right) \sin\theta \right|$$

$$\simeq -\frac{ct}{\rho}\epsilon_1 + \theta\epsilon_2 . \tag{5-25}$$

One now substitutes these approximations into Eq. (5-22) to obtain for the intensity of radiation per unit solid angle per unit frequency interval

$$\frac{dI(\omega)}{d\Omega} \simeq \frac{e^2\omega^2}{4\pi^2 c} \left| -\epsilon_1 F_1(\omega) + \epsilon_2 F_2(\omega) \right|^2 , \tag{5-26}$$

where

$$F_1(\omega) = (c/\rho) \int t e^{i\phi}\, dt , \tag{5-27a}$$

$$F_2(\omega) = \theta \int e^{i\phi}\, dt , \tag{5-27b}$$

and

$$\phi = \tfrac{1}{2}\omega[(\gamma^{-2} + \theta^2)t + (c^2t^3/3\rho^2)] . \tag{5-28}$$

The limits of integration in Eqs. (5-27) have purposely been omitted because the physical assumption of a very short radiation time implies some uncertainty as to the values of these limits.

The integrals in Eqs. (5-27) have the general form

$$\int_a^b g(t) e^{i\omega f(t)}\, dt ,$$

where g and f are real functions. Such integrals were first studied by Stokes[8] for large values of ω, and Kelvin[9] later formulated their behavior explicitly for adaptation to physical problems. The essential point is that for large values of ω the integrand oscillates very rapidly and the dominant contributions to the integral come from neighborhoods about points for which $f(t)$ has an extremum, or is stationary. Thus, this behavior has come to be known as Kelvin's *principle of stationary phase*, and a formal mathematical proof of the principle was first given by Watson.[10] The reader is referred to the papers by Copson[11] and Erdélyi[12] for further details of the analysis.

In applying the principle of stationary phase to the integrals of Eqs. (5-27), we first note that the frequency distribution for ultrarelativistic charged particles will be appreciable only for high frequencies. One verifies this from the general relation $\Delta\omega \sim 1/\Delta t$ obtained for Fourier transform pairs, and in the present situation only times close to $t = 0$ are pertinent. Since ω is then large, the dominant contribution arises from those times for which ϕ is stationary. From Eq. (5-28) we see that the requirement

$\beta \sim 1$ implies that these points are very close to $t = 0$. Hence, the limits of integration in Eqs. (5-27) really have no influence on the value of the integral and we can extend the latter from $(-\infty)$ to $(+\infty)$ with small fear of error.

Now change the integration variable to

$$x = (ct/\rho)(\gamma^{-2} + \theta^2)^{1/2},$$

and set

$$\alpha = (\omega\rho/3c)(\gamma^{-2} + \theta^2)^{3/2}. \tag{5-29}$$

Upon making these substitutions and exploiting the symmetry of the integrands in Eqs. (5-27), we find that

$$F_1(\omega) \simeq 2i\frac{\rho}{c}(\gamma^{-2} + \theta^2)\int_0^\infty x \sin\left[\frac{3}{2}\alpha\left(x + \frac{1}{3}x^3\right)\right] dx, \tag{5-30a}$$

$$F_2(\omega) \simeq 2\frac{\rho}{c}(\gamma^{-2} + \theta^2)^{1/2}\int_0^\infty \cos\left[\frac{3}{2}\alpha\left(x + \frac{1}{3}x^3\right)\right] dx. \tag{5-30b}$$

These integrals are recognized as representations of the Airy functions, which in turn can be related to modified Bessel functions of fractional order. The needed relations are[13]

$$Ai[\pm(3a)^{1/3}x] = \frac{(3a)^{1/3}}{\pi}\int_0^\infty \cos(at^3 \pm xt)\, dt, \tag{5-31a}$$

$$Ai(z) = \frac{1}{\pi}\left(\frac{z}{3}\right)^{1/2} K_{1/3}\left(\frac{2}{3}z^{3/2}\right), \tag{5-31b}$$

$$A'i(z) = \frac{1}{\pi}\left(\frac{z}{3}\right)^{1/2} K_{2/3}\left(\frac{2}{3}z^{3/2}\right), \tag{5-31c}$$

where the last expression involving the derivative of Ai with respect to its argument is needed to write F_1 in terms of F_2.

Referring again to Fig. 14, one sees that the cross term in Eq. (5-26) will vanish for ultrarelativistic particles, because in that region ϵ_1 is very closely orthogonal to ϵ_2. Then, using Eqs. (5-30) and (5-31) in Eq. (5-22), it is now an easy matter to write for the intensity of radiation per unit frequency interval per unit solid angle

$$\frac{dI(\omega)}{d\Omega} \simeq \frac{e^2}{3\pi^2 c}\left(\frac{\omega\rho}{c}\right)^2(\gamma^{-2} + \theta^2)^2\left[K_{2/3}^2(\alpha) + \frac{\theta^2}{\gamma^{-2} + \theta^2}K_{1/3}^2(\alpha)\right], \tag{5-32}$$

where the first term corresponds to polarization in the plane of the orbit and the second to polarization perpendicular to that plane. This equation was first obtained by Schwinger.[6]

Exercise. Verify the algebra leading to Eq. (5-32).

The expression of Eq. (5-32) is a fairly complicated result and, therefore, it is necessary to analyze and approximate the right-hand side with some care in order to extract its physical information. The three measurable quantities which can be obtained from this equation are the angular distribution of the total radiated energy, the frequency distribution of the total radiated energy, and the total radiated energy itself. Thus, the first quantity can be calculated by substituting into Eq. (5-16) from Eq. (5-32). The necessary integral is given in most integral tables, but its region of validity has been corrected by Armstrong,[14] who finds that

$$\int_0^\infty K_\nu^2(t)t^n\,dt = \frac{2^{n-2}}{\Gamma(n+1)}\Gamma\left(\frac{1}{2}+\nu+\frac{n}{2}\right)\Gamma\left(\frac{n+1}{2}-\nu\right)\Gamma^2\left(\frac{n+1}{2}\right),$$
$$\text{for} \quad \mathrm{Re}\,(n \pm 2\nu + 1) > 0\,. \tag{5-33}$$

Applying this formula, along with the analytic continuations $z\Gamma(z) = \Gamma(z+1)$ and $\Gamma(z)\Gamma(1-z) = \pi/\sin(\pi z)$, we find for the integral of interest

$$\frac{dW}{d\Omega} = \int_0^\infty \frac{dI(\omega)}{d\Omega}\,d\omega$$
$$\simeq \frac{7}{16}\frac{e^2}{\rho}(\gamma^{-2}+\theta^2)^{-5/2}\left[1+\frac{5}{7}\frac{\theta^2}{\gamma^{-2}+\theta^2}\right], \tag{5-34}$$

again remembering that the first term corresponds to polarization in the plane of the orbit.

It is instructive at this point to make some observations about this angular distribution. First, it reflects our assumptions about small angles being important, in that the region $\theta \ll |\pi/2|$ dominates the distribution. Second, one can now integrate over all angles and, since the dominant contribution to the integral comes from the region around $\theta \sim 0$, the limits can be taken as $(-\pi/2, \pi/2)$. Remembering that θ is the colatitude, we find that seven times more energy is radiated with polarization in the plane of the orbit than with polarization perpendicular to that plane.

Exercise. Verify this last statement.

Let us now return to Eq. (5-32), and note that the asymptotic behavior of the modified Bessel functions (for $\nu \neq 0$) is[15]

$$K_\nu(z) \xrightarrow[z\to 0]{} \tfrac{1}{2}\Gamma(\nu)(z/2)^{-\nu} \tag{5-35a}$$

$$\xrightarrow[z\to\infty]{} (\pi/2z)^{1/2}e^{-z}\,. \tag{5-35b}$$

Therefore, for large values of α, Eq. (5-32) indicates that the radiation is negligible. At a fixed frequency, Eq. (5-29) implies that this will occur for large angles, verifying the observation made above. From the same equation it can be concluded that, as the frequency is increased, the angular spread in which significant radiation will be detected becomes smaller. Thus, there will exist a critical frequency beyond which there is negligible radiation, even at $\theta = 0$, and this can be defined by setting $\alpha = 1$ at $\theta = 0$, the point at which the intensity has dropped off by $1/e^2$. From Eq. (5-29),

$$\omega_c = 3\gamma^3(c/\rho)\,. \tag{5-36}$$

In order to render these qualitative remarks more precise, we now calculate the frequency distribution by integrating Eq. (5-32) over all angles. Because the significant radiation is confined to a neighborhood $\theta \sim 0$, the integration over the colatitude θ can again be extended to $(-\pi/2, \pi/2)$ and the factor $\cos\theta$ appearing in the surface element taken to be unity. In fact, these limits can actually be extended to $(-\infty, \infty)$, since their actual values are irrelevant. Then, we write

$$\begin{aligned} I(\omega) &\simeq 2\pi \int_{-\infty}^{\infty} \frac{dI(\omega)}{d\Omega}\, d\theta \\ &= \frac{4e^2}{3\pi c}\left(\frac{\omega\rho}{c}\right)^2 \left\{\int_0^\infty (\gamma^{-2} + \theta^2)^2 K_{2/3}^2[a(\gamma^{-2} + \theta^2)^{3/2}]\, d\theta \right. \\ &\quad \left. + \int_0^\infty \theta^2(\gamma^{-2} + \theta^2) K_{1/3}^2[a(\gamma^{-2} + \theta^2)^{3/2}]\, d\theta \right\}, \end{aligned} \tag{5-37}$$

where $a = \omega\rho/3c$. Now make the following successive changes of variable: $\theta^2 = x$, $t = \gamma^{-2} + x$, $z = at^{3/2}$, so that Eq. (5-37) becomes

$$I(\omega) \simeq \frac{4e^2}{\pi c}\, \gamma \left(\frac{\omega}{\omega_c}\right)^{2/3} J\left(\frac{\omega}{\omega_c}\right), \tag{5-38}$$

where

$$J(x) = \int_x^\infty z^{1/3}[(z/x)^{2/3} - 1]^{1/2} \left[\frac{(z/x)^{2/3}}{(z/x)^{2/3} - 1} K_{2/3}^2(z) + K_{1/3}^2(z)\right] dz\,. \tag{5-39}$$

This complicated integral cannot be evaluated much further analytically unless some assumptions are made regarding the values of $x = \omega/\omega_c$ [although see Eq. (5-44) below]. In order to get a qualitative idea of the frequency spectrum, we can evaluate $J(x)$ in the limits of large and small x. In the latter case it is an easy matter to show that

$$\begin{aligned} J(x) &\underset{x\to 0}{\longrightarrow} x^{-1/3} \int_0^\infty z^{2/3}[K_{2/3}^2(z) + K_{1/3}^2(z)]\, dz \\ &= \tfrac{1}{2}\pi\sqrt{3}\,\Gamma(2/3) x^{-1/3}, \end{aligned} \tag{5-40}$$

and therefore

$$I(\omega) \underset{\omega/\omega_c \ll 1}{\longrightarrow} 2\sqrt{3}\,\Gamma(2/3)\gamma(e^2/c)(\omega/\omega_c)^{1/3}\,. \tag{5-41}$$

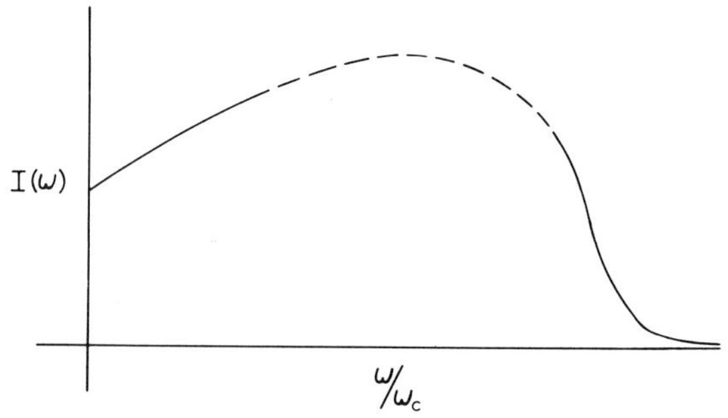

Fig. 15. Frequency spectrum of synchrotron radiation described by Eq. (5-37). The solid portions of the curve refer to the asymptotic forms (5-41) and (5-43).

Exercise. Evaluate the integrals in Eq. (5-40) by means of Eq. (5-33).

In the opposite limit we change the integration variable to $t = z - x$ and employ the asymptotic form (5-35b) to write

$$\begin{aligned} J(x) \underset{x\to\infty}{\longrightarrow} & \frac{\pi}{2} e^{-2x} \int_0^\infty e^{-2t} \frac{2(1 + t/x)^{2/3} - 1}{(t + x)^{2/3}[(1 + t/x)^{2/3} - 1]^{1/2}} \, dx \\ & \sim \frac{\pi}{2} e^{-2x} \left(\frac{3}{2}\right)^{1/2} x^{-1/6} \int_0^\infty \frac{e^{-2t}}{t^{1/2}} \, dt \\ & = \frac{\pi}{2} \left(\frac{3\pi}{2}\right)^{1/2} x^{-1/6} e^{-2x} , \end{aligned} \qquad (5\text{-}42)$$

where the exponential cutoff has been used to justify the expansion of the integrand in going from the first line to the second. Thus,

$$I(\omega) \underset{\omega/\omega_c \gg 1}{\longrightarrow} (3\pi)^{1/2} \frac{e^2}{c} \gamma \left(\frac{\omega}{\omega_c}\right)^{1/2} \exp\left(-\frac{2\omega}{\omega_c}\right). \qquad (5\text{-}43)$$

In Fig. 15 the intensity per unit frequency interval is plotted as a function of ω/ω_c, and one observes that the behavior at both ends of the spectrum is that given by Eqs. (5-41) and (5-43). The entire curve has been constructed by actually performing a numerical integration of the integral $J(x)$. For this purpose it is more convenient to use the alternative form of $J(x)$ developed by Schwinger[6]:

$$J(x) = \frac{\sqrt{3}}{8} x^{1/3} \int_x^\infty K_{5/3}(z) \, dz . \qquad (5\text{-}44)$$

The radiation described by Eq. (5-32) and Fig. 15 is called *synchrotron radiation*, since it was first observed in an electron synchrotron.[16] In the visible region it is bluish white in color, and for an 80-MeV machine ω_c is of the order $10^{16}\,\text{sec}^{-1}$. As mentioned previously, the details as presented here were worked out by Schwinger for synchrotrons,[6] but the fundamental ideas were first given by Schott[17] long ago. Explicit comparisons with the theory have been made from synchrotron measurements by Elder *et al.*,[18] and by Tomboulain and Hartman,[19] and the agreement is excellent.

As is well known, a charged particle undergoes transverse accelerations during motion in a magnetic field, and Shklovsky[20] has proposed that the ensuing synchrotron radiation may be responsible for much of the optical and radio emission from the galaxy and elsewhere. In particular, the light from the Crab Nebula seems to be strongly polarized, and analyses by Dombrovskii[21] and Oort and Walraven[22] suggest that the synchrotron mechanism discussed in this chapter is the dominant process contributing to the continuous spectrum. The latter authors have applied the above formulas to the radiation data from the Crab Nebula and find that it corresponds to electrons with a median energy of about 2×10^{11} eV moving in a magnetic field strength of approximately 10^{-3} G. The origin of the **B** field and energy source of the elecrons is still not clear, for the observed kinetic energy of the Crab Nebula is apparently insufficient to provide the necessary energy.

In another area of interest in astrophysics, Baldwin[23] has given evidence for a homogeneous galactic corona of radiation at a 3.7-m wavelength. Since the possibility of attributing this corona to discrete sources is ruled out by other observations, Shklovsky[24] has also suggested the above mechanism to be responsible in this case. Spitzer[25] argues that the corona is probably due to a gas at $T \approx 10^{6}\,{}^\circ\text{K}$ in which there is a small number of radiation-producing relativistic electrons. Using Baldwin's data, one can construct a model with $B \approx 10^{-6}$ gauss and $\omega_c \approx 8 \times 10^{7}\,\text{sec}^{-1}$, which leads to electron energies on the order of 2.2×10^{9} eV, and densities of relativistic electrons on the order of $4 \times 10^{-12}/\text{cm}^3$. Again, the nature of the electron energy source is in doubt.

Gardner and Shain[26] have observed a continuous emission from the planet Jupiter which is strongly polarized and corresponds to a critical frequency of about $10^{7}\,\text{sec}^{-1}$. It is quite possible that this is due to synchrotron radiation from an ionized layer high above the planet, but the evidence is not really clear.

The indication that synchrotron radiation is a strongly contributing factor to the radiation phenomena in the universe is fairly convincing. As the techniques of radio astronomy improve, the fundamental radiation processes should continue to be an excellent tool for understanding many of the physical phenomena in the universe. A particularly noteworty example is the recent discovery of pulsars and current attempts to discern their structure.

C. ČERENKOV RADIATION

In 1937 Čerenkov[27] observed blue light being radiated from charged particles moving with high *uniform velocity* through a material medium. This is, at first thought, rather perplexing, for the criterion governing radiation from a charged particle requires the particle to have a nonzero acceleration [see the discussion following Eq. (5-5)]. However, the phenomenon becomes less mysterious when one notes that the particle velocity can actually exceed the *phase velocity* of light in a material medium.

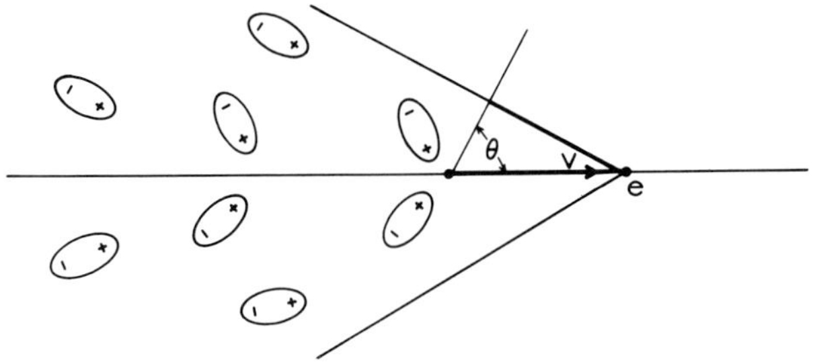

Fig. 16. Polarization of the medium by a very high-speed charged particle, leading to Čerenkov radiation.

Let us recall the second term of Eq. (4-63), which implies that the radiation field is zero if $\dot{\beta} = 0$. If, however, the denominator factor also vanishes, then this conclusion does not necessarily follow. Denoting the index of refraction of the medium by n, we can write

$$K = 1 - (vn/c)\cos\theta\,, \tag{5-45}$$

so that it is possible to have $vn > c$. The behavior of the field is not clear in a cone defined by

$$\cos\theta = c/nv\,, \tag{5-46}$$

known as the Čerenkov condition. Thus, when this condition is satisfied one might expect a physical disturbance to accompany the particle, in the form of an electromagnetic shock wave.

Physically, we have seen that the observed radiation cannot come from the particle, and so it must be due to the medium itself. As the particle moves through the material at high speed, it polarizes the nearby atoms, and, as long as $vn < c$, the resulting polarization field is essentially symmetric about the charge. On the other hand, if $vn > c$, the field is symmetric only about the path of the charge, because the disturbance cannot propagate forward fast enough (see Fig. 16). Consequently, there is a total pulsating dipole field in the wake of the particle, thereby setting up a radiation field. Therefore, it is actually the medium, excited in an asymmetric way by the passage of the charge, which produces the Čerenkov radiation.

Quantitatively, the radiation can be described by calculating the fields due to the current represented by the moving charge, or the expression (5-22) can be modified to give the frequency and angular distribution. The first method was used by Frank and Tamm[28] in developing the original theory of Čerenkov radiation, but we shall find it more convenient to pursue the second course. Hence, in Eq. (5-22) we make the replacements

$$c \rightarrow c/n\,, \qquad e \rightarrow e/n\,, \tag{5-47}$$

the first being obvious, and the second because the displacement vector **D** must be used in the medium, rather than **E**. Then,

$$\frac{dI(\omega)}{d\Omega} = \frac{e^2\omega^2 n}{4\pi^2 c^3}\left|\int_{-\infty}^{\infty} \mathbf{z} \times (\mathbf{z} \times \mathbf{v}) \exp\left[i\omega\left(t - \frac{\mathbf{z}\cdot\mathbf{r}}{c/n}\right)\right] dt\right|^2, \tag{5-48}$$

where the unit vector locating the observer with respect to the particle has been denoted by **z** to avoid confusion with the index of refraction. For uniform motion in a straight line we can set $\mathbf{r} = \mathbf{v}t$. Furthermore, the integrand in Eq. (5-48) should be defined to be zero outside of some time interval $2T$, since the particle only moves through the medium for a finite time. Although we have used similar arguments in the past to extend such integration limits to infinity, this procedure is to be avoided in the present situation in order to obtain a finite result. One readily finds that

$$\frac{dI(\omega)}{d\Omega} = \frac{e^2\beta^2}{c/n}\sin^2\theta\left|\frac{\omega T}{\pi}\frac{\sin[\omega T(1 - n\beta\cos\theta)]}{\omega T(1 - n\beta\cos\theta)}\right|^2. \tag{5-49}$$

The frequency distribution is now calculated by integrating over all angles:

$$I(\omega) = \frac{e^2\beta^2}{c/n}\left(\frac{\omega T}{\pi}\right)^2 2\pi \int_{-1}^{+1} \frac{\sin^2[\omega T(1 - n\beta s)]}{[\omega T(1 - n\beta s)]^2}(1 - s^2)\, ds\,, \tag{5-50}$$

where we have set $s = \cos\theta$. For large ω the integrand behaves like a δ function times $(1 - s^2)$. The observation that Čerenkov radiation is in the blue end of the spectrum, and that the particles have very high velocities, indicates that the physical region of interest is that of very large ω. Hence,

$$I(\omega) \simeq (2vT)\frac{e^2\omega}{c^2}\left[1 - \frac{c^2}{v^2 n^2}\right]. \tag{5-51}$$

The measurable quantity of interest is taken to be the energy radiated per unit frequency interval per unit length of path. Hence,

$$\frac{dI(\omega)}{dx} \simeq \frac{e^2}{c^2}\left[1 - \frac{c^2}{v^2 n^2}\right]\omega , \qquad (5\text{-}52)$$

which is a positive quantity according to Eq. (5-46).

It is seen that the radiated energy increases monotonically with frequency, verifying our assumption that the high-frequency radiation is dominant. If one now attempts to integrate over all frequencies to obtain an expression for the total energy radiated per unit path length, it is evident that the integral diverges. This is only an apparent divergence, however, because the dispersive nature of the medium, reflected in the frequency dependence of the index of refraction, will cut off the spectrum, prohibiting propagation beyond some critical frequency. Thus, one must actually know the dispersion relation for the particular medium in order to calculate the total energy radiated.

One cannot infer from the above discussion that Čerenkov radiation is confined to high frequencies, although, according to Eq. (5-52), the intensity will fall off rapidly at long wavelengths. In fact, the radiation has been observed and identified unambiguously at microwave frequencies,[29] and considerable ingenuity has gone into the construction of generating devices in this region.

Čerenkov radiation has probably proved most valuable as a tool in high-energy physics. Čerenkov counters are employed as instruments for measuring particle velocities and as velocity selectors for eliminating unwanted slow particles. For a more detailed discussion of Čerenkov radiation, the reader is referred to the comprehensive work by Jelley.[30] An excellent survey of Čerenkov counters and their use in detecting relativistic particles has been presented by Yuan.[31]

D. TRANSITION RADIATION

The outstanding physical feature of the Čerenkov effect is the conclusion that electromagnetic radiation can indeed be associated with a charged particle in uniform motion, under certain conditions. As we have seen, polarization of a medium by a high-velocity charged particle is an essential aspect of this radiation phenomenon, and it seems worthwhile to explore further the possible existence of similar effects which also may depend strongly on the optical properties of the medium. Frank and Ginzberg[32] investigated this idea further and discovered that radiation should also be emitted at the interface of two different media when traversed by a uniformly moving charged particle. They named the effect *transition radiation,* and it is the purpose of this section to discuss very briefly the most significant characteristics of this phenomenon.

Consider a charged particle in vacuum approaching the ideally sharp surface of another medium with uniform velocity. For simplicity, assume the particle trajectory to be normal to the surface, let $v \ll c$, and suppose the particle to enter the medium without any change in its velocity. We can now imagine the particle to be stopped suddenly at the boundary, and then accelerated just as suddenly to resume its path through the medium at the same velocity. If we pursue this *gedanken* experiment, it becomes necessary to calculate the radiation emitted when a particle is suddenly stopped, which follows easily from substitution of Eq. (5-21) into (5-17):

$$\frac{dI(\omega)}{d\Omega} = \frac{e^2}{4\pi^2 c}\left|\int_{-\infty}^{\infty} \frac{\exp[i\omega(t' - \mathbf{n}\cdot\mathbf{r}/c)]}{(1 - \mathbf{n}\cdot\boldsymbol{\beta})^2}\, \mathbf{n} \times [(\mathbf{n} - \boldsymbol{\beta}) \times \dot{\boldsymbol{\beta}}]\, dt'\right|^2. \tag{5-53}$$

An approximate evaluation of this quantity begins with the nonrelativistic condition $|\boldsymbol{\beta}| \ll 1$, and the observation that the acceleration is antiparallel to the velocity in this problem. Let $\langle \mathbf{v} \rangle$ be the average velocity of the particle during the very short period of deceleration, τ. Then $\mathbf{r} \approx \langle \mathbf{v} \rangle \tau$ and the second terms in both the exponential and denominator can be neglected as $O(v/c)$. Thus,

$$\frac{dI(\omega)}{d\Omega} \simeq \frac{e^2}{4\pi^2 c}\left|\int \mathbf{n} \times (\mathbf{n} \times \dot{\boldsymbol{\beta}}) e^{i\omega t'}\, dt'\right|^2. \tag{5-54}$$

Now recall the previous discussion of the principle of stationary phase, which implies that the exponential is essentially unity for $\omega\tau \ll 1$, and oscillates very rapidly otherwise, leading to a zero contribution at high frequencies due to destructive interference. Except for an angular factor, the integrand reduces to $\dot{\boldsymbol{\beta}}(t')$, and the integral therefore is just the initial velocity, up to a sign factor. Consequently, one verifies that the frequency spectrum becomes

$$\frac{dI(\omega)}{d\Omega} \simeq \frac{e^2 v^2}{4\pi^2 c^3} \sin^2\theta \tag{5-55}$$

for low frequencies, and is cut off at $\omega \sim \tau^{-1}$ at the high end. The angle of observation θ is measured from the normal to the surface, so that the angular distribution is similar to that of Fig. 10. Note that Eq. (5-55) is the same as Eq. (5-3) if, in the latter, $\dot{v}^2$ is replaced by v^2/π.

We can now return to our original problem and note that stopping the particle suddenly at the boundary will create a radiation field leading to Eq. (5-55). Sudden acceleration of the particle in the medium to the initial velocity will then produce a field of the *same* intensity but with *opposite* sign. The two fields together represent that of a charged particle in uniform motion. The total field is obtained by adding the two amplitudes, and, if the medium has the constitutive properties of the vacuum, the amplitudes cancel and we have the familiar result that a particle moving uniformly in vacuum does not radiate. Note that to be perfectly correct we must consider the stop and start of the particle to be infinitely rapid.

Suppose, however, that the second medium differs from the vacuum and is described by an index of refraction n. The cancellation of field amplitudes will then be incomplete and one can expect to observe electromagnetic radiation. Moreover, in order to properly account for the total field, one must introduce the electrical image of the incident particle, moving from the medium toward the vacuum and stopping suddenly at the surface. We then obtain for the spectral density of the radiation per unit solid angle in vacuum[32]

$$\frac{dI(\omega)}{d\Omega} = \frac{e^2v^2}{4\pi^2c^3} \sin^2\theta \,|\, 1 + r - f/n \,|^2 , \tag{5-56}$$

where r and f are the Fresnel coefficients for the reflected and refracted waves, respectively. From classical electromagnetic theory[33] we have the relations

$$r = \frac{n^2 \cos\theta - [n^2 - \sin^2\theta]^{1/2}}{n^2 \cos\theta + [n^2 - \sin^2\theta]^{1/2}} , \tag{5-57a}$$

$$f = \frac{2n\cos\theta}{n^2 \cos\theta + [n^2 - \sin^2\theta]^{1/2}} , \tag{5-57b}$$

$$1 + r = fn . \tag{5-58}$$

The prediction of transition radiation, Eq. (5-56), contains several physical features which are distinctly different from Čerenkov radiation. For instance, it is to be observed that there is no threshold velocity for emission of transition radiation, so that one expects to observe the effect at any speed. Nevertheless, at low speeds the radiation is generally masked by *Bremsstrahlung* (see the following chapter), particularly for electrons, and at high enough speeds Čerenkov radiation is a much larger effect when the transition is between two similar media. (In this respect, one should notice that passage of a charged particle at a uniform high velocity through a thin slab will be described by precisely the same equations as for a particle accelerated rapidly to the same speed within the medium, emitting Čerenkov radiation over a distance equal to the thickness of the slab, and then being decelerated rapidly, still within the medium.) On the other hand, an important characteristic of transition radiation is that it is expected to be completely polarized with the electric vector in the plane defined by the normal to the surface and the incident beam, as can be verified from the discussion preceding Eq. (5-3). This polarization is helpful in experimentally distinguishing this radiation from other types, and is some compensation for the masking problem noted above.

When the medium is actually the vacuum ($n^2 \to 1$), Eq. (5-58) implies that the radiation intensity vanishes, whereas in the limit of a perfect conductor ($n^2 \to \infty$), Eqs. (5-57) show that Eq. (5-56) becomes

$$\frac{dI(\omega)}{d\Omega} = \frac{e^2 v^2}{\pi^2 c^3} \sin^2\theta\,. \tag{5-59}$$

Note that this is the same as the radiation emitted when two opposite charges e meet with relative velocity v (see Problem 5-6). This suggests the following physical interpretation of transition radiation when a charge approaches a metal conductor in a direction normal to the surface: the image charge within the conductor simultaneously approaches the surface with velocity $-\mathbf{v}$ and the two annihilate, emitting radiation. What one actually seems to observe is the disappearance of the charge within the metal.

The angular distribution when the medium is an ideal conductor is given by Eq. (5-59) and Fig. 10, so that in this case the maximum intensity occurs at $\theta = \pi/2$. In a real medium the index of refraction affects the distribution considerably, and the maximum for metals such as silver, tungsten, aluminum, and nickel occurs very close to $\theta = \pi/3$. The angular distribution is tilted forward much like that of Fig. 11. Finite indices of refraction also control the frequency dependence of the transition radiation, so that Eq. (5-56) does not lead to an infinite result when integrated over all frequencies.

Equation (5-56) was derived on the basis of the simplest possible assumptions, and it can be generalized in several ways. For instance, Pafomov[34] has shown that this result also holds when the particle is obliquely incident upon the surface if one merely replaces v by the component of v normal to the surface. This simple substitution is only valid nonrelativistically and cannot be made in the relativistic generalization of (5-56):

$$\frac{dI(\omega)}{d\Omega} = \frac{e^2 v^2}{4\pi^2 c^3} \sin^2\theta \left| \frac{1}{1+\beta\cos\theta} + \frac{r}{1-\beta\cos\theta} - \frac{f}{n^2}\frac{1}{1+n\beta\cos\phi} \right|^2, \tag{5-60}$$

where

$$n\cos\phi = [n^2 - \sin^2\theta]^{1/2}\,. \tag{5-61}$$

For oblique incidence the expression becomes much more complicated.[34]

The relativistic formula is interesting in several respects, particularly in connection with the possible experimental observation of transition radiation. First, the presence of β introduces a directivity not seen nonrelativistically, in that β changes sign when the particle moves from the medium into the vacuum. In the latter case we note that the third term of Eq. (5-60) also includes Čerenkov radiation when n is greater than unity. Also, in this case ($\beta \rightarrow -\beta$), the first term in Eq. (5-60) completely dominates the other two when: (1) θ is very small, (2) β approaches unity, (3) $n \neq 1$, and (4) ϕ differs considerably from the Čerenkov angle. Then the frequency spectrum intergrated over all angles increases logarithmically as the particle energy, and is independent of the refractive index. Consequently, one can sum the radiation from many successive surfaces and therefore enhance the efficiency of particle detection. Similar considerations suggest that transition radiation can also be extended into the X-ray and γ-ray regions for high-energy particles. Thus, it is quite possible that effective detectors for very-high-energy particles utilizing transition radiation will become feasible.[35]

Unambiguous detection of transition radiation has only recently become practicable. Tanaka and Katayama[36] have observed the polarized visible radiation from 6–19-keV electrons on NiO, and the angular distribution agrees almost perfectly with the theory. Oostens *et al.*[37] have also made observations in the GeV range, and have observed the logarithmic increase of the radiation with particle energy. Recently, the same group has detected transition radiation in the X-ray region,[38] lending further support to the possibility of using the phenomenon in high-energy particle detection. A more detailed discussion of the experimental situation and the general theory can be found in the review article by Frank.[39]

PROBLEMS

5-1. Consider a Lorentz frame in which the angular distribution of radiation is $I(\theta)$. In a frame moving with velocity $\mathbf{v}$ with respect to the first frame and along their common x axis (see Fig. 1) the intensity is $\bar{I}(\bar{\theta})$. Show that the two distributions are related by

$$I(\theta) = \gamma^2(1 + \beta \cos \bar{\theta})^2 \bar{I}(\bar{\theta}) .$$

5-2. Demonstrate that the invariant instantaneous rate of radiation from an arbitrarily accelerated relativistic particle can be written in the from

$$\begin{aligned} W(t') &= -\tfrac{2}{3}(e^2/c^3)a^\mu a_\mu \\ &= \tfrac{2}{3}(e^2/c^3)\gamma^4[\mathbf{a}^2 + \gamma^2(\boldsymbol{\beta}\cdot\mathbf{a})^2] , \end{aligned} \tag{5-62}$$

where a^μ is defined in Eq. (3-35). Show that this result is equivalent to that of Eq. (5-5).

5-3. Verify Eqs. (5-8) and (5-10) by performing the indicated angular integrations in Eqs. (5-7) and (5-9), respectively.

5-4. Consider a relativistic electron moving in a constant, uniform magnetic field **B**.

(a) Calculate the total instantaneous power radiated by the electron.
(b) Suppose the particle is trapped in the magnetic dipole field of the earth, and is spiraling back and forth along a magnetic field line. Decide whether the particle radiates more energy near the equator or near the poles, being as quantitative as possible. Discuss a possible experimental observation of your conclusions.

5-5. An electron enters a gas with a very high, but constant velocity. It is found that the medium has an index of refraction

$$n = 1 + \tfrac{1}{2}a^2/(\omega_0{}^2 - \omega^2),$$

where a and ω_0 are characteristic of the gas. What is the total energy radiated per unit length of path?

5-6. Consider an electron and a proton approaching each other at speed $v \ll c$. Show that the ensuing radiation is described by Eq. (5-59).

REFERENCES

1. See, e.g., W. Heitler, "The Quantum Theory of Radiation," p. 17. Oxford Univ. Press, London and New York, 1936.
2. T. Fulton and F. Rohrlich, *Ann. Phys.* (*N. Y.*) **9,** 499 (1960).
3. J. Larmor, *Phil. Mag* [5] **44,** 503 (1897).
4. A. Liénard, *Eclairage Elec.* **16,** 5, 53, 106 (1898).
5. R. A. Mould, *Ann. Phys.* (*N. Y.*) **27,** 1 (1964).
6. J. Schwinger, *Phys. Rev.* **75,** 1912 (1949).
7. P. M. Morse and H. Feshbach, "Methods of Theoretical Physics," Chapter 4. McGraw-Hill, New York, 1953.
8. G. G. Stokes, *Cambridge Phil. Trans.* **9,** 166 (1856).
9. Lord Kelvin, *Phil. Mag.* [5] **23,** 252 (1887).
10. G. N. Watson, *Proc. Cambridge Phil. Soc.* **19,** 49 (1918).
11. E. T. Copson, "The Asymptotic Expansion of a Function Defined By a Definite Integral or a Contour Integral." Admirality Comput. Serv., London, 1946.
12. A. Erdélyi, *J. Soc. Ind. Appl. Math.* **3,** 17, (1955).
13. M. Abramowitz and I. Stegun, eds., "Handbook of Mathematical Functions," p. 447. Natl. Bur. Std., AMS 55, Washington, D. C., 1964.
14. B. H. Armstrong, *Phys. Rev.* **130,** 2506 (1963).
15. M. Abramowitz and I. Stegun, eds., "Handbook of Mathematical Functions," pp. 377–378. Natl. Bur. Std., AMS 55, Washington, D. C., 1964.
16. F. R. Elder, A. M. Gurewitsch, R. V. Langmuir, and H. C. Pollock, *Phys. Rev.* **71,** 839 (1947).
17. G. A. Schott, "Electromagnetic Radiation." Cambridge Univ. Press, London and New York, 1912.
18. F. R. Elder, R. V. Langmuir, and H. H. Pollock, *Phys. Rev.* **74,** 52 (1948).
19. D. H. Tomboulain and P. L. Hartman, *Phys. Rev.* **102,** 1423 (1956).
20. I. S. Shklovsky, *Astron. Zh.* **29,** 418 (1952).
21. V. A. Dombrovskii, *Dokl. Akad. Nauk SSSR* **94,** 1021 (1954).
22. J. H. Oort and T. Walraven, *Bull. Astron. Inst. Neth.* **12,** 285 (1956).
23. J. E. Baldwin, *Monthly Notices Roy. Astron. Soc.* **115,** 690 (1955).
24. I. S. Shklovsky, *Astron. Zh.* **30,** 15 (1953).

25. L. Spitzer, Jr., *Astrophys. J.*, **124,** 20 (1956).
26. F. F. Gardner and C. A. Shain, *Australian J. Phys.* **11,** 55 (1958).
27. P. A. Čerenkov, *Dokl. Akad. Nauk SSSR* **8,** 451 (1934); *Phys. Rev.* **52,** 378 (1937); S. I. Vavilov, *Dokl. Akad. Nauk SSSR* **8,** 457 (1934).
28. I. M. Frank and I. Tamm, *Dokl. Akad. Nauk SSSR* **14,** 109 (1937)
29. M. Danos, S. Geschwind, H. Lashinsky, and A. van Trier, *Phys. Rev.* **92,** 828 (1953); see also M. Danos and H. Lashinsky, *IRE Trans.* **MTT-2,** 21 (1954).
30. J. V. Jelley, "Čerenkov Radiation." Pergamon, New York, 1958.
31. L. C. L. Yuan, *Science* **154,** 124 (1966).
32. I. M. Frank, Izv. *Akad. Nauk SSSR Ser. Fiz.* **6,** 3 (1942); V. L. Ginzberg and I. M. Frank, *Zh. Eksperim. i Teor. Fiz.* **16,** 1 (1946).
33. W. K. H. Panofsky and M. Phillips, "Classical Electricity and Magnetism," 2nd ed., Section 11-5. Addison-Wesley, Reading, Massachusetts, 1962.
34. V. E. Pafomov, *Radiofizika* (*USSR*) **5,** 485 (1962).
35. A. I. Alikhanian, *Proc. Intern. Conf. Instrumentation High Energy Phys. Stanford, 1966*, p. 419. IUPAP and USAEC, Washington, D. C. 1966.
36. S. Tanaka and Y. Katayama, *J. Phys. Soc. Japan* **19,** 40 (1964).
37. J. Oostens, S. Prünster, C. L. Wang, and L. C. L. Yuan, *Phys. Rev. Letters* **19,** 541 (1967).
38. L. C. L. Yuan, C. L. Wang, and S. Prünster, *Phys, Rev, Letters* **23,** 496 (1969); see also *Phys. Today* **22,** No. 11, 59 (1969).
39. I. M. Frank, *Soviet Phys. Usp.* (*English Transl.*) **8,** 729 (1966); see also R. M. Lewis and J. K. Cohen, *J. Math. Phys.* **11,** 296 (1970).

VI SCATTERING PROCESSES

In the preceding chapter the radiation from accelerated charged particles was studied and a series of useful formulas describing the type of radiation to be expected in varying circumstances was derived. However, an accurate description of an experimental situation can only be given by actually examining the source of the accelerating force, for this will generally influence the nature of the observations.

An important class of processes in which the nature of the force producing the acceleration of the charged particle is significant involves scattering mechanisms. As is well known from a study of classical mechanics,[1] the measurable quantity of interest is the *differential scattering cross section,* defined as*

$$\frac{d\sigma}{d\Omega} = \frac{\text{scattered energy/unit time/unit solid angle}}{\text{incident energy flux/unit area/unit time}}\,. \tag{6-1}$$

The energy flux is used to include the situation where the incident energy consists of a beam of particles, or light. One can then obtain the *total scattering cross section* by integrating over all angles:

$$\sigma = \int \left(\frac{d\sigma}{d\Omega}\right) d\Omega\,. \tag{6-2}$$

Figure 17 exhibits a schematic diagram of the process for particle–particle scattering, and the distance b is called the *impact parameter*. Since b is closely related to the scattering angle, it is obvious that the total cross section can also be obtained from the differential cross section by integrating

* For time-dependent processes one should use the appropriate time averages in Eq. (6-1).

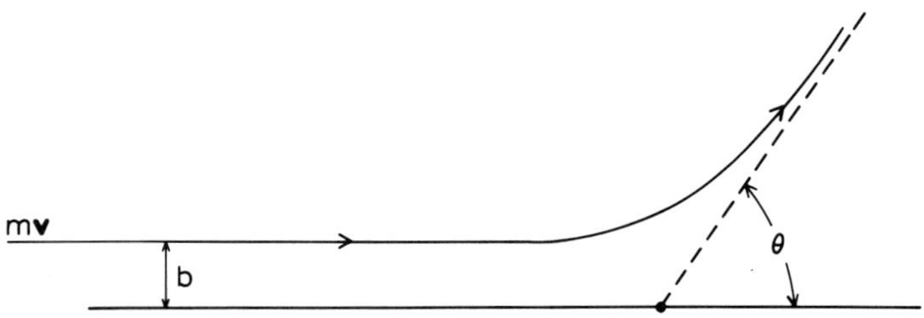

Fig. 17. Relation of the impact parameter to the scattring angle in the classical scattering process.

over all impact parameters. In this chapter we shall calculate these quantities for two very important physical processes.

A. *BREMSSTRAHLUNG*

The equations describing the radiation from a charged particle accelerated in a direction parallel or antiparallel to its velocity were derived in the last chapter. A common example of this type of process, and one which is important in the production of X-rays, is the deceleration of electrons in the Coulomb field of a nucleus, and the resulting radiation is referred to as *Bremsstrahlung*.

When the electron energies in an incident beam impinging on a material sample are very low the collisions with nuclei are rare. In this case the electron–electron interactions are dominant, and the overall conservation laws can only be satisfied by considering collective effects due to all the atomic electrons. A rather exact theory of these processes has been given by Sommerfeld,[2] and the reader is referred to that source for further details.

As the energy of the incident electrons is increased, the radiation due to deflection of the electrons in the Coulomb field of the nucleus becomes dominant, and the physics can be described by single collisions. The resulting "soft" photons give rise to the continuous X-ray spectrum. We shall now derive the expected cross section for this process.

Consider an electron of velocity v passing a nucleus of charge Ze at a distance b, as shown in Fig. 18. The time of closest approach is taken as $t = 0$. The magnitude of the Coulomb force on the electron is then

$$F = m\dot{v} = \frac{Ze^2}{b^2 + v^2t^2}. \tag{6-3}$$

Since the dominant contribution to the radiation comes from the trans-

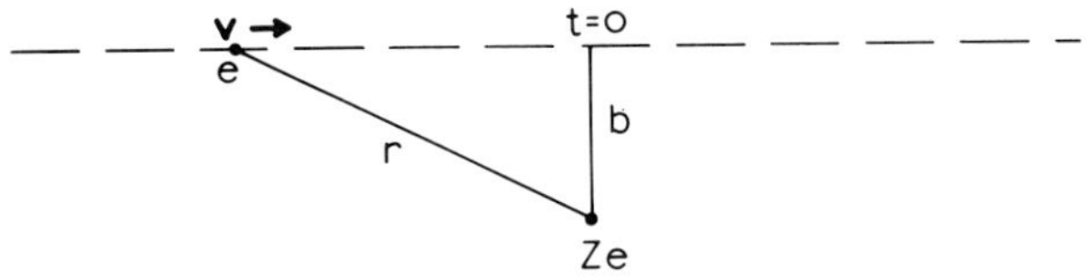

Fig. 18. Closest approach of an electron to the scattering center in the *Bremsstrahlung* process.

verse component of the acceleration,* we see from the figure that the appropriate acceleration to consider is

$$\dot{v} = \frac{Ze^2}{m} \frac{b}{(b^2 + v^2t^2)^{3/2}} . \tag{6-4}$$

The total instantaneous power radiated can then be obtained immediately by substitution into the Larmor formula, Eq. (5-4).

It is the frequency distribution, however, that we wish to measure, and so it is necessary to first Fourier analyze $\dot{v}$. With reference to Eq. (5-14b),

$$\begin{aligned} \dot{v}(\omega) &= \frac{1}{(2\pi)^{1/2}} \int_{-\infty}^{\infty} \dot{v}(t) e^{i\omega t}\, dt \\ &= \frac{2Ze^2 b}{m_0 (2\pi)^{1/2}} \int_0^{\infty} \frac{\cos \omega t}{(b^2 + v^2t^2)^{3/2}}\, dt . \end{aligned} \tag{6-5}$$

This integral can be identified as a modified Bessel function of order one,[3] so that

$$\dot{v}(\omega) = \frac{2Ze^2\omega}{m_0 v^2 (2\pi)^{1/2}} K_1\left(\frac{\omega b}{v}\right). \tag{6-6}$$

It is the soft X-ray region in which we are interested, and so the asymptotic forms (5-35) can be employed to write

$$\dot{v}(\omega) \simeq \begin{cases} \dfrac{2Ze^2}{m_0 v b (2\pi)^{1/2}}, & \omega < v/b \\ 0, & \omega > v/b . \end{cases} \tag{6-7}$$

Finally, by referring to Eq. (5-17), the Larmor formula can be converted to a frequency-dependent expression to give for the intensity of

* One cannot really infer this dominance from the relativistic discussion of the last chapter, but, rather, the argument is that the screening by atomic electrons actually cuts off the Coulombic force until the electron is close in. At this point the interaction is essentially Coulombic and the acceleration is almost completely in the transverse direction.

radiation

$$I(\omega, b) = \frac{8}{3\pi}\left(\frac{e^2}{m_0c^2}\right)^2 \frac{Z^2e^2}{c}\left(\frac{c}{v}\right)^2 \frac{1}{b^2}, \qquad \omega < \frac{v}{b}, \tag{6-8}$$

where the expression on the right-hand side is implicitly assumed to vanish for higher frequencies. The total cross section as a function of frequency, called the *radiation cross section*, is now obtained by integrating over all impact parameters*:

$$\sigma(\omega) = \int_{b_-}^{b_+} I(\omega, b) 2\pi b \, db = \frac{16}{3} r_0^2 \frac{Z^2e^2}{c}\left(\frac{c}{v}\right)^2 \ln\left(\frac{b_+}{b_-}\right), \tag{6-9}$$

where we have introduced the *classical electron radius* (terminology justified in Chapter VII)

$$r_0 = e^2/m_0c^2, \tag{6-10}$$

and designated the maximum and mimimum impact parameters by b_+ and b_-, respectively.

The maximum impact parameter is easily obtained from the inequality associated with Eq. (6-8), $b_+ = v/\omega$. The calculation can be completed in a strictly classical manner by determining b_- from the requirement that the energy transfer be bounded above by the kinetic energy of the incoming electron, and then equating this to the potential energy Ze^2/b. The resulting cross section is grossly incompatible with the data, however, so that the derived formula is of little value. Actually, the minimum impact parameter must be provided by quantum mechanics, and the distance of closest approach is governed by the uncertainty principle. Thus,

$$b_- \simeq \hbar/m_0 v, \tag{6-11}$$

and Eq. (6-9) becomes

$$\sigma(\omega) \simeq \frac{16}{3} r_0^2 \frac{Z^2e^2}{c}\left(\frac{c}{v}\right)^2 \ln\left(\frac{mv^2}{\hbar\omega}\right). \tag{6-12}$$

The semiclassical cross section of Eq. (6-12) and that which would have been obtained completely classically are compared with experimental results in Chapter XII (see Fig. 37), and we shall see that it is quite necessary to introduce some quantum mechanical behavior into the theory in order to obtain a reasonable fit to the data. At low frequencies Eq. (6-12) gives a fairly accurate fit, but the curve falls off too slowly in the region of hard photons. The reason for this is clear: at high frequencies the photon spec-

* Note that we have *not* normalized this quantity $\sigma(\omega)$ by the incident energy. Therefore, its dimensions are $(area)(energy)/(frequency)$.

trum must reflect the true quantum behavior of the system. In this case it is necessary to perform a complete quantum mechanical calculation, which yields the correct Bethe–Heitler formula. We shall derive this expression in Chapter XII. One also notes that small corrections to Eq. (6-12) due to the effects of atomic electrons should be included. Thus, a calculation incorporating the electronic screening of the Coulomb interaction between the incident electron and the nucleus provides a somewhat closer fit to the data. This correction is examined in the problems (see Problem 6-1).

It is also possible to derive a cross section similar to Eq. (6-12) when the electron is incident with relativistic velocities. There is little point in actually making this calculation, however, because it is very difficult to present a convincing derivation in the present context. The mixture of special relativity and only a little bit of quantum mechanics leads to a rather mysterious brew, presenting more questions than it answers. One must really investigate in full the relativistic quantum dynamics of the *Bremsstrahlung* process in order to obtain equations compatible with high-energy measurements, and even when this program is carried out one finds very few experimental results for making comparisons.

B. RADIATION SCATTERING FROM ELECTRONS

Another important radiation process occurs when monochromatic plane waves are scattered by free electrons. (The scattering from bound electrons represents a slightly more complicated physical process and will be taken up in the next chapter.) If the frequency of the incident wave is small, then the problem can be considered as nonrelativistic, the recoil of the electron can be neglected, and the scattered wave will have the same frequency as that of the incident wave. At higher frequencies the particle will have significant recoil and there will be a frequency shift of the scattered wave according to the previous discussion of the Compton effect (see Problem 3-1), so that the problem becomes inherently relativistic. We shall restrict our study to the low-frequency case for the moment.

The electric field vector of the incident plane wave can be written

$$\mathbf{E}(\mathbf{r}, t) = \boldsymbol{\epsilon} E_0 \exp[i(\mathbf{k} \cdot \mathbf{r} - \omega t)], \tag{6-13}$$

where $\boldsymbol{\epsilon}$ is the unit polarization vector of the incident wave, and the subsequent (nonrelativistic) force on the electron is given by Newton's second law:

$$\mathbf{F} = m_0 \dot{\mathbf{v}}(t) = e\mathbf{E}\,. \tag{6-14}$$

Combining these two equations, we have

$$\dot{\mathbf{v}}(t) = \boldsymbol{\epsilon}(e/m_0) E_0 \exp[i(\mathbf{k} \cdot \mathbf{r} - \omega t)]\,. \tag{6-15}$$

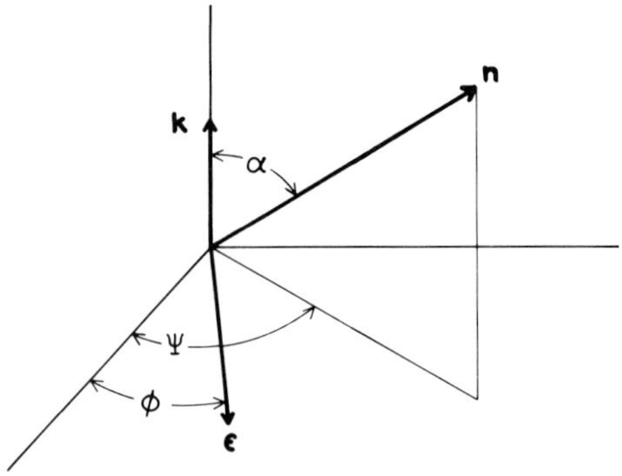

Fig. 19. Orientation of angles as described by Eq. (6-19).

One must recall that the time average of the square of the real part of a harmonic function of time[4] is $\frac{1}{2}(\mathbf{f}\cdot\mathbf{f}^*)$, so that substitution of Eq. (6-15) into Eq. (5-3) yields for the time-average power radiated per unit solid angle

$$\frac{dW}{d\Omega} = \frac{c}{8\pi}\,|E_0|^2\left(\frac{e^2}{m_0c^2}\right)^2 \sin^2\theta\,. \tag{6-16}$$

The differential scattering cross section is then obtained from Eq. (6-1) as

$$\frac{d\sigma}{d\Omega} = r_0^2 \sin^2\theta\,, \tag{6-17}$$

where r_0 is defined by Eq. (6-10). The total cross section is obtained by integrating over all angles, and one easily finds

$$\sigma_{\mathrm{T}} = (8\pi/3)r_0^2\,, \tag{6-18}$$

called the *Thomson cross section.*[5]

While Eq. (6-17) gives the angular distribution of radiation, it does not exhibit explicitly the polarization, and this can be an important aspect of the comparison with experiment. If the polarization vector is oriented as in Fig. 19, then the angle θ between the acceleration vector and the observer is related to the other angles by the spherical harmonic addition theorem:

$$\sin^2\theta = 1 - \sin^2\alpha\cos^2(\psi - \phi)\,. \tag{6-19}$$

Since the polarization cannot generally be specified, it is customary to average over all polarization angles ϕ.

Exercise. Show that

$$\langle\sin^2\theta\rangle_\phi = \tfrac{1}{2}(1 + \cos^2\alpha)\,. \tag{6-20}$$

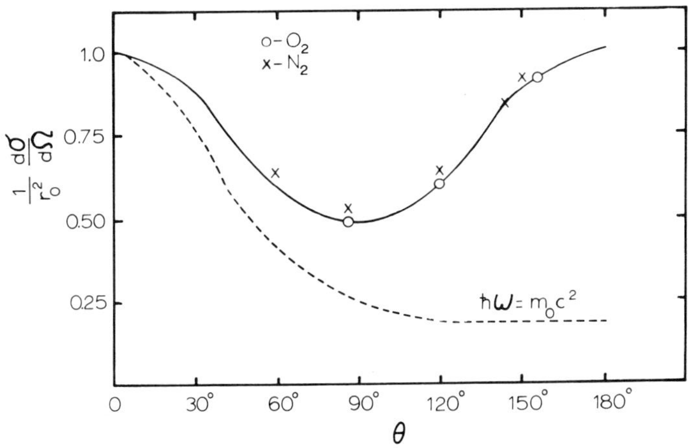

Fig. 20. Comparison of the Thomson formula, (6-21), with the relativistic cross section (dashed curve) and with experimental results of Barrett[7] for O_2 and N_2 at incident energies of 45 and 63 keV, respectively. The measured intensities are only relative, and so have been adjusted to the solid curve.

Hence, in terms of the angle between the wave vector of the incident wave and the observer, the differential cross section becomes the *Thomson formula*[5]

$$d\sigma/d\Omega = \tfrac{1}{2}r_0^2(1 + \cos^2 \alpha)\,. \qquad (6\text{-}21)$$

As mentioned previously, the differential scattering cross section given by Eq. (6-21) and the total cross section of Eq. (6-18) are valid only at low frequencies of the incident wave, or for low-energy photons. When the photon energy $\hbar\omega$ becomes comparable with the rest energy m_0c^2 of the electron, then the process must be treated both relativistically and quantum mechanically. This, of course, happens when the wavelength of the incident wave is of the order of an electron Compton wavelength, $\lambda_c = h/m_0c$. The relativistic kinematics of the problem are easy enough to handle, the frequency shift being obtained as in problem 3-1. However, to calculate the Compton cross section, the quantum mechanical Dirac equation for the electron must be used to obtain the correct Klein–Nishina formula.[6] This expression reduces to Eq. (6-18) in the low-frequency limit, and for high frequencies has the limiting form

$$\sigma_{KN} \simeq \pi r_0^2 \frac{m_0c^2}{\hbar\omega}\left[\frac{1}{2} + \ln\left(\frac{2\hbar\omega}{m_0c^2}\right)\right], \qquad \hbar\omega \gg m_0c^2\,. \qquad (6\text{-}22)$$

We shall return to the discussion of the quantum-mechanical behavior in Chapter XII. However, the Thomson theory is compared with a few experimental results[7] in Fig. 20.

Finally, one further point should be made with respect to comparing the Thomson formulas with experiment. The agreement is quite good for the scattering of soft X-rays from electrons or γ-rays from light atoms,[7] but if the experiment involves scattering from a distribution of electrons, such as in an electron gas or a heavy atom, then the observed scattering is much larger in the forward direction than indicated by Eq. (6-21). The reason for this behavior is that there will be a coherent addition of the radiation amplitudes of all the electrons, the effects being dominant at wavelengths large compared to the electron spacing. In these cases it is perhaps easier to treat the problem as one of wave propagation in a dispersive medium.[8]

C. SUMMARY

The object of this chapter has been to study only two of the very basic scattering processes leading to electromagnetic radiation. This brief discussion was motivated not only by the desire to understand some important physical problems, but also to provide some insight into the types of radiation cross section one can calculate in the classical theory. In subsequent chapters we shall want to look at these problems from the standpoint of quantum mechanics and take a detailed look at any new physical predictions the quantum theory may make. In this sense, the Thomson and *Bremsstrahlung* processes are of fundamental interest for the sake of comparison.

There are, of course, many other phenomena which could yet be studied with the classical theory as developed thus far. For instance, we have not discussed energy-loss formulas in charged-particle collisions, multiple scattering, radiative decays, and other processes of interest in atomic physics. While important, these problems are of little concern to us in studying the basis on which electrodynamics is built, and the interested reader can find them treated quite nicely in a book such as that of Jackson.[9] On the other hand, we have also not considered other single-particle processes, such as radiation scattering from bound electrons and absorption phenomena, which are important to a fundamental understanding of electrodynamics. However, their correct treatment depends on a deeper investigation into the structure of charged particles, and we shall initiate just such an investigation in the next chapter.

PROBLEMS

6-1. In order to account for screening effects in the *Bremsstrahlung* process, consider the electron to be deflected by a screened-Coulomb potential

$$V(r) = Ze^2 \frac{\exp(-\alpha r)}{r}.$$

Assume that the particle moves in a straight-line trajectory past the center of force, and calculate the cross section for radiation as a function of the limiting impact parameters b_- and b_+. Compare the result with that of Eq. (6-12) in the limit $\alpha \to 0$.

6-2. Describe the behavior of the radiation cross section $\sigma(\omega)$ under Lorentz transformations.

6-3. Consider a nonrelativistic electron scattered from a very heavy particle of charge Ze. Calculate the differential scattering cross section in the center-of-mass system, in terms of the initial electron kinetic energy.

6-4. If in Problem 6-3 the electron is initially at rest and the heavy particle incident, calculate the energy transferred to the electron as a function of the impact parameter.

REFERENCES

1. H. Goldstein, "Classical Mechanics," Section 3-7. Addison-Wesley, Reading, Massachusetts, 1959.
2. A. Sommerfeld, "Atombau und Spektrallinien," Vol. II. Vieweg, Braunschweig, 1939.
3. M. Abramowitz and I. Stegun, eds., "Handbook of Mathematical Functions," p. 376. Natl. Bur. Std., AMS-55, Washington, D. C., 1964.
4. W. K. H. Panofsky and M. Phillips, "Classical Electricity and Magnetism," p. 190. Addison-Wesley, Reading, Massachusetts, 1962.
5. J. J. Thomson, "Conduction of Electricity through Gases," 3rd ed., Vol. II, p. 256. Cambridge Univ. Press, London and New York, 1933.
6. O. Klein and Y. Nishina, *Z. Physik* **52,** 853 (1929).
7. C. S. Barrett, *Phys. Rev.* **32,** 22 (1928).
8. See, e.g., J. A. Stratton, "Electromagnetic Theory," Chapter 5. McGraw-Hill, New York, 1941.
9. J. D. Jackson, "Classical Electrodynamics," in particular, Chapters 13 and 15. Wiley, New York, 1962.

VII THE CLASSICAL ELECTRON

One might well suspect at this point that the theory as developed in the preceding chapters is quite complete, in that we have studied in some detail the relativistic motion of charged particles and the subsequent radiated energy. Hopefully, further refinements of the theory and more complicated applications would merely be a matter of grinding the mathematical crank more diligently. Unfortunately, nature is rarely so accommodating, and there are several logical objections to the theory as developed thus far. It is incumbent upon us at this time to formulate these questions and attempt to answer them.

Given the motion of a charged particle in an external field and the distribution of the associated radiation, if any, why should we assume these to be consistent results? Clearly, the radiation field carries off momentum and energy, so how can the energy of the particle remain constant under a uniform external force? Does the radiation process alter the internal structure and energy of a charged particle? Why do the fields of the particle itself not react back on their source and alter the motion of the particle? Finally, and probably most important, in view of these very serious questions, why do the calculations of the preceding chapters agree so well with experiment in a great many cases?

It is well to begin with the last question, which has an obvious answer, that the reaction of the fields on the particle itself is in most instances negligible. In order to see this more clearly, consider an external field to accelerate a charged particle uniformly for a period of time T. From Larmor's formula, Eq. (5-4), the energy radiated is

$$\tfrac{2}{3}(e^2/c^3)a^2T\,. \tag{7-1}$$

If this energy is much smaller than other energies involved in the problem, then reaction effects must be negligible. On the other hand, if the kinetic energy after the acceleration, $\sim m(aT)^2$, is of the same order as the radiated energy, Eq. (7-1), the reaction effects are important, and equating the two energies yields a characteristic time

$$\tau_0 = \tfrac{2}{3}e^2/mc^3 . \tag{7-2}$$

For an electron, $\tau_0 \approx 10^{-24}$ sec, in which time light would travel 10^{-13} cm. Consequently, only for situations involving such times or distances will the reaction become an important concept, which indicates why the above questions had no bearing on the problems considered previously.*

Regardless of when these questions become significant, however, the important point is that the theory is in some sense incomplete, and in order to have a comprehensive understanding of classical electrodynamics it is necessary to study in some detail the structure of a charged particle, and to determine its actual equations of motion. We shall find that such a detailed investigation will be of great value in understanding the quantum problem later.

A. CLASSICAL THEORY OF CHARGED PARTICLES

It has been said that those who do not understand the past are doomed to repeat it. Therefore, in order to obtain some perspective for trying to understand the current problems with elementary charged particles, it will be the objective of this section to trace briefly the historical efforts toward construction of a model of the electron.

Historically, the attempts at developing a theory of charged particles centered around the idea of a completely electromagnetic description. The dominant thought was that the mass of the electron was completely electromagnetic, and that electrons represented "atoms of electricity." As we shall see, the development of this point of view was primarily due to Lorentz[1] and Abraham.[2] Thomson[3] discovered the electron in 1897, but many years prior to this discovery he had been the first to calculate an expression for the electromagnetic mass of the particle.[4] Thomson ascribed a spherical charge distribution to an electron in uniform motion and calculated the

* The characteristic time defined by Eq. (7-2) is actually more significant than indicated here. Although there is no direct evidence for a quantization of time, τ_0 for an electron appears to be the smallest time interval encountered in nature, being the time in which the strong interactions take place. One can, of course, infer even smaller characteristic times for more massive particles from Eq. (7-2), but these do not appear to have been observed. Moreover, the age of the universe as calculated from Hubble's expansion constant in units of τ_0 is $\sim 10^{40}$, which is also the ratio of the electron-proton electromagnetic interaction to their gravitational interaction. There is not yet any real understanding of these strange relationships, but it would be truly amazing if they were only coincidental.

kinetic energy of its electromagnetic field. Let us assume a surface charge density of radius r_0. Then, Thomson identified an electromagnetic mass of the electron as

$$m = \tfrac{2}{3}e^2/r_0c^2 . \tag{7-3}$$

It is worthwhile to digress a moment and note that if one requires the electron mass to be entirely electromagnetic in origin, then this mass should be identifiable by equating the rest energy to the electrostatic energy. For the surface charge distribution one obtains

$$m_e c^2 = (1/8\pi) \int |\mathbf{E}|^2 \, d^3r = e^2/2r_0 ,$$

or

$$m_e = \tfrac{1}{2}e^2/r_0c^2 , \tag{7-4}$$

which we shall call the *electromagnetic mass*. For different types of spherical charge distribution the numerical factor in Eq. (7-4) changes, but remains of order unity. Hence, it is customary to define the *classical electron radius* as

$$r_0 = e^2/m_0c^2 , \tag{7-5}$$

as in Eq. (6-10).

Comparison of Eqs. (7-3) and (7-4) shows that the mass identified by Thomson is

$$m = \tfrac{4}{3}m_e ,$$

a result which will appear to be more confounding shortly.

At about the same time Lorentz constructed a complete theory of charged-particle dynamics in an attempt to replace Maxwell's macroscopic theory with a microscopic description.[1] The major ideas were that charged particles interact via the electromagnetic field, rather than directly, and that the particle itself can be described by a charge density ρ. When in motion the charge defines a current density

$$\mathbf{j} = \rho\mathbf{v} , \tag{7-6a}$$

and its total charge is

$$e = \int \rho \, d^3r . \tag{7-6b}$$

These equations, along with the Lorentz force, Eq. (4-34), are sufficient to determine the theory.

Lorentz also realized the need for considering the effect of the charge's field on the motion of the charge, and the necessity for investigating energy conservation in the radiation process. He was, in fact, able to derive an equation of motion for the electron which includes a force due to the effects of radiation reaction, a detailed derivation of which will be given below.

It was clear even at this time, however, that there had to exist something for the electron fields to react onto, so that a model for the structure of the electron had to be developed.

The major contributions to understanding the structure of the electron were due to Abraham,[2] who took as the model a uniform, spherically symmetric charge distribution. In terms of the Poynting vector, Eq. (4-39), he assumed the electron momentum due to its own fields to be

$$\mathbf{p}_e = (1/c^2)\int \mathbf{S}\, d^3r\,. \tag{7-7}$$

For a charge in motion one must substitute the fields of Eqs. (4-63) and (4-64) into the integral, and a short calculation yields

$$\mathbf{p}_e = \tfrac{4}{3}m_e\mathbf{v}\,, \tag{7-8}$$

with m_e given by Eq. (7-4). Thus, the usual momentum–velocity relation for particles does not seem to be valid for the electron, and the anomalous factor of $\frac{4}{3}$ remained a bothersome quantity for many years.

Exercise. Verify Eq. (7-8).

In order to account for the reaction of the fields on the source itself, it is possible to derive an expression for the self-force needed to ensure a consistent equation of motion. This should be obtainable by integrating the Lorentz force density over the electron volume and using the charge's own fields. Hence

$$\mathbf{F}_s = \int \rho\left(\mathbf{E}_s + \frac{1}{c}\mathbf{v}\times\mathbf{B}_s\right) d^3r\,, \tag{7-9}$$

where the subscript s refers to the self-fields. The spherical charge distribution of radius r_0 is assumed to be rigid; although this is not a relativistically invariant assumption, we remedy this by performing the calculation in the rest frame of the particle. The generalization to a moving electron is then made by performing a Lorentz transformation. With this assumption the second term in the integrand can be dropped.

The evaluation of the integral in Eq. (7-9) must be done very carefully. The field $\mathbf{E}_s$ at the location of a charge element de' due to an element de is, of course, a retarded field. Because the two charge elements are rigidly connected, de' moves along with de, and the retarded field, Eq. (4-63), must be modified, since it refers to a *fixed* observer. This modification can be accomplished by setting $\mathbf{v} = 0$ in the expression for the retarded field, consistent with our assumption of an instantaneous rest frame. The dimensions of the electron are very small, so that $t' = t - R/c$ differs from t by a quantity of order r_0/c. Thus, the integrand in Eq. (7-9) can be expanded in a Taylor series about $t' = t$. The calculation is straightforward, but tedious, and is done very nicely by Heitler.[5] One finds

$$\mathbf{F}_s = -\frac{2}{3c^2}\,\dot{\mathbf{v}} \iint \frac{\rho(\mathbf{r})\rho(\mathbf{r}')}{|\mathbf{r}-\mathbf{r}'|}\,d^3r\,d^3r' + \frac{2}{3}\frac{e^2}{c^3}\,\ddot{\mathbf{v}} + O(r_0)\,, \tag{7-10}$$

where we have omitted all terms depending on r_0. The form of these omitted terms will be examined in the problems. The integral in Eq. (7-10) can be identified with the self-energy of the particle,[6]

$$W_s = \frac{1}{2}\iint \frac{\rho(\mathbf{r})\rho(\mathbf{r}')}{|\mathbf{r}-\mathbf{r}'|}\,d^3r\,d^3r'\,, \tag{7-11}$$

so that

$$\mathbf{F}_s = \frac{2}{3}\frac{e^2}{c^3}\,\ddot{\mathbf{v}} - \frac{4}{3c^2}\,W_s\dot{\mathbf{v}} + O(r_0)\,. \tag{7-12}$$

The first term is just the self-force obtained by Lorentz.[7]

If one considers a closed system of particle plus external forces, then the total time rate-of-change of momentum must vanish. Thus, the external force must be the negative of $\mathbf{F}_s$, which will yield an equation of motion. Such an equation will bear some resemblance to Newton's equation of motion if in Eq. (7-12) we define a "self-mass" by*

$$m_s = W_s/c^2 = m_e\,, \tag{7-13}$$

from Eq. (7-4). Hence,

$$\tfrac{4}{3}m_e\dot{\mathbf{v}} - \tfrac{2}{3}(e^2/c^3)\ddot{\mathbf{v}} = \mathbf{F}_{\text{ext}}\,, \tag{7-14}$$

which is the *Abraham–Lorentz equation of motion* when structure-dependent terms are ignored.[7]

Until the advent of relativity in the period 1904–1905 the derivation of the above equation of motion was thought to constitute a successful theory of the purely electromagnetic charged particle. True, there were some difficulties, but these were generally attributed to poor mathematical methods. The first drawback of the theory is that Eq. (7-14) is not really an equation of motion in the Newtonian sense, because it contains in general derivatives of $\mathbf{v}$ to all orders, and, even if the structure-dependent terms are dropped, it still remains a third-order differential equation for the particle position vector. Thus, the initial position and velocity of the particle are not sufficient to specify its motion. The difficulty with such an equation can be seen by solving Eq. (7-14) when $\mathbf{F}_{\text{ext}} = 0$, in which case there are

* The identification of an energy with a mass times c^2 is an interesting one at this point in the theory, but was by no means recognized as having the deep meaning later brought out by special relativity.

two possible solutions:

$$\dot{\mathbf{v}}(t) = 0\,, \tag{7-15a}$$

$$\dot{\mathbf{v}}(t) = \mathbf{a}\exp(4t/3\tau_0)\,, \tag{7-15b}$$

where τ_0 is defined by Eq. (7-2). Only the first solution can be considered physically reasonable, the second being known as the *runaway solution*. Results such as (7-15b) can be discarded as meaningless, but that they should appear at all is an unhealthy sign.

The second difficulty with Eq. (7-14) is that one cannot rigorously eliminate the structure terms by letting $r_0 \to 0$, because then the self-energy diverges [see Eqs. (7-4) and (7-13)]. Thus, the electron must have structure. Nevertheless, these structure terms can be considered formally small, and it was thought that the theory was well in hand.

B. THE RELATIVISTIC ELECTRON

With the advent of the special theory of relativity several new, more severe criticisms emerged against the theory of the electron. Perhaps the most striking difficulty was due to the relations among energy, momentum, and velocity required by Lorentz invariance. Referring to Eqs. (3-18) and (3-22), we observe that

$$\mathbf{p} = (E/c^2)\mathbf{v}\,, \tag{7-16}$$

which must be true for *all* particles regardless of the value taken by $|\mathbf{v}|$. Nevertheless, a comparison of Eqs. (7-8) and (7-13) indicates that for a purely electromagnetic electron

$$\mathbf{p}_e = \tfrac{4}{3}(W_s/c^2)\mathbf{v}\,. \tag{7-17}$$

This immediately suggests that the electron cannot be purely electromagnetic, but that the self-energy must be partly due to something else.

The conclusion that classical electron theory is not in agreement with special relativity actually turns out to be fallacious, for the difference between Eqs. (7-16) and (7-17) arises from an incorrect application of the Lorentz transformations. In fact, not only does a correct treatment of the transformation properties invalidate the argument against the electromagnetic electron, but it also explains the anomalous factor of $\frac{4}{3}$ associated with m_e. This was first pointed out* in 1922 by Fermi.[8]

* It is historically interesting that Fermi's work went largely unnoticed, and that the correct form for $\mathbf{p}_e$ was rediscovered by Wilson[9] in 1936, again unnoticed, again found by Kwal[10] in 1949, and once more forgotten. Finally, Rohrlich[11] discovered the treatment for a fourth time in 1960. It is somewhat fortunate that all theoretical discoveries do not require 38 years in which to enter the mainstream of physical thought!

The error lies with the definition of momentum assumed in Eq. (7-7), which is relativistically incorrect when applied to other than free fields. The 4-momentum of the self-field must be defined as in Eq. (4-45), and this can also be written in the form

$$p^\mu = (\mathbf{p}, W/c)\,. \tag{7-18}$$

Now, the timelike vector $d\sigma_\nu$ in Eq. (4-45) has the form $(\mathbf{0}, d^3r)$ in the rest frame of the particle. This spacelike plane can be described with a unit vector n^μ by the equation $n^\mu x_\mu + c\tau = 0$, so that in general

$$d\sigma_\nu = \gamma(v_\nu/c)\, d^3r\,. \tag{7-19}$$

Hence, for the self-fields of the electron p^μ has components*

$$W_s = \gamma^2 \int u\, d^3r - \frac{\gamma^2}{c^2} \int \mathbf{S} \cdot \mathbf{v}\, d^3r\,, \tag{7-20}$$

$$\mathbf{p}_e = \frac{\gamma^2}{c^2} \int \mathbf{S}\, d^3r + \frac{\gamma^2}{c^2} \int \mathsf{T} \cdot \mathbf{v}\, d^3r\,, \tag{7-21}$$

where the integrals extend over all space outside of the electron, and the Maxwell stress tensor has been written in dyadic form.

These last two equations can be evaluated in the nonrelativistic limit in order to extract their essential physical content. In this limit $\mathbf{B}_s \sim (\mathbf{v} \times \mathbf{E}_s)/c$, and the second term in Eq. (7-20) is of higher order. Thus, in first approximation the self-energy is that given by Eq. (7-11). However, the nonrelativistic limit of Eq. (7-21) which follows from (4-38) is

$$\begin{aligned}\mathbf{p}_e &\simeq \frac{1}{4\pi c^2} \int [\mathbf{E}_s^2 \mathbf{v} - (\mathbf{E}_s \cdot \mathbf{v})\mathbf{E}_s]\, d^3r \\ &\quad + \frac{1}{4\pi c^2} \int \left[(\mathbf{E}_s \cdot \mathbf{v})\mathbf{E}_s - \frac{1}{2}\mathbf{E}_s^2 \mathbf{v}\right] d^3r \\ &= \frac{W_s}{c^2}\mathbf{v}\,,\end{aligned} \tag{7-22}$$

in agreement with the relativistic requirement of Eq. (7-16), and in disagreement with Eq. (7-8).

Exercise. Show explicitly that for a spherically symmetric field the first term in the first line of Eq. (7-22) is responsible for the factor of $\frac{4}{3}$ in the previous work.

Although the previous calculation demonstrated that it is possible to formulate the theory of the electron in agreement with special relativity,

* See Eqs. (4-37) and (4-45).

the desire to have a purely electromagnetic electron was still frustrated by the relativistic facts of life. As we have seen in Chapter I, the concept of rigidity is not relativistically invariant, so that the model of a rigid charge distribution had to be abandoned. Lorentz[7] adopted the model of a compressible electron which could be deformed by the Lorentz contraction, and he worked out a very complicated function of (v/c) for its charge distribution. Unfortunately, this function is not an invariant under Lorentz transformations, and it is still not clear whether such a construction is useful.

At about the same time another, more serious question was raised regarding the stability of the electron. If the particle has structure, then the various parts of the charged sphere should repel one another via Coulomb's law. This gives rise to an inherent instability which was quite puzzling, particularly since it was not observed. A brilliant solution was suggested by Poincaré, who postulated the existence of nonelectromagnetic forces within the electron which held it together. The nature of these forces is outside the realm of classical theory, to be sure, but it is more than a little informative to investigate Poincaré's solution.[12]

Consider once more Eqs. (4-37) and (4-45), imagining that the self-fields of the electron are to be inserted in the former. In the rest system of the spherically symmetric charge distribution it is clear that the right-hand side of Eq. (4-45) vanishes for $\mu \neq \nu$. Moreover, from the exercise following Eq. (4-40) the trace of $\mathsf{W}_{\mu\nu}$ vanishes, so that

$$\int \overline{\mathsf{W}}_{11} d^3r = \int \overline{\mathsf{W}}_{22} d^3r = \int \overline{\mathsf{W}}_{33} d^3r = \tfrac{1}{3} \int \overline{\mathsf{W}}_{44} d^3r = \tfrac{1}{3} W_s \,, \tag{7-23}$$

where the superposed bar indicates quantities in the rest frame of the particle. The components $\overline{\mathsf{W}}_{11}, \overline{\mathsf{W}}_{22}, \overline{\mathsf{W}}_{33}$, are called the *self-stresses* of the particle, and, since $W_s \neq 0$, they are nonzero. This nonvanishing of the self-stresses therefore renders the electron unstable and induces a positive internal pressure tending to tear it apart. We shall examine this instability more quantitatively in the problems (see Problem 7-2).

Exercise. From Eq. (7-23) show that it is just the self-stresses that contribute the extra amount of $\frac{1}{3}m_e$ to the momentum $\mathbf{p}_e$, Eq. (7-17), leading to the disagreement with Newton's laws and the special theory of relativity.

Poincaré suggested, then, that the existence of a stable electron implied that there are "cohesive" forces of a nonelectromagnetic nature holding the charge distribution together, and that these forces must be such as to render the overall self-stresses in the electron rest frame zero.

Formally, the cohesive forces can be introduced by defining a nonelectromagnetic symmetric tensor $\mathsf{S}^{\mu\nu}$, such that $\mathsf{W}^{\mu\nu}$ of Eq. (4-37) is to be replaced in the theory by

$$\mathsf{T}^{\mu\nu} = \mathsf{W}^{\mu\nu} + \mathsf{S}^{\mu\nu} , \tag{7-24}$$

under the condition that

$$\partial_\mu \mathsf{T}^{\mu\nu} = 0 . \tag{7-25}$$

That is, Eq. (4-45) is replaced by

$$P^\mu = \frac{1}{c} \int \mathsf{T}^{\mu\nu} \, d\sigma_\nu . \tag{7-26}$$

The components of $\mathsf{S}^{\mu\nu}$ are then chosen so as to make the self-stresses vanish in the proper frame of the electron:

$$\bar{\mathsf{T}}^{kk} = 0 , \tag{7-27}$$

and one then has

$$m_0 c^2 = W_s = \int (\mathsf{W}^{44} + \mathsf{S}^{44}) \, d^3r . \tag{7-28}$$

Hence, by explicitly accounting for the structure of the charged particle, Poincaré was able to bring the theory into accord with special relativity, and also to render the electron stable. The price paid was rather high, however, because the concept of a purely electromagnetic charged particle was no longer possible. The electron self-energy and mass now contained nonelectromagnetic parts, so that the quantity to be used in the equations was the *experimentally observed rest mass*

$$m_0 = m_c + m_e , \tag{7-29}$$

where m_c is the contribution from the cohesive forces. It is only the sum in Eq. (7-29) that has physical meaning, and the contributions m_c and m_e cannot even in principle be measured separately.* We have here the first inkling of the *mass renormalization* which will play such an important role in the subsequent discussion of charged particles.

It is well to point out in passing that the introduction of cohesive forces within the charged particle, or the requirement of manifestly covariant expressions for energy and momentum, Eqs. (7-20) and (7-21), really amount to the same thing, for one can identify the velocity-dependent terms in the

* The Newtonian relation $\mathbf{F} = e\mathbf{E} = m\mathbf{a}$ is, of course, well verified experimentally, demonstrating the classical coupling of the electron to the electromagnetic field. However, to the author's knowledge, there is no clear-cut experimental evidence that the electron has a coupling to the gravitational field in the same way as a charge-neutral particle. While recent experimental[13] and theoretical[14] work has begun to shed some light on the question, the situation is still not unequivocably clear.

latter two equations as the negative of the work done by the fields and the momentum carried by the fields, respectively. The conservation laws demand that these be canceled by some corresponding quantities which must be of nonelectromagnetic origin. Clearly, the same physical result obtains, irrespective of the choice of either Eq. (7-18) or Eq. (7-26).

By the early part of the twentieth century the theory of charged particles was considered in somewhat satisfactory condition in that it described a structured charged particle which was stable and behaved in accordance with the special theory of relativity, and for which an equation of motion was available: Eq. (7-14) without the factor of $\frac{4}{3}$. The concept of a purely electromagnetic electron had to be abandoned and the structure-dependent terms of Eq. (7-12) taken seriously, but they were assumed to be small and could hopefully be neglected in any calculation. The mass used in the equations was the observed quantity of Eq. (7-29).

Nevertheless, the necessity of a structure-dependent theory and the concomitant inability of classical physics to describe this structure remained a weak spot in theoretical physics. In the meantime, the development of quantum mechanics had placed emphasis on point-particle theories, and one might hope that a well-defined point-charge ($r_0 \to 0$) limit of the Abraham–Lorentz theory existed. The structure of an elementary particle, after all, was not within the domain of classical physics, so that there ought to exist a theory of classical point charges. This point of view received concrete support when in 1933 Wentzel[15] discovered that the fields were not necessarily divergent at the position of a point charge, nor were the associated energy and momentum, as long as one pursued a very careful limiting procedure.

Finally, in 1938 Dirac[16] succeeded in deriving an equation of motion for the point electron which was relativistically covariant, and the form of which had been indicated earlier by Abraham[17] and von Laue[18] in attempts at generalizing the Lorentz equation, Eq. (7-14). It is instructive, in fact, to return to this equation and attempt to put it into covariant form. The first term on the left-hand side is easily generalized by omitting the factor of $\frac{4}{3}$ as noted in the relativistic treatment discussed above, and $\dot{\mathbf{v}}$ is replaced by $dv^\mu/d\tau$. The external force is merely a Minkowski 4-force. The remaining term, called the *radiative reaction* because it corresponds to the energy loss due to radiation, can be generalized by using the acceleration 4-vector, Eq. (3-35). This generalization of the radiative reaction force, Γ^μ, must be done very carefully, because it should have the same covariant *form* as the Minkowski force, Eq. (3-34). However, one readily sees that this last equation, along with the definition of 4-velocity, Eq. (3-16), requires the 4-force to be orthogonal to the 4-velocity. Consequently, we must impose the condition

$$\Gamma^\mu v_\mu = 0\,. \tag{7-30}$$

Therefore, Γ^μ is defined only up to a term proportional to the 4-velocity, and we write

$$\Gamma^\mu = \frac{2}{3}\frac{e^2}{c^3}\left(\frac{d^2 v^\mu}{d\tau^2} + \frac{1}{c^2}\,Bv^\mu\right), \tag{7-31}$$

where B is an invariant. Applying the condition (7-30), we find

$$\begin{aligned} B &= -v_\mu \frac{d^2 v^\mu}{d\tau^2} \\ &= \frac{dv_\mu}{d\tau}\frac{dv^\mu}{d\tau} = a_\mu a^\mu\,. \end{aligned} \tag{7-32}$$

Exercise. Verify the second line of Eq. (7-32).

Thus, the proper generalization of the radiative reaction term is

$$\Gamma^\mu = \frac{2}{3}\frac{e^2}{c^3}\left(\dot{a}^\mu + \frac{1}{c^3}\,a^\nu a_\nu v^\mu\right), \tag{7-33}$$

known as the *Abraham 4-vector* of radiation reaction. The Lorentz equation, Eq. (7-14), then has the covariant generalization

$$m_0 \dot{v}^\mu = F^\mu_{\text{ext}} + \Gamma^\mu\,, \tag{7-34}$$

where m_0 is the observed mass. This equation is called the *Lorentz–Dirac equation*, and it should be noted that we derived it by ignoring the structure-dependent terms implicit in Eq. (7-14).

Dirac derived Eq. (7-34) in a more fundamental way from the Maxwell–Lorentz equations and the conservation laws of energy and momentum. Moreover, and this is most important, he based his derivation on the assumption of point charges *ab initio*, so that the Lorentz–Dirac equation in no way reflects the structure of the electron. In Dirac's theory the electromagnetic mass is still infinite, but it is combined with any nonelectromagnetic mass, as in Eq. (7-29), and m_0 is interpreted as the observed rest mass. Because m_0 is a relativistic invariant, this *renormalization procedure* yields a completely covariant theory.

An important feature of Dirac's theory was his observation that it is very advantageous to consider combinations of *both* advanced and retarded fields, Eqs. (4-63) and (4-64), by writing for the field-strength tensor

$$\mathsf{F}^{\mu\nu}_{\pm} = \tfrac{1}{2}(\mathsf{F}^{\mu\nu}_{\text{ret}} \pm \mathsf{F}^{\mu\nu}_{\text{adv}})\,. \tag{7-35}$$

He found that in the limit $r_0 \rightarrow 0$ one obtains

$$\Gamma^\mu = (e/c)\mathsf{F}_-^{\mu\nu} v_\nu \,, \tag{7-36}$$

completely independent of the charged-particle structure. This remarkable observation thereby laid a cornerstone for a sound point-particle theory. We shall not pursue here the algebraic details of the ensuing derivation of Eq. (7-34), but instead refer the reader to the work of Dirac[16] or Rohrlich.[19]

The combinations (7-35) led to another, essentially equivalent, formulation of classical electrodynamics by Wheeler and Feynman[20] which has some intrinsic interest. They assumed that an electron really interacts with all the other electrons in the universe and that these particles completely absorb the radiation from the initial electron. This many-particle, action-at-a-distance theory adopts the point of view that the particles interact only with one another, *without* the presence of a mediating field. Thus, all of the radiation emitted is reabsorbed, and radiation appears as a secondary effect when there is incomplete absorption.

The interesting physical point here is that in our previous work we have always considered the physical solutions of the field equations as the retarded quantities. However, both the equations of electromagnetic theory and classical mechanics are time-reversal invariant, and we have *ourselves* destroyed the symmetry of the solution by discarding the advanced fields. Since there is really no formal objection to these solutions, it may be of interest to examine the time-symmetric solution in the form $\mathsf{F}_+^{\mu\nu}$, Eq. (7-35), which in the absorber model is taken as the correct form of the field-strength tensor. The assumption that the absorbing cloud contains the fields then leads to the equation for the single (jth) electron:

$$\sum_{\substack{i=0\\ i\neq j}}^{n-1} \mathsf{F}_+^{\mu\nu}(i) = \sum_{\substack{i=0\\ i\neq j}}^{n-1} [\mathsf{F}_{\text{ret}}^{\mu\nu}(i) + \mathsf{F}_-^{\mu\nu}(j)] \,, \tag{7-37}$$

where $\mathsf{F}_-^{\mu\nu}(j)$ is the radiative reaction. Thus, not only do we obtain results equivalent to those of Dirac, but the expression $\mathsf{F}_-^{\mu\nu}$ for the radiative reaction results in a mutual cancellation of the divergent contributions to the mass. This *subtraction procedure* is equivalent to a mass renormalization and yields a convergent, Lorentz-covariant theory. The procedure had actually been considered earlier by Fokker,[21] but he was unable to introduce radiation, which the absorber model handles quite nicely.

While the theory of Wheeler and Feynman gives consistent results, one might object to its physical interpretation. The elimination of *all* electromagnetic fields seems conceptually contradictory with the observed independence of radiation fields and the successful notion of photons in the quantum theory. The model *does* resolve the outstanding inconsistencies of classical electrodynamics, but we shall now see that these can be solved in a somewhat more transparent and satisfying way.*

One must return to the Lorentz–Dirac equation, Eq. (7-34), and ask two questions: (1) what, if any, are the difficulties with this equation? and (2) can it be considered an equation of motion? In the remainder of this section we shall show that the answers to these questions lead to the conclusion that Eq. (7-34) does indeed form a sound basis for the classical theory of charged particles.

Two of the major difficulties with the Lorentz–Dirac equation are the necessary mass renormalization due to an infinite electromagnetic contribution, and the existence of runaway solutions of the type indicated by Eq. (7-15b). Concerning the latter, Dirac himself noticed that they could be eliminated by requiring the acceleration to vanish asymptotically. In 1955 Haag[23] pointed out the importance of an asymptotic quantum field theory, and these studies resulted in a better understanding of the importance of asymptotic conditions.

One must inquire as to how the behavior of an electron is to be measured in connection with a physical process in which it takes part. In any given interaction the electron must become free of the force as $|\tau| \to \infty$, and, in the same limit, the momentum must be the free-particle momentum, and the field surrounding the particle in its rest frame must be a pure Coulomb field. With these observations it is then the rest mass m_0 which is measured, and this is independent of the kinematic changes and dynamic effects taking place due to the interaction. Also, we see that the Lorentz–Dirac equation is *not* the equation of motion, but that coupled to it is the *asymptotic condition*

$$\lim_{|\tau| \to \infty} a^{\mu}(\tau) = 0 \, . \tag{7-38}$$

The point of view to be taken in classical electrodynamics, then, is that Eqs. (7-34) and (7-38) *together* constitute the equations of motion for a charged particle, and that the mass which appears is always the *phenomenologically* determined quantity appearing in the region in which measurements are made. The classical theory cannot hope to account for the structure of a charged particle, because the electron size is not within the domain of competence of classical physics. The theory describes the motion of a point particle, which is defined as having dimensions incapable of being determined by classical methods. Thus, the classical theory which emerges is entirely satisfactory, in that it can describe all possible behaviors of charged

* Quite recently Hoyle and Narlikar[22] have resurrected the time-symmetric theory as a method for deciding among various cosmological models. They have had some success with quantizing the theory.

particles within the domain of its applicability. It cannot, of course, account for the charge, mass, or structure of an electron, but this is of no consequence, since it cannot be expected to accomplish such objectives. These latter goals are properly the goal of quantum mechanics. Unfortunately, we shall see in the sequel that the quantum theory has not been able to live up to its expectations in this respect, and that an adequate theory of elementary particles remains to be found. In the classical theory, however, these difficulties are of no concern.

We now derive the classical equations of motion for a charged particle in covariant from, which were first developed by Rohrlich.[24] Let us first rewrite Eq. (7-34) as

$$m_0(a^\mu - \tau_0 \dot{a}^\mu) = K^\mu\,, \tag{7-39}$$

where τ_0 is defined by Eq. (7-2), and

$$K^\mu = F^\mu_{\text{ext}} + (m_0\tau_0/c^2)a^\nu a_\nu v^\mu\,. \tag{7-40}$$

This rearrangement is quite useful because it now allows us to eliminate $\dot{a}^\mu$ by multiplying Eq. (7-39) by an integrating factor $\exp(-\tau/\tau_0)$. Thus,

$$-\frac{d}{d\tau}\left\{\left[\exp\left(-\frac{\tau}{\tau_0}\right)\right]a^\mu(\tau)\right\} = \frac{1}{m_0\tau_0}\left[\exp\left(-\frac{\tau}{\tau_0}\right)\right]K^\mu(\tau)\,,$$

and integrating between τ and infinity (due to the asymptotic condition) yields

$$a^\mu(\tau) = \frac{\exp(\tau/\tau_0)}{m_0\tau_0}\int_\tau^\infty \left[\exp\left(\frac{-\tau'}{\tau_0}\right)\right] K^\mu(\tau')\,d\tau'\,. \tag{7-41}$$

According to Eq. (7-40), this is a nonlinear integral equation for the 4-acceleration of the particle. However, one must demand that the condition (7-38) be satisfied, so that the restriction on the right-hand side of Eq. (7-41) is that it vanish in the limit $|\tau| \to \infty$. This is in fact a necessary condition that two more integrals of the equation exist in order to obtain a solution for the world line of the particle. We shall not, however, integrate Eq. (7-41) further, but only consider the *integral equation of motion*

$$a^\mu(\tau) = \int_\tau^\infty \left[\exp\frac{\tau-\tau'}{\tau_0}\right] \times \left[\frac{1}{m_0\tau_0}F^\mu_{\text{ext}}(\tau') + \frac{1}{c^2}a^\nu(\tau')a_\nu(\tau')v^\mu(\tau')\right]d\tau'\,. \tag{7-42}$$

In the remainder of this section we shall examine the mathematical properties and physical content of this equation, which must always be taken in conjunction with Eq. (7-38).

Mathematically, the first observation of interest is that Eqs. (7-38) and (7-42) do not admit the so-called runaway solutions of the type (7-15b). Thus, if solutions to the equations of motion exist, then one can have some confidence that they are physical. Regarding the question of existence, Hale and Stokes[25] have solved this problem under very weak conditions, and it appears that solutions exist for all cases of physical interest, including most reasonable initial conditions. The mathematical problem of the uniqueness of these solutions, however, has not been solved and therefore remains one of the unsolved problems of the theory. Nonetheless, in the absence of clear-cut physical reasons for such ambiguity, one must conclude on intuitive grounds that the solutions are indeed unique.

Prior to recognition of the important role played by the asymptotic conditions, Eliezer[26] presented a series of proofs that no physical solution of the equations of motion existed for certain important problems. Actually, he overlooked the physical solutions to these problems, and Plass[27] later discussed them in detail. It is just the asymptotic condition, Eq. (7-38), which eliminates the unphysical results. An important problem in this class is that of a charged particle moving in one dimension under an attractive Coulomb force. This problem has no solutions at the origin in terms of ordinary functions, but Clavier[28] has obtained satisfactory solutions which are distributions.[29] Nevertheless, the physical situation at the center of force is really outside the realm of classical physics so that we must actually take the point of view that such effects are classically unobservable.

Physically, the second square-bracketed expression in the integrand of Eq. (7-42) can be interpreted as an effective force, which is the external force minus the radiation reaction force. That is, according to Problem 5-2, the second term in this expression is the negative rate at which radiation is emitted. Notice that this radiation rate also occurs in the Lorentz–Dirac equation (7-34) as *part* of the Abraham 4-vector Γ^μ, Eq. (7-33). Thus, the interpretation of Γ^μ as the radiation reaction force is incorrect, because radiation can be emitted regardless of whether $\Gamma^\mu = 0$ or not (recall the Lorentz invariant criterion of Problem 5-2).

Exercise. If it is the case that $\Gamma^\mu = 0$, what type of motion does the particle undergo?

Further physical properties of Eq. (7-42) can be observed most readily by changing variables in Eq. (7-41) to

$$x = (\tau' - \tau)/\tau_0 . \tag{7-43}$$

Then

$$m_0 a^\mu(\tau) = \int_0^\infty K^\mu(\tau + x\tau_0)\, e^{-x}\, dx . \tag{7-44}$$

This is seen to be a very *nonlocal* equation, in that the acceleration depends on the force at time τ as well as all other times. The origin of this nonlocal behavior can be easily pinpointed by noting that the integral in Eq. (7-44) is cut off exponentially for times greater than $x = 1$. Hence, for the case in which K^μ remains constant over such time intervals we can write

$$m_0 a^\mu(\tau) \simeq K^\mu(\tau + x\tau_0) \simeq K^\mu(\tau)\,, \tag{7-45}$$

because of Eq. (7-2). The difference between this equation and the Lorentz–Dirac equation, Eq. (7-34), is just the quantity

$$\tfrac{2}{3}(e^2/c^3)\dot{a}^\mu = m_0\tau_0\dot{a}^\mu\,, \tag{7-46}$$

called the *Schott term.* We see that it is just the absence of this term which is responsible for the nonlocal time dependence, and this term was eliminated from the equations of motion by the integrating factor $\exp(-\tau/\tau_0)$. The Schott term actually has further physical significance, and we shall examine it in more detail below.

The nonlocal time behavior discussed above has been used to infer a possible violation of causality in the equations of motion. While it is clear that these equations predict the future motion and ensure signal velocities less than c (relativistic invariance), it is possible that a nonzero acceleration could occur prior to the action of the force. From Eq. (7-44) one sees that for a force which vanishes outside a time interval $(\tau_1 - \tau_2)$ there can be a finite acceleration at time $\tau < \tau_1$, since the integral goes over *all* future times. However, it is obvious that such *preaccelerations* must take place within time intervals on the order of τ_0, and there are no classical measurements which can possibly be made with time resolutions of this accuracy. As discussed at the beginning of this chapter, times of the order of magnitude τ_0 are really related to questions concerning particle structure, and therefore are necessarily quantum mechanical. At this time it would seem that not even quantum mechanics is capable of making such observations.*

An interesting physical question often asked centers around the principle of energy conservation. That is, from where is the energy supplied which the electron radiates when accelerated? The question can be examined most readily by referring to the Lorentz–Dirac equation, Eq. (7-34), and observing the fourth component. One finds for the rate at which the external force

* It might also be added that the concept of causality is maintained intact as long as one can always distinguish between cause and effect, irrespective of which of these is observed first on some time scale. Thus, one cannot *logically* rule out precognition, even though no firm evidence for such an effect in nature has been found. There is great danger in taking too dogmatic a stance when discussing time intervals on the order of τ_0.

does work

$$\frac{dW_{\text{ext}}}{c\,dt} = \frac{dT}{dt}\frac{\gamma^3}{c^2} - \frac{2}{3}\frac{e^2}{c^3}\left(c\frac{da^4}{dt} + \frac{1}{c}a^\mu a_\mu\right), \tag{7-47}$$

where T is the kinetic energy of the particle. Referring again to Problem 5-2, we see that the energy lost by radiation is entirely accounted for by the work done by radiation reaction. This would be an excellent situation were it not for the remaining term in Eq. (7-47), which is just the fourth component of the Schott term, Eq. (7-46). The quantity

$$Q = \tfrac{2}{3}(e^2/c^3)a^4 = m_0\tau_0 a^4 \tag{7-48}$$

has been called acceleration energy by Schott,[30] and may be positive or negative. (Actually cQ has the dimensions of energy.)

The second term on the right-hand side of Eq. (7-47) has been interpreted as the change in the particle's internal energy, which can contribute to the radiation. This cannot affect the rest mass of the particle, since it is easy to see that $a^4 = 0$ in the rest frame. Most probably, this energy is accounted for by the velocity field, which remains associated with the particle and does not detach itself as does the radiation field. However, the complete understanding of this term will probably remain unsatisfactory within the scope of classical electrodynamics, and should be considered quantum mechanically along with the electron structure.

Finally, it should be mentioned that along with the foregoing discussion of classical electrodynamics there exist still other formulations of the theory which are essentially equivalent. Rohrlich[31] has given an action principle which yields the previous equations of motion, along with the conservation laws, and which is free of mathematical difficulties such as divergent self-energy terms. Another approach which appears to have some merit is that taken by Bradbury,[32] in which he performs the calculations in the accelerated (noninertial) rest frame of the particle and then finds the transformation to an inertial frame. Although the equations derived are essentially those obtained by other formulations, the calculation for three-dimensional motion is quite tedious. At any rate, the preceding discussion would seem to indicate that with Eqs. (7-38) and (7-42) we have a quite satisfactory classical theory of charged particles. We shall close this chapter by examining some actual problems falling within the domain of competence of the theory.

C. APPLICATIONS OF THE THEORY

In conclusion it is appropriate to examine some typical physical situations so as to have available some actual predictions of the theory. It is sufficient to restrict our considerations to the nonrelativistic limit, since this simplifies the analysis and allows one to focus on the effects of radiation reaction rather than on the relativistic complications.

Presumably, the nonrelativistic limit ($v \ll c$) of the Lorentz–Dirac equation (7-34) is the Abraham–Lorentz equation (7-14) but with the factor of $\frac{4}{3}m_e$ replaced by m_0. This is indeed true if the time rate-of-change of the 3-acceleration is in some sense greater than (va^2/c^2).

Exercise. Examine in detail the conditions under which the modified Abraham–Lorentz equation is the nonrelativistic limit of the Lorentz–Dirac equation.

It must also be kept in mind that the nonrelativistic limit of the asymptotic condition (7-38) is an integral part of the ensuing equations of motion. Applying techniques similar to those leading to Eq. (7-41), we easily derive from the Abraham–Lorentz equation the integral equation

$$m_0\mathbf{a}(t) = \frac{1}{\tau_0}\int_t^\infty \left[\exp\left(\frac{t-t'}{\tau_0}\right)\right]\mathbf{F}_{\text{ext}}(t')\,dt' . \tag{7-49}$$

This is to be compared with the nonrelativistic limit of Eq. (7-42),

$$m_0\mathbf{a}(t) = \frac{1}{\tau_0}\int_t^\infty \left[\exp\left(\frac{t-t'}{\tau_0}\right)\right]\left[\mathbf{F}_{\text{ext}}(t') + m_0\tau_0\frac{a^2(t')}{c^2}\,\mathbf{v}(t')\right]dt' . \tag{7-50}$$

Equations (7-49) and (7-50) differ by the force of radiative reaction, although the former does contain the energy loss by radiation. The importance of the extra term in Eq. (7-50) is again measured by the magnitude (a^2v/c^2).

Exercise. Verify the derivations of Eqs. (7-49) and (7-50). Also show that the former can be rewritten as

$$m_0\mathbf{a}(t) = \int_0^\infty \mathbf{F}_{\text{ext}}(t + x\tau_0)\,e^{-x}\,dx . \tag{7-51}$$

Let us consider an electron moving nonrelativistically in a static and uniform magnetic field $\mathbf{B}$, taken in the z direction.[27] The solution is obtained most readily from the modified Abraham–Lorentz equation, where the external force is just the Lorentz force. Thus, the equations of motion in component form are

$$\dot{v}_x - \tau_0\ddot{v}_x = \omega_L v_y , \tag{7-52a}$$

$$\dot{v}_y - \tau_0\ddot{v}_y = -\omega_L v_x , \tag{7-52b}$$

$$\dot{v}_z - \tau_0\ddot{v}_z = 0 , \tag{7-52c}$$

where the dots indicate total derivatives with respect to t, and

$$\omega_L = eB/m_0c \tag{7-53}$$

is the Larmor frequency. For simplicity we invoke the initial conditions $v_x(0) = v$ and $v_y(0) = 0 = v_z(0)$. The velocity component along the field is then zero, and application of the asymptotic condition leads to the following solutions of Eqs. (7-52):

$$\begin{aligned} v_x &= v\,e^{-\alpha t}\cos\beta t\,, \\ v_y &= -v\,e^{-\alpha t}\sin\beta t\,. \end{aligned} \tag{7-54}$$

The constants α and β are determined by substitution into the original differential equation, giving

$$\alpha = (2\tau_0)^{-1}\{[\tfrac{1}{2} + \tfrac{1}{2}(1 + 16\tau_0^2\omega_L^2)^{1/2}]^{1/2} - 1\}\,, \tag{7-55a}$$

$$\beta = (2\tau_0)^{-1}[-\tfrac{1}{2} + \tfrac{1}{2}(1 + 16\tau_0^2\omega_L^2)^{1/2}]^{1/2}\,. \tag{7-55b}$$

In order to extract some physics from these results, one should note that in almost all physical situations, $\tau_0\omega_L \ll 1$.

Exercise. How strong a field is necessary in order to have $\tau_0\omega_L \sim 1$? Compare this figure with uniform laboratory fields available today.

Thus, the approximations

$$\alpha = \tau_0\omega_L^2(1 - 5\tau_0^2\omega_L^2 + \cdots)\,, \tag{7-56a}$$

$$\beta = \omega_L(1 - 2\tau_0^2\omega_L^2 + \cdots)\,, \tag{7-56b}$$

are generally quite good. The motion of the particle is essentially circular at the Larmor frequency, departing from this motion slightly only at very high field strengths. From Eq. (7-54), however, we see that the motion will decay slightly as time goes on, the particle eventually spiraling toward the center of the motion as a result of the energy lost due to radiation. For high enough field strengths the particle will radiate most of its energy in a single revolution, as can be seen by considering the limit $\tau_0\omega_L \gg 1$, in which case

$$\alpha = \beta = (\omega_L/2\tau_0)^{1/2}\,. \tag{7-57}$$

The frequency of the motion and the decay constant are then equal. A possible trajectory for the intermediate case $\omega_L \approx \tau_0$ is shown in Fig. 21.

An important physical situation to which we can apply the non-relativistic equations is that of the forced vibration of a harmonically bound electron with a spherically symmetric, radial restoring force $m_0\omega_0^2\mathbf{x}$. The

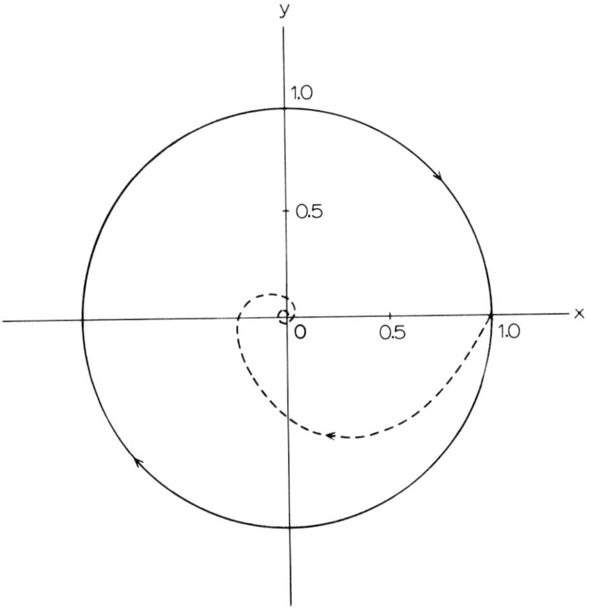

Fig. 21. Trajectory of a charged particle moving nonrelativistically in a uniform magnetic field in the z direction. The solid circular path neglects radiation reaction, whereas inclusion of this effect for $\omega_L \sim \tau_0$ results in the dashed-line trajectory in which the particle loses a considerable fraction of its energy in one revolution.

results of such an analysis form the basis for the classical theory of spectral emission, absorption, and scattering. The external force is composed of the restoring force plus an incident plane wave, so that the Abraham–Lorentz equation becomes

$$\ddot{\mathbf{x}} - \tau_0 \dddot{\mathbf{x}} + \omega_0^2 \mathbf{x} = (e/m_0)\boldsymbol{\epsilon} E_0 \, e^{-i\omega t} \, , \tag{7-58}$$

where $\boldsymbol{\epsilon}$ is the polarization vector of the incident wave. We shall be interested only in the steady-state solution, which is therefore proportional to $e^{-i\omega t}$. Thus, a reasonably good approximation is

$$\dddot{\mathbf{x}} \simeq -\omega^2 \dot{\mathbf{x}} \, . \tag{7-59}$$

The steady-state solution of the ensuing equation is then easily found to be

$$\mathbf{x} = \frac{e}{m_0} \boldsymbol{\epsilon} \frac{E_0 \, e^{-i\omega t}}{\omega_0^2 - \omega^2 - i\omega^3 \tau_0} \, . \tag{7-60}$$

Equation (7-60) can now describe the accelerated motion of the charge which gives rise to a radiation field. The radiated electric field for nonrelativistic excitation is given by Eq. (5-1), so that the component with polarization ϵ' is

$$\epsilon' \cdot \mathbf{E} = \omega^2 r_0 \left(\frac{\epsilon \cdot \epsilon'}{r}\right) \frac{E_0\, e^{i(kr-\omega t)}}{\omega_0^2 - \omega^2 - i\omega^3\tau_0}. \tag{7-61}$$

The differential scattering cross section now follows, using Eq. (6-1) and time averages:

$$\frac{d\sigma(\omega, \epsilon')}{d\Omega} = r_0^2(\epsilon \cdot \epsilon')^2 \frac{\omega^4}{(\omega_0^2 - \omega^2)^2 + \omega^6\tau_0^2}. \tag{7-62}$$

Note that the first two factors comprise the Thomson differential cross section, Eq. (6-17).

An important limiting case of Eq. (7-62) arises when the frequency of the incident wave is much lower than the binding frequency ($\omega \ll \omega_0$), so that

$$\frac{d\sigma(\omega, \epsilon')}{d\Omega} \simeq r_0^2(\epsilon \cdot \epsilon')^2 \left(\frac{\omega}{\omega_0}\right)^4, \tag{7-63}$$

independent of τ_0. This is the famous *Rayleigh scattering law*, which was originally derived assuming that the molecules in a sample scatter in a random fashion, so that one can add the individual intensities. Rayleigh[33] employed the fourth-power law to explain the blueness of the sky, the random-phase assumption being justified later by Einstein[34] and von Smoluchowski,[35] who showed that this assumption follows from the smallness of density fluctuations in the atmosphere. One sees that the above scattering processes contribute most at the high-frequency, or blue, end of the spectrum.

For small binding frequencies ($\omega_0 \ll \omega$) Eq. (7-62) becomes

$$\frac{d\sigma(\omega, \epsilon')}{d\Omega} \simeq r_0^2(\epsilon \cdot \epsilon')^2 \frac{1}{1 + \tau_0^2\omega^2}, \tag{7-64}$$

which exhibits a *radiative correction* to the Thomson cross section. Such a correction is generally negligible, but exhibits the effects of radiative reaction.*

The scattering shows a resonance in the region $\omega \simeq \omega_0$, and in this case Eq. (7-62) becomes

$$\frac{d\sigma(\omega, \epsilon')}{d\Omega} \simeq \frac{9}{16}\left(\frac{c}{\omega_0}\right)^2 \frac{\Gamma^2}{(\omega - \omega_0)^2 + (\Gamma/2)^2} (\epsilon \cdot \epsilon')^2, \tag{7-65}$$

where

$$\Gamma = \omega_0^2\tau_0 \tag{7-66}$$

* This correction factor should really be set equal to unity classically, because, if it was significant, it would correspond to high-energy photons.

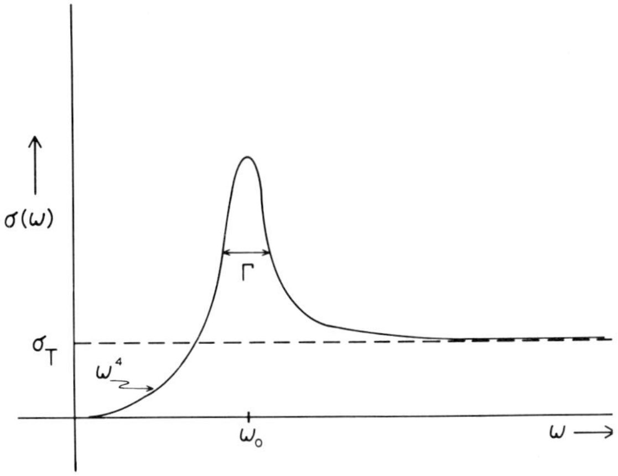

Fig. 22. Frequency distribution of radiation due to the forced vibration of a harmonically bound electron, exhibiting Rayleigh and Thomson scattering at the extreme ends of the spectrum.

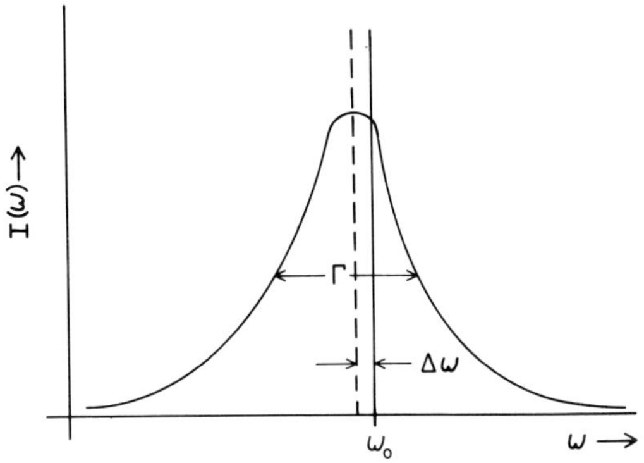

Fig. 23. Lorentzian line shape due to the natural vibrations of a harmonically bound charged particle.

is called the *half-width* (that is, the width of the frequency distribution at half-maximum intensity).

Exercise. Verify Eq. (7-65).

The total scattering cross section as a function of frequency is obtained by integrating over all angles, and we find

$$\sigma(\omega) \simeq \frac{3\pi}{2}\left(\frac{c}{\omega_0}\right)^2 \frac{\Gamma^2}{(\omega - \omega_0)^2 + (\Gamma/2)^2}\,. \tag{7-67}$$

In Fig. 22 the scattering cross section is plotted as a function of frequency, the shape of which is a common phenomenon. We have been treating here the forced vibrations of a harmonically bound electron, but it is possible to also study the natural vibrations without the external force. In Problem 7-4, we find that the frequency distribution of radiation in this case is

$$I(\omega) = I_0 \frac{\Gamma}{2\pi} \frac{1}{(\omega - \omega_0 - \Delta\omega)^2 + (\Gamma/2)^2}, \tag{7-68}$$

normalized to the total energy radiated, I_0. The half-width Γ is given by Eq. (7-66), and the *level shift* $\Delta\omega$ is defined as

$$\Delta\omega = -\tfrac{5}{8}\omega_0{}^3\tau_0{}^2 . \tag{7-69}$$

The shape of $I(\omega)$, plotted in Fig. 23, is called a *Lorentzian line shape*, and the fact that it is not a sharp line is directly attributable to the broadening caused by radiative reaction. As expected, then, the scattering cross section $\sigma(\omega)$ exhibits a similar structure, the sharp maximum at $\omega = \omega_0$ being referred to as *resonance fluorescence*.

In a similar manner the cross sections for absorption and emission of radiation can be calculated. Rather than pursue their development at this time, however, we will defer their discussion until we come to study quantum electrodynamics.

PROBLEMS

7-1. Verify Eq. (7-10) for the self-force on the electron, and obtain the next term in the expansion. That is, calculate the lowest-order, structure-dependent term.

7-2. Assume as a model of the electron a spherical distribution of uniform surface charge density, and calculate the self-force in the rest frame of the particle. Show that the internal pressure (self-force per unit surface element) is in the radial direction and of magnitude

$$\mathscr{P} = \tfrac{2}{3} W_s / V , \tag{7-70}$$

where V is the electronic volume. Comment on this form of expression for the pressure, which arises in other physical problems as well.

7-3. Assuming unpolarized light from the sun and single scattering processes, how much polarization is to be expected in sky light scattered at an angle θ?

7-4. Consider a charged particle bound by a one-dimensional linear restoring force with elastic constant $k = m_0\omega_0{}^2$. Assume a solution of the form $x(t) = x_0 e^{-\alpha t}$, and evaluate α in the limit $\omega_0\tau_0 \ll 1$. Finally, derive Eq. (7-68).

7-5. It has been claimed[36] that the solutions (7-54) to the problem of an electron in a magnetic field give rise to a paradox. That is, if one calculates the total energy radiated from $t = 0$ to $t = \infty$ by integrating Eq. (5-62), one finds that the energy radiated is greater than the initial kinetic energy of the electron. In particular, for $\omega\tau_0 = 0.968$, we find that $W_{\text{rad}} = 1.57mv^2/2$. Resolve this "paradox."

REFERENCES

1. H. A. Lorentz, *Arch. Nederl. Sci.* **25,** 363 (1892).
2. M. Abraham, *Ann. Physik* **10,** 105 (1903).
3. J. J. Thomson, *Phil. Mag.* **44,** 298 (1897).
4. J. J. Thomson, *Phil. Mag.* [5] **11,** 229 (1881).
5. W. Heitler, "The Quantum Theory of Radiation," 3rd ed., pp. 27–29. Oxford Univ. Press, London and New York, 1954.
6. See, e.g., J. D. Jackson, "Classical Electrodynamics," p. 21. Wiley, New York, 1962.
7. H. A. Lorentz, "The Theory of Electrons," 2nd ed. Dover, New York, 1952. These lectures were first given at Columbia University in 1906.
8. E. Fermi, *Physik. Z.* **23,** 340 (1922).
9. W. Wilson, *Proc. Phys. Soc.* (*London*) **48,** 736 (1936).
10. B. Kwal, *J. Phys. Radium* **10,** 103 (1949).
11. F. Rohrlich, *Am. J. Phys.* **28,** 639 (1960).
12. H. Poincaré, *Compt. Rend.* **140,** 1504 (1905); *Rend. Circ. Mat. Palermo* **21,** 129 (1906).
13. F. C. Witteborn and W. M. Fairbank, *Phys. Rev. Letters* **19,** 1049, 1123 (1967).
14. L. I. Schiff and M. V. Barnhill, *Phys. Rev.* **151,** 1067 (1966).
15. G. Wentzel, *Z. Physik* **86,** 479, 635 (1933); **87,** 726 (1934).
16. P. A. M. Dirac, *Proc. Roy. Soc.* (*London*) **A167,** 148 (1938).
17. M. Abraham, "Theorie der Elektrizität," Vol. II. Springer, Leipzig, 1905.
18. M. von Laue, *Ann. Physik* **28,** 436 (1909).
19. F. Rohrlich, "Classical Charged Particles," in particular, Section 6.5. Addison-Wesley, Reading, Massachusetts, 1965.
20. J. A. Wheeler and R. P. Feynman, *Rev. Mod. Phys.* **17,** 157 (1945); **21,** 425 (1949).
21. A. D. Fokker, *Z. Physik* **58,** 386 (1929).
22. F. Hoyle and J. V. Narlikar, *Ann. Phys.* (*New York*) **54,** 207 (1969).
23. R. Haag, *Z. Naturforsch.* **10a,** 752 (1955).
24. F. Rohrlich, *Ann. Phys.* (*New York*) **13,** 93 (1961).
25. J. K. Hale and A. P. Stokes, *J. Math. Phys.* **3,** 70 (1962).
26. C. J. Eliezer, *Rev. Mod. Phys.* **19,** 147, (1947).
27. G. N. Plass, *Rev. Mod. Phys.* **33,** 37 (1961).
28. P. A. Clavier, *Phys. Rev.* **124,** 616 (1961).
29. L. Schwartz, "Theorie des Distributions." Hermann, Paris, 1950.
30. G. A. Schott, "Electromagnetic Radiation." Cambridge Univ. Press, London and New York, 1912.
31. F. Rohrlich, "Classical Charged Particles," Section 6.9. Addison-Wesley, Reading, Massachusetts, 1965.
32. T. C. Bradbury, *Ann. Phys.* (*New York*) **19,** 323 (1962).
33. Lord Rayleigh (J. W. Strutt), *Phil. Mag.* **47,** 375 (1899).
34. A. Einstein, *Ann. Physik* **33,** 1294 (1910).
35. M. von Smoluchowski, *Ann. Physik* **25,** 205 (1908).
36. W. J. M. Cloetens, B. Castillo, E. DeEscobar, and R. Merlos, *Nuovo Cimento* **62A,** 247 (1969).

VIII CANONICAL FORMULATION OF CLASSICAL ELECTRODYNAMICS*

The radical changes in microscopic physical theory generated by the discovery of quantum mechanics made it immediately clear that the classical theory of electrodynamics, including the physical picture of the electron, was very incomplete. From this point on, it is our purpose to study the essentially more correct description of electromagnetic phenomena in the microscopic domain known as quantum electrodynamics (QED). In order to pursue this study, we begin with the well-known fact that quantum theory in its formal aspects is intimately related to the canonical Hamiltonian formalism of classical mechanics. Therefore, as a first step in developing QED we shall reformulate the entire classical theory of radiation in canonical form. This procedure introduces no new physics, but merely serves to express the equations of the field, and its interaction with matter, in a form eminently well suited for quantization.

It should be emphasized at the outset that the following chapters will, for the most part, be concerned with nonrelativistic QED, or the noncovariant theory. This will lead to an understanding of a great many physical phenomena, with a minimum of mathematical complications, and take us to the point where divergences in the theory require the notion of Lorentz invariance to play an explicit role in formulation of the equations and calculational procedures. As simple limit on the objectives of the next several chapters, we shall describe the motion of charged particles by the Schroedinger equation, and the relativistic Dirac equation will not be considered.

* The formulation presented in this chapter was first developed by Fermi.[1]

A. THE PURE RADIATION FIELD

As a first step in the program outlined above, let us consider only the free electromagnetic field. In classical mechanics it is well known that a system is understood in principle if it can be described in terms of normal modes. Furthermore, a system with a finite number of degrees of freedom undergoing small oscillations from a configuration of stable equilibrium is always solvable in terms of simple harmonic oscillations, and the energy of such a system is*

$$W = \tfrac{1}{2} \sum_i (\dot{q}_i^2 + \omega_i^2 q_i^2) . \tag{8-1}$$

The numbers q_i, of course, are the generalized coordinates appropriate to the particular problem. Classically, the vacuum ($\mathbf{E} = \mathbf{B} = 0$) is a state of stable equilibrium, so that one might expect a normal-mode description to apply to the free electromagnetic field, but with an infinite number of degrees of freedom. This expectation is strengthened by recalling that such a description indeed exists for electromagnetic cavities,[2] and with a *countably* infinite number of modes. In fact, these systems suggest the means for generalizing the description to encompass the free field. The existence of this generalization in a manner such that the field energy takes the form of Eq. (8-1) is known as *Jeans' theorem.*[3]

In order to derive the canonical equations, it is assumed that the field is enclosed in a very large cube of volume $V = L^3$, where L is extremely large compared with the dimensions of any system considered. For the free field, **E** and **B** are transverse so that they can be derived from a vector potential alone. The result of Problem 4-7 suggests that a convenient choice of gauge in this case is the Coulomb gauge, and much stronger support for this choice will be found in the next section. The vector potential **A** then satisfies the homogeneous wave equation

$$\nabla^2 \mathbf{A} - (1/c^2)\, \partial_t^2 \mathbf{A} = 0 , \tag{8-2}$$

subject to the subsidiary condition

$$\nabla \cdot \mathbf{A} = 0 . \tag{8-3}$$

The choice of boundary conditions is somewhat arbitrary, but it will be found convenient to require **A** to be periodic on the surface of V.

The situation is now that of the usual eigenvalue problem for the wave equation, and we assume that **A** can be expanded in terms of its eigenfunctions $\mathbf{A}_i(\mathbf{r})$:

$$\mathbf{A}(\mathbf{r}, t) = \sum_i q_i(t) \mathbf{A}_i(\mathbf{r}) . \tag{8-4}$$

* Superposed dots will always indicate total time derivatives.

Substitution into Eq. (8-2) allows a separation of variables in a familiar manner, yielding the equations determining q_i and $\mathbf{A}_i$:

$$\nabla^2 \mathbf{A}_i(\mathbf{r}) + (\omega_i^2/c^2)\mathbf{A}_i(\mathbf{r}) = 0\,, \tag{8-5}$$

$$\ddot{q}_i(t) + \omega_i^2 q_i(t) = 0\,. \tag{8-6}$$

Thus, the coefficients $q_i(t)$ satisfy the harmonic oscillator equation, and, according to Eqs. (8-3) and (8-4), the eigenfunctions of $\mathbf{A}$ satisfy the equation

$$\nabla \cdot \mathbf{A}_i(\mathbf{r}) = 0\,. \tag{8-7}$$

It must be emphasized that the functions $\mathbf{A}_i$ *do not* in general form a complete set of orthogonal vector functions, because no longitudinal vector field can be expanded in terms of functions satisfying Eq. (8-7). However, they are sufficient for representing the free electromagnetic field, because in this case $\mathbf{E}$ and $\mathbf{B}$ are transverse.

The numbers ω_i are, of course, separation constants and can be identified as frequencies of the eigenvibrations if we introduce a *propagation vector* $\mathbf{k}_i$ with magnitude

$$k_i = |\mathbf{k}_i| = \omega_i/c\,. \tag{8-8}$$

Due to the periodic boundary conditions, $\mathbf{k}_i$ can assume only a discrete set of values:

$$\begin{aligned} k_{ix} &= (2\pi/L)n_{ix}\,, \\ k_{iy} &= (2\pi/L)n_{iy}\,, \\ k_{iz} &= (2\pi/L)n_{iz}\,, \end{aligned} \tag{8-9}$$

where n_{ix}, n_{iy}, and n_{iz} are integers. This in turn requires the $\mathbf{A}_i$ to be simple harmonic functions, such as

$$(8\pi c^2/V)^{1/2}\epsilon_i \cos(\mathbf{k}_i \cdot \mathbf{r})\,, \qquad (8\pi c^2/V)^{1/2}\epsilon_i \sin(\mathbf{k}_i \cdot \mathbf{r})\,, \tag{8-10}$$

where ϵ_i is a unit polarization vector. We have chosen the normalization of the orthogonal set $\{\mathbf{A}_i\}$ such that

$$\int \mathbf{A}_i \cdot \mathbf{A}_j \, d^3r = 4\pi c^2 \delta_{ij} \tag{8-11}$$

The reason for this choice will be examined more closely later.

For each value of $\mathbf{k}_i$ there are two independent directions of polarization. The transverse nature of the eigenwaves is demonstrated by substituting functions like those exhibited in Eq. (8-10) into Eq. (8-7). In any case one finds that

$$\mathbf{k}_i \cdot \epsilon_i = 0\,, \tag{8-12}$$

the *transversality condition.*

Exercise. Verify that the functions of Eq. (8-10) satisfy Eq. (8-11). (Remember the two independent directions of polarization.)

Since the $\mathbf{A}_i$ are given functions of the coordinates, it follows from Eq. (8-4) that the electromagnetic field is completely specified by the functions $q_i(t)$ which satisfy Eq. (8-6). Therefore, the field has indeed been represented by an infinite number of harmonic oscillators, and the canonical equations now follow from the Hamiltonian function for an oscillator:

$$H_i = \tfrac{1}{2}(p_i^2 + \omega_i^2 q_i^2) . \tag{8-13}$$

The p_i and q_i satisfy Hamilton's equations,

$$\begin{aligned} \partial H_i/\partial q_i &= -\dot{p}_i , \\ \partial H_i/\partial p_i &= \dot{q}_i = p_i . \end{aligned} \tag{8-14}$$

Exercise. Derive Eq. (8-6) from Eqs. (8-13) and (8-14).

It is apparent from the preceding discussion that the total free field can be described by an infinite set of canonical variables with a total Hamiltonian

$$H = \sum_i H_i . \tag{8-15}$$

In classical dynamics H_i in Eq. (8-13) is the energy of a harmonic oscillator, and it is now necessary to show that H is the total energy of the field. From electromagnetic theory

$$W = (1/8\pi) \int (\mathbf{E}^2 + \mathbf{B}^2)\, d^3r , \tag{8-16}$$

with field strengths

$$\mathbf{E} = -(1/c)\,\partial_t \mathbf{A} = -(1/c) \sum_i p_i \mathbf{A}_i , \tag{8-17a}$$

$$\mathbf{B} = \nabla \times \mathbf{A} = \sum_i q_i \nabla \times \mathbf{A}_i , \tag{8-17b}$$

where Eqs. (4-2) and (8-4) have been used. Substitution of Eqs. (8-17) into (8-16) gives for the energy

$$\begin{aligned} W = (1/8\pi c^2) \sum_{i,j} \int [&p_i p_j \mathbf{A}_i \cdot \mathbf{A}_j \\ &+ c^2 q_i q_j (\nabla \times \mathbf{A}_i) \cdot (\nabla \times \mathbf{A}_j)]\, d^3r . \end{aligned} \tag{8-18}$$

The second term in the integrand can be evaluated by employing a standard vector identity† and introducing Gauss' divergence theorem:

$$\int (\nabla \times \mathbf{A}_i) \cdot (\nabla \times \mathbf{A}_j)\, d^3r$$
$$= \int \mathbf{A}_i \times (\nabla \times \mathbf{A}_j) \cdot d\mathbf{S} + \int \mathbf{A}_i \cdot \nabla \times \nabla \times \mathbf{A}_j\, d^3r\,.$$

The first integral on the right-hand side vanishes due to the periodic boundary conditions on the surface of V, and the second integral can be developed using the identity $\nabla \times \nabla \times \mathbf{A} = \nabla(\nabla \cdot \mathbf{A}) - \nabla^2\mathbf{A}$, along with Eq. (8-7). This bit of algebra leads to

$$\int (\nabla \times \mathbf{A}_i) \cdot (\nabla \times \mathbf{A}_j)\, d^3r = -\int \mathbf{A}_i \cdot \nabla^2 \mathbf{A}_j\, d^3r\,. \tag{8-19}$$

Using Eq. (8-5) in this last result, we now find for the total energy

$$W = (1/8\pi c^2) \sum_{i,j} \int [p_i p_j + \omega_j^2 q_i q_j] \mathbf{A}_i \cdot \mathbf{A}_j\, d^3r\,. \tag{8-20}$$

The orthogonality relation (8-11) now gives Eq. (8-1) immediately, and the proof of Jeans' theorem is complete—the total field energy is the sum of the energies of the countably infinite set of oscillators. Moreover, Eq. (8-20) now explains the choice of normalization constant in Eq. (8-11).

As a final matter for this section, it is of interest to note that the harmonic solutions to the wave equation exhibited in Eq. (8-10) can be taken in linear combination as complex exponential functions. The vector potential is a real function, and we can write it as the sum of a complex quantity and its complex conjugate:

$$\mathbf{A}(\mathbf{r}, t) = \sum_i [q_i(t)\mathbf{A}_i(\mathbf{r}) + q_i^*(t)\mathbf{A}_i^*(\mathbf{r})]\,. \tag{8-21}$$

In this complex representation the respective solutions to Eqs. (8-5) and (8-6) are

$$\mathbf{A}_i(\mathbf{r}) = (4\pi c^2/V)^{1/2} \epsilon_i \exp(i\mathbf{k}_i \cdot \mathbf{r})\,, \tag{8-22}$$
$$q_i(t) = |q_i| \exp(-i\omega_i t)\,. \tag{8-23}$$

The values of $\mathbf{k}_i$ can be proportional to positive *or* negative integers, and the orthogonality condition becomes

$$\int \mathbf{A}_i^* \cdot \mathbf{A}_j\, d^3r = 4\pi c^2\, \delta_{ij}\,. \tag{8-24}$$

Note that these complex q_i are *not* canonical variables, but that a set of

† Namely, $\nabla \cdot (\mathbf{A} \times \nabla \times \mathbf{B}) = (\nabla \times \mathbf{A}) \cdot (\nabla \times \mathbf{B}) - \mathbf{A} \cdot \nabla \times \nabla \times \mathbf{B}$.

real canonical variables can be constructed from them as follows:

$$Q_i = q_i + q_i^*\,, \tag{8-25a}$$

$$P_i = -i\omega_i(q_i - q_i^*) = \dot{Q}_i\,. \tag{8-25b}$$

Hamilton's equations can now be written

$$H_i = \tfrac{1}{2}(P_i^2 + \omega_i^2 Q_i^2) = 2\omega_i^2 q_i q_i^*\,, \tag{8-26a}$$

$$\partial H_i/\partial Q_i = -\dot{P}_i\,, \qquad \partial H_i/\partial P_i = \dot{Q}_i\,. \tag{8-26b}$$

Exercise. Demonstrate explicitly that for these new variables $W = \sum_i H_i$.

The primary motivation for introducing this complex representation is due to the fact that in this form the theory is most easily transcribed into a quantum mechanical formalism. The complex exponentials have the same functional form as the plane-wave momentum eigenfunctions of quantum mechanics.

Irrespective of the choice of eigenfunctions, however, we obtain the same number of radiation oscillators in the volume V with a given direction of propagation and polarization within a solid angle element $d\Omega$ and frequency interval $d\omega$. If, as we assume, the linear dimensions of the volume are large compared with the wavelength, Eqs. (8-8) and (8-9) imply that

$$\omega_i^2 = (2\pi c/L)^2(n_{ix}^2 + n_{iy}^2 + n_{iz}^2)\,. \tag{8-27}$$

The number we wish to calculate can be defined in terms of ρ_ω, the *density of light waves* in the volume V.†

Exercise. Show that

$$\rho_\omega\, d\omega\, d\Omega = (\omega^2\, d\omega\, d\Omega)/(2\pi c)^3\,. \tag{8-28}$$

B. FIELD AND PARTICLES COMBINED

In order to generalize the discussion to the case when both particles and fields are present, it is first necessary to find the Hamiltonian for a charged particle in the presence of an electromagnetic field. Nonrelativistically, the particle is under the influence of the Lorentz force, Eq. (4-34). By replacing the fields with the potentials, Eq. (4-2), we can write for the force on the particle

$$\mathbf{f} = e[-\nabla\phi - (1/c)\,\partial_t\mathbf{A} + (1/c)\,\mathbf{v} \times (\nabla \times \mathbf{A})]\,. \tag{8-29}$$

† See, e.g., Problem 13-8 of Panofsky and Phillips.[4]

Exercise. Show that

$$(\mathbf{v} \times \nabla \times \mathbf{A})_x = \partial_t A_x + \partial_x(\mathbf{v} \cdot \mathbf{A}) - (dA_x/dt)\,. \tag{8-30}$$

With this identity, and an alternative notation for the total derivative, one can now write

$$\begin{aligned} f_x &= e\left[-\partial_x\left(\phi - \frac{1}{c}\,\mathbf{v} \cdot \mathbf{A}\right) - \frac{1}{c}\frac{d}{dt}\,(\partial_{v_x}\mathbf{A} \cdot \mathbf{v})\right] \\ &= -\partial_x U + \frac{d}{dt}\,(\partial_{v_x} U)\,, \end{aligned} \tag{8-31}$$

where

$$U \equiv e\phi - \frac{e}{c}\,\mathbf{A} \cdot \mathbf{v}\,. \tag{8-32}$$

Equation (8-31) is precisely the Euler–Lagrange equation of the motion, so that the generalization to three dimensions immediately yields the Lagrangian

$$L = \frac{p^2}{2m} - e\phi + \frac{e}{c}\,\mathbf{A} \cdot \mathbf{v}\,. \tag{8-33}$$

In the usual manner, then, we find for the Hamiltonian function describing a charged particle with charge e and mass m in the presence of an electromagnetic field

$$H = e\phi + \left(\mathbf{p} - \frac{e}{c}\,\mathbf{A}\right)^2 \Big/ 2m\,. \tag{8-34}$$

This equation, in fact, provides a general prescription for converting the equations of classical mechanics into forms incorporating electromagnetic forces.[4]

The generality of Eq. (8-34) cannot be emphasized enough, for the vector potential $\mathbf{A}$ and scalar potential ϕ contain more than just the external fields in which the particle moves. They also refer to the particle's own fields, which means that the force of radiative reaction is automatically incorporated into the canonical formulation. Later we will see that this observation forms an essential part of the treatment.

In order to complete the canonical formulation, it is now necessary to write down the equations for the coupled system of charged particles and fields. If there is only one charged particle to be considered, then the task reduces to that of simply writing down the sum of Eqs. (8-15) and

(8-34) for the total Hamiltonian. When one wishes to consider an entire system of charged particles, the problem is not so trivial. Therefore, to remain as general as possible, we shall consider a number of charged particles with total Hamiltonian

$$H_p = \sum_k H_p(k) , \tag{8-35}$$

where the sum is over all particles, and

$$H_p(k) = e_k\phi(k) + \frac{1}{2m_k}\left[\mathbf{p}_k - \frac{e_k}{c}\mathbf{A}(k)\right]^2 . \tag{8-36}$$

Here $\phi(k)$ and $\mathbf{A}(k)$ are the potentials at the position of the kth particle due to the presence of all particles, including the kth particle, as well as any external fields. The tremendous coupling in the system is now apparent.

There are two different directions in which we can go at this point. The coupled system could be described by a Lagrangian density and the canonical equations obtained from the variational principle, or we can proceed in the manner of the last section. The former approach is essentially field theoretic in nature and is described quite well, for instance, in the books by Goldstein[5] and Barut.[6] For reasons of convenience in the subsequent chapters we shall follow the latter method and assume the system to be enclosed in a volume V such that $\mathbf{A}$ and ϕ are periodic on the surface. It is, however, important to consider the question of gauge before proceeding further.

In the last section the transverse nature of the free fields essentially dictated the choice of Coulomb gauge because the condition (8-3) is a natural expression of transversality. When particles are present, however, the fields are not necessarily transverse and one must admit the possibility of longitudinal waves. In this case, the choice of gauge should be made on possibly different grounds of convenience, and the result of Problem 4-7 offers a strong suggestion here. In the Coulomb gauge the scalar potential reduces to the instantaneous, static Coulomb interaction between the charged particles, and we receive an added bonus in that only transverse eigenvibrations will be present. Were we to consider here the relativistic motion of the charges, then the Lorentz gauge would be a better choice, since this represents a covariant gauge condition. But we shall not consider relativistic QED in what follows, and, moreover, we will want to have a transparent view of the effects of particle–particle interations. Thus, our objectives demand the choice of Coulomb gauge, Eq. (8-3).

One finds, then, that the equations satisfied by the potentials are*

$$-\left[\nabla^2\mathbf{A} - \frac{1}{c^2}\,\partial_t{}^2\mathbf{A}\right] + \frac{1}{c}\,\nabla\dot{\phi} = \frac{4\pi}{c}\,\rho\mathbf{v}\,, \tag{8-37}$$

$$\nabla^2\phi = -4\pi\rho\,. \tag{8-38}$$

The solution for ϕ is immediate:

$$\phi(\mathbf{x}) = \sum_k e_k/|\,\mathbf{r}_k - \mathbf{x}\,|\,, \tag{8-39}$$

as expected. In arriving at the form (8-39), we have assumed the system to consist of point charges, so that the charge density can be written

$$\rho = \sum_k e_k\,\delta(\mathbf{x} - \mathbf{r}_k)\,. \tag{8-40}$$

It is again possible to expand the vector potential as in Eq. (8-4), because the set of functions $\mathbf{A}_i$ still represents a complete set for the transverse vector field $\mathbf{A}$. Moreover, the $\mathbf{A}_i$ satisfy the differential equation (8-5), but $\mathbf{A}$ *does not*. [It satisfies Eq. (8-37).] In a like manner, the scalar potential ϕ can be expanded as

$$\phi(\mathbf{x}) = \sum_j b_j(t)\phi_j(\mathbf{x})\,, \tag{8-41}$$

in terms of a complete set of orthogonal functions satisfying

$$\nabla^2\phi_j + (\omega_j{}^2/c^2)\phi_j = 0\,, \tag{8-42}$$

and periodic boundary conditions on the surface of V. We shall normalize the ϕ_j as in Eq. (8-24).

The differential equations determining the canonical coordinates are now obtained by substituting Eqs. (8-4) and (8-41) into Eq. (8-37), and then taking the scalar product with $\mathbf{A}_j$, followed by an integration over the volume. The result for point charges is

$$-\sum_i q_i \int \mathbf{A}_j \cdot \nabla^2\mathbf{A}_i\, d^3r + \frac{1}{c^2}\sum_i \ddot{q}_i \int \mathbf{A}_j \cdot \mathbf{A}_i\, d^3r$$
$$+ \frac{1}{c}\sum_i \dot{b}_i \int \mathbf{A}_j \cdot \nabla\phi_i\, d^3r = \frac{4\pi}{cV}\sum_k e_k \int \mathbf{A}_j \cdot \mathbf{v}_k\, d^3r\,. \tag{8-43}$$

* It should be recalled that Maxwell's equations in terms of the potentials are, prior to the removal of the degree of freedom in choosing the gauge,

$$\nabla^2\mathbf{A} - (\partial_t{}^2\mathbf{A})/c^2 - \nabla[\nabla\cdot\mathbf{A} + (\partial_t\phi)/c] = -4\pi\mathbf{J}/c\,,$$
$$\nabla^2\phi + \partial_t(\nabla\cdot\mathbf{A})/c = -4\pi\rho\,.$$

Exercise. By means of the vector formula

$$\int [(\nabla \times \mathbf{A}) \cdot (\nabla \times \mathbf{B}) + (\nabla \cdot \mathbf{A})(\nabla \cdot \mathbf{B}) + \mathbf{A} \cdot \nabla^2 \mathbf{B}]\, d^3r$$
$$= \oint [\mathbf{A} \times (\nabla \times \mathbf{B}) + \mathbf{A}(\nabla \cdot \mathbf{B})] \cdot d\mathbf{S}\,,$$

show that the third integral on the left-hand side of Eq. (8-43) vanishes.

The result of this exercise, along with Eqs. (8-5) and (8-11), then yields

$$\ddot{q}_i + \omega_i^2 q_i = (1/cV) \sum_k e_k \int \mathbf{v}_k \cdot \mathbf{A}_i\, d^3r\,. \tag{8-44}$$

Thus, as might have been expected, the field variables represent the *forced* vibrations of harmonic oscillators, and the driving force is due to the motion of the charged particles. The canonical equations now follow by noting that the right-hand side of Eq. (8-44) must come from a term in the Hamiltonian containing particle variables, and that term is precisely H_{p}, Eq. (8-35).

Equation (8-38) can be treated in a similar manner, and a procedure analogous to the above algebra yields

$$\omega_j^2 b_j = \sum_k e_k \phi_j(k)\,. \tag{8-45}$$

This expression contains no new information, however, because the right-hand side is a well-defined function of the particle coordinates at a given time t, and the b_j are *not* independent variables. All of the information concerning the scalar potential is contained in Eq. (8-39), and if this is substituted into (8-36), we find that the total effect of ϕ in the Hamiltonian is a term

$$\tfrac{1}{2} \sum_{i \neq j} (e_i e_j / r_{ij})\,. \tag{8-46}$$

That is, the *static*, instantaneous Coulomb interactions among the particles are actually contained in the fields when the latter also include those due to the particles. Equation (8-46) has been written so as to only count each interaction once, and, of course, the self-energies occurring for $i = j$ are excluded.

The total Hamiltonian function for the combined system of particles and field can now be written as

$$H = \sum_k H_k + \sum_i H_i + \tfrac{1}{2} \sum_{i \neq j} (e_i e_j / r_{ij})\,, \tag{8-47}$$

where

$$H_k = (1/2m_k)[\mathbf{p}_k - (e_k/c)\mathbf{A}]^2 , \tag{8-48}$$

$$H_i = \tfrac{1}{2}(p_i^2 + \omega_i^2 q_i^2) . \tag{8-49}$$

If external fields are to be included, then a term $e_k \phi_{ex}(k)$ is to be added to H_k, and a term $\mathbf{A}_{ex}$ is to be added to $\mathbf{A}$.

Exercise. Show that Hamilton's equations of motion for the particle and field variables lead to Eq. (8-44) for the field and to the Lorentz equation for the particles.

Equations (8-47)–(8-49) represent the canonical formulation of classical electrodynamics for a system of charged particles interacting with the field and with each other. Physically, the acceleration of a charged particle will excite the field oscillators into one of a countably infinite number of states, and the radiation is described by transverse waves. That the particle–particle interaction is the static Coulomb potential is deceiving, because the Hamiltonian actually does contain the retarded interaction. The effects of retardation are contained in the transverse vector potential $\mathbf{A}$ in Eq. (8-48), and appear as a mutual emission and absorption of light waves by the particles. This mutual exchange of light waves will become physically more transparent in the quantum theory.

PROBLEMS

8-1. Assuming eigenvibrations of the sinusoidal type, Eq. (8-10), use the canonical formalism to derive Eq. (5-52) for the radiative energy loss due to Čerenkov radiation. The medium within V is nonpermeable and the dielectric constant is related to the index of refraction by $\varepsilon = n^2 > 1$.

8-2. Find a Lagrangian from which the field equations (8-6) can be derived, and find a Lagrangian density from which this Lagrangian can be derived.

REFERENCES

1. E. Fermi, *Rev. Mod. Phys.* **4,** 87 (1932).
2. J. D. Jackson, "Classical Electrodynamics," Section 8.6. Wiley, New York, 1962.
3. J. H. Jeans, *Phil. Mag.* **10,** 91 (1905).
4. W. K. H. Panofsky and M. Phillips, "Classical Electricity and Magnetism," 2nd ed. Addison-Wesley, Reading, Massachusetts, 1962.
5. H. Goldstein, "Classical Mechanics," Section 7.3. Addison-Wesley, Reading, Massachusetts, 1959.
6. A. O. Barut, "Electrodynamics and Classical Theory of Fields and Particles." Macmillan, New York, 1964.

IX QUANTIZATION OF THE RADIATION FIELD

With the canonical formulation of classical electrodynamics achieved in the last chapter it is now possible to begin a study of quantum electrodynamics. Following the previous procedure, we first concern ourselves with the quantization of the radiation field.

The necessity of quantizing the radiation field stems directly from Planck's quantum hypothesis, made in order to remove the ultraviolet divergence from the theory of blackbody radiation. The simple assumption made by Planck[1] was that the energy of a monochromatic plane wave with frequency ν can assume only the values

$$E_n = nh\nu = n\hbar\omega \,, \tag{9-1}$$

where n is an integer, h is a universal constant, and $\hbar = h/2\pi$. This assumption not only means that the classical theory of the preceding chapters has to be reexamined (hopefully, not completely abandoned), but it also leads to the entire quantum theory of particle kinematics and dynamics. It is interesting to note, in fact, that if the radiation field were not to be quantized in the manner indicated by Eq. (9-1), the entire quantum mechanical theory of material particles would be invalid: One could then measure the position of an electron, say, with as "soft" a photon as desired, thereby transferring arbitrarily small momentum in the process.

In this chapter we shall follow the original method of obtaining quantum equations from the classical theory by using the correspondence principle as a guide. That is, we follow Bohr in replacing the usual canonical variables of the system by linear operators which satisfy a noncommutative algebra, the noncommutativity being measured by the nonzero value of Planck's constant. The Hamiltonian method of the last chapter will be followed, although it should be noted that one could begin with the Lagrangian formulation and obtain the equations describing the system from Hamilton's principle. The latter approach is quite useful in deriving a field description of physics, and the ensuing development will tend to stress more the particle aspects of the theory. Complementarity, of course, allows the two views to be interchangeable.

A. DIRECT QUANTIZATION

Quantization of the free field begins with the plane wave expansion of the vector potential, Eqs. (8-21)–(8-23). As noted previously, the q_i are *not* canonical variables, and thus are unsuitable for quantization purposes. However, it was demonstrated in the last chapter that canonical variables could be defined by the linear combinations of Eq. (8-25), so that the energy of a light wave could be written

$$H_i = \tfrac{1}{2}(P_i^2 + \omega_i^2 Q_i^2) . \tag{9-2}$$

The quantization procedure now requires these canonical variables to be interpreted as linear operators satisfying the commutation relations

$$[P_i , Q_j] = P_i Q_j - Q_j P_i = -i\hbar\, \delta_{ij} , \tag{9-3a}$$

$$[P_i , P_j] = [Q_i , Q_j] = 0 . \tag{9-3b}$$

The solution to the harmonic oscillator problem with Hamiltonian (9-2) is well known in quantum mechanics, so that one can immediately write down the energy levels:

$$E_i = (n_i + \tfrac{1}{2})\hbar\omega_i , \tag{9-4}$$

where the integer $n_i \geqslant 0$. The total Hamiltonian of the system is

$$\mathsf{H} = \tfrac{1}{2} \sum_i (P_i^2 + \omega_i^2 Q_i^2) , \tag{9-5a}$$

with total energy

$$E = \sum_i \hbar\omega_i (n_i + \tfrac{1}{2}) . \tag{9-5b}$$

The wave function describing the state of the system is then obtained from the Schroedinger equation

$$\mathsf{H}(P, Q)\psi = i\hbar\, \partial_t \psi . \tag{9-6}$$

We shall work almost exclusively in the Schroedinger picture, so that all the operators are time independent and the wave functions evolve in time. Although the converse is true in the Heisenberg picture, the two can be made to coincide at $t = 0$, a convention to which we will adhere.

It is at times convenient to explicitly indicate the linear operator nature of H, P, and Q by exhibiting their infinite matrix forms in the representation of the Hamiltonian, Eq. (9-5a), in which H is diagonal. Thus, for $t = 0$ and *one particular state* of the system

$$\mathsf{H}_i = \frac{1}{2}\begin{bmatrix} \hbar\omega_i & 0 & 0 & \cdots \\ 0 & 3\hbar\omega_i & 0 & \cdots \\ 0 & 0 & 5\hbar\omega_i & \cdots \\ \vdots & \vdots & \vdots & \end{bmatrix}, \tag{9-7}$$

$$Q_i = \left(\frac{\hbar}{2\omega_i}\right)^{1/2}\begin{bmatrix} 0 & \sqrt{1} & 0 & 0 & \cdots \\ \sqrt{1} & 0 & \sqrt{2} & 0 & \cdots \\ 0 & \sqrt{2} & 0 & \sqrt{3} & \cdots \\ \vdots & \vdots & \vdots & \vdots & \end{bmatrix}, \tag{9-8}$$

$$P_i = i\left(\frac{\hbar\omega_i}{2}\right)^{1/2}\begin{bmatrix} 0 & -\sqrt{1} & 0 & 0 & \cdots \\ \sqrt{1} & 0 & -\sqrt{2} & 0 & \cdots \\ 0 & \sqrt{2} & 0 & -\sqrt{3} & \cdots \\ \vdots & \vdots & \vdots & \vdots & \end{bmatrix}. \tag{9-9}$$

At $t = 0$ these operators are defined on a vector space spanned by state vectors described by the number of excitations of the oscillator for the ith state. We represent these (orthonormal) vectors by

$$|N_i\rangle = \begin{bmatrix} 0 \\ 0 \\ \vdots \\ 0 \\ 1 \\ 0 \\ \vdots \end{bmatrix}, \tag{9-10}$$

where the "1" corresponds to the n_ith position. We also define a *number operator* for the ith state:

$$\mathsf{N}_i = \begin{bmatrix} 0 & 0 & 0 & 0 & \cdots \\ 0 & 1 & 0 & 0 & \cdots \\ 0 & 0 & 2 & 0 & \cdots \\ 0 & 0 & 0 & 3 & \cdots \\ \vdots & \vdots & \vdots & \vdots & \end{bmatrix}. \tag{9-11}$$

Exercise. Verify Eq. (9-2) in the matrix representation, as well as the operator equations

$$\mathsf{N}_i \,|\, N_i\rangle = n_i \,|\, N_i\rangle\,, \tag{9-12a}$$

$$\mathsf{H}_i \,|\, N_i\rangle = (n_i + \tfrac{1}{2})\hbar\omega_i \,|\, N_i\rangle\,. \tag{9-12b}$$

It is seen, therefore, that the basis vectors $|\,N_i\rangle$ are eigenstates of both H_i and N_i.

We can also deduce matrix representations of the q_i from the foregoing, but it is convenient to consider instead the operators obtained by the replacement

$$q_i \to (\hbar/2\omega_i)^{1/2} a_i\,. \tag{9-13}$$

One then verifies, by means, of Eqs. (8-25), that

$$a_i = \begin{bmatrix} 0 & \sqrt{1} & 0 & 0 & \cdots \\ 0 & 0 & \sqrt{2} & 0 & \cdots \\ 0 & 0 & 0 & \sqrt{3} & \cdots \\ \vdots & \vdots & \vdots & \vdots & \end{bmatrix}, \tag{9-14a}$$

$$a_i^\dagger = \begin{bmatrix} 0 & 0 & 0 & 0 & \cdots \\ \sqrt{1} & 0 & 0 & 0 & \cdots \\ 0 & \sqrt{2} & 0 & 0 & \cdots \\ \vdots & \vdots & \vdots & \vdots & \end{bmatrix}, \tag{9-14b}$$

where the dagger, of course, indicates the Hermitian conjugate (transposed complex conjugate). Referring to Eq. (8-23), we see that the time dependence of a_i is

$$a_i(t) = a_i e^{-i\omega_i t}\,. \tag{9-15}$$

Although we shall use the operators only at time $t = 0$, Eq. (9-15) will be needed whenever we wish to examine the time derivatives at $t = 0$.

Now, the canonical variables P_i and Q_i were originally introduced because they formed canonically conjugate pairs, whereas p_i and q_i did not. In quantum mechanical parlance, P_i and Q_i are Hermitian operators, as is easily verified from Eqs. (9-8) and (9-9). The operators q_i and p_i are *not* Hermitian, and so neither are a_i and $a_i^\dagger$, and therefore Eqs. (9-14) cannot represent physical observables. Nevertheless, we shall see below that a_i and $a_i^\dagger$ are the eminently useful operators with which to develop the quantum theory of radiation. The two operators are related through the commutation relations

$$[a_i, a_i^\dagger] = 1\,, \tag{9-16a}$$

$$[a_i, a_i] = [a_i^\dagger, a_i^\dagger] = 0\,, \tag{9-16b}$$

as is verified trivially from the above matrix forms.

It is possible, of course, to define the operators a_i and $a_i^\dagger$ for all possible states of the field oscillators, but this is rather cumbersome in the present context, and the next chapter will be specifically devoted to the development of a calculus governing the behavior of these operators. However, it is possible to obtain a generalization of Eqs. (9-16) to include different states by referring to the definition (9-13) and the commutation relations (9-3).

Exercise. Derive the relations

$$[a_i, a_j^\dagger] = \delta_{ij}\,, \tag{9-17a}$$

$$[a_i, a_j] = [a_i^\dagger, a_j^\dagger] = 0\,. \tag{9-17b}$$

The number operator and Hamiltonian can now be written

$$\mathsf{N}_i = a_i^\dagger a_i\,, \tag{9-18}$$

$$\mathsf{H}_i = \hbar\omega_i(a_i^\dagger a_i + \tfrac{1}{2})\,, \tag{9-19}$$

which are both Hermitian. A physical interpretation of a_i and $a_i^\dagger$ follows from observing their effect on a state vector $|N_i\rangle$:

$$a_i\,|N_i\rangle = \sqrt{n_i}\,|(N_i - 1)\rangle\,, \tag{9-20a}$$

$$a_i^\dagger\,|N_i\rangle = (n_i + 1)^{1/2}\,|(N_i + 1)\rangle\,, \tag{9-20b}$$

as is easily verified from the preceding matrix representations.* Since a_i reduces the number of eigenwaves in the state by one, it is called an *annihilation operator*; because $a_i^\dagger$ increases this number by one, it is called a *creation operator*. The appearance of such operators is, of course, a com-

* The annihilation operator is defined with the additional restriction that $a_i\,|0\rangle = 0$, where $|0\rangle$ is the "no-particle," or *vacuum state*.

pletely new concept, and their very fundamental significance cannot yet be fully appreciated. This significance will take form in the subsequent discussion.

It is of interest at this point to exhibit the explicit Fourier representation of **A** in terms of creation and annihilation operators. Substituting Eqs. (9-13) and (8-22) into Eq. (8-21), we find

$$\mathbf{A} = \sum_{\mathbf{k},i} (2\pi\hbar c/kV)^{1/2}\boldsymbol{\epsilon}_{\mathbf{k}}^{(i)}[a_{\mathbf{k}i}\,e^{i\mathbf{k}\cdot\mathbf{r}} + a_{\mathbf{k}i}^{\dagger}\,e^{-i\mathbf{k}\cdot\mathbf{r}}]\,, \tag{9-21}$$

where the sum over **k** goes over all the discrete states of Eq. (8-9), and the sum over i is over the two linearly independent directions of polarization.* We have also used the classical relation $\omega = ck$, the quantum-physical interpretation of which will be discussed below. In the same way one finds similar expressions for the fields **E** and **B** by means of Eqs. (4-2):

$$\mathbf{E} = i\sum_{\mathbf{k}} (2\pi\hbar ck/V)^{1/2}\boldsymbol{\epsilon}_{\mathbf{k}}[a_{\mathbf{k}}\,e^{i\mathbf{k}\cdot\mathbf{r}} - a_{\mathbf{k}}^{\dagger}\,e^{-i\mathbf{k}\cdot\mathbf{r}}]\,, \tag{9-22}$$

$$\mathbf{B} = i\sum_{\mathbf{k}} (2\pi\hbar c/kV)^{1/2}(\mathbf{k}\times\boldsymbol{\epsilon}_{\mathbf{k}})[a_{\mathbf{k}}\,e^{i\mathbf{k}\cdot\mathbf{r}} - a_{\mathbf{k}}^{\dagger}\,e^{-i\mathbf{k}\cdot\mathbf{r}}]\,. \tag{9-23}$$

One of the most important means of describing the behavior of a quantum system is through a study of the commutation relations satisfied by the observables. In this sense, the commutators of the field strengths are every bit as important for QED as the corresponding relations are for quantum particle mechanics. In order to derive such expressions at general points in space–time, it is necessary to momentarily reintroduce the explicit time dependence of the operators, Eq. (9-15). For two components of the electric field evaluated at different points in space–time, say, we have

$$\begin{aligned}[E_i(\mathbf{x},t),\, E_j(\mathbf{x}',t')] &= -\sum_{\mathbf{k},\mathbf{k}'} [2\pi\hbar c(kk')^{1/2}/V](\boldsymbol{\epsilon}_{\mathbf{k}})_i(\boldsymbol{\epsilon}_{\mathbf{k}'})_j \\ &\quad\times [a_{\mathbf{k}}(t)\,e^{i\mathbf{k}\cdot\mathbf{x}} - a_{\mathbf{k}}^{\dagger}\,e^{-i\mathbf{k}\cdot\mathbf{x}},\, a_{\mathbf{k}'}(t')\,e^{i\mathbf{k}'\cdot\mathbf{x}'} - a_{\mathbf{k}'}^{\dagger}(t')\,e^{-i\mathbf{k}'\cdot\mathbf{x}'}]\,. \end{aligned} \tag{9-24}$$

The implicit sum over polarization indices is performed using the result of Problem 9-1. With the aid of this formula and the commutation relation (9-17), one obtains

$$[E_i(\mathbf{x},t),\, E_j(\mathbf{x}',t')] = \frac{4\pi i\hbar c}{V}\sum_{\mathbf{k}} (\mathbf{k}^2\delta_{ij} - k_ik_j)\,\frac{\sin[\mathbf{k}\cdot(\mathbf{x}-\mathbf{x}') - kc(t-t')]}{k}\,. \tag{9-25}$$

* From now on we shall suppress the polarization index and sum, so that it is always understood that sums over states implicitly include sums over polarization indices.

We shall use the notation $\mathbf{r} = \mathbf{x} - \mathbf{x}'$ and $\tau = t - t'$ in that which follows. Referring to Eqs. (8-8) and (8-28), we see that the number of light waves in the volume with wave vectors in the element d^3k is just $V d^3k/(2\pi)^3$. Thus, when the dimensions of V are very large compared to the wavelengths considered, it is possible to replace sums over $\mathbf{k}$ states by integrals, using the following prescription:

$$\sum_{\mathbf{k}} \to [V/(2\pi)^3] \int d^3k \,. \tag{9-26}$$

This recipe is actually very important, and transcends its immediate application here. In the subsequent chapters we shall also apply Eq. (9-26) to sums over particle states, making similar arguments regarding the particle de Broglie wavelengths.*

Upon changing the sum to an integral in Eq. (9-25) we find that the right-hand side involves the function

$$\begin{aligned} \Delta(\mathbf{r}, \tau) &\equiv \frac{1}{(2\pi)^3} \int \frac{\sin(\mathbf{k}\cdot\mathbf{r} - kc\tau)}{k}\, d^3k \\ &= \frac{\delta(r + c\tau) - \delta(r - c\tau)}{4\pi r}\,, \end{aligned} \tag{9-27}$$

as is easily verified, and we have finally

$$[E_i(\mathbf{x}, t), E_j(\mathbf{x}', t')] = i\hbar c[\delta_{ij}(1/c^2)\,\partial^2_{tt'} - \partial^2_{x_i x_j'}]\,\Delta(\mathbf{r}, \tau)\,. \tag{9-28}$$

Thus, the components of the electric field evaluated at two different points in space–time commute only if the points cannot be connected by a light ray.

Exercise. In the same manner as above, verify the commutation relations

$$[B_i(\mathbf{x}, t), B_j(\mathbf{x}', t')] = [E_i(\mathbf{x}, t), E_j(\mathbf{x}', t')]\,, \tag{9-29}$$

$$[E_i(\mathbf{x}, t), B_j(\mathbf{x}', t')] = i\hbar\, \partial^2_{x_k' t}\, \Delta(\mathbf{r}, \tau)\,, \tag{9-30}$$

where (i, j, k) can be an even permutation of (x, y, z). Equations (9-28)–(9-30) were first derived by Jordan and Pauli.[3]

An important result of elementary quantum mechanics is the conclusion that if two operators representing observables do not commute,

$$[\mathsf{A}, \mathsf{B}] = \mathsf{C}\,, \tag{9-31a}$$

* In statistical mechanics the replacement (9-26) corresponds to the infinite-volume, or thermodynamic, limit. See, for example, the discussion by Huang.[2]

then the corresponding uncertainties satisfy the Heisenberg inequality, or uncertainty relation,

$$\Delta \mathsf{A}\, \Delta \mathsf{B} \gtrsim |\mathsf{C}| . \tag{9-31b}$$

The customary mathematical definition of uncertainty will be adopted here, namely,

$$\Delta \mathsf{A} = [\langle \mathsf{A}^2 \rangle - \langle \mathsf{A} \rangle^2]^{1/2} . \tag{9-32}$$

The uncertainty relations corresponding to the above commutation relations can be obtained by considering average values, which are the only quantities one can measure. These values are obtained by averaging over the two regions of space–time indicated by the two sets of variables appearing in Eqs. (9-28)–(9-30). By uncertainty, we generally have in mind the physical problem of performing simultaneous measurements, and this concept requires considerable analysis for electromagnetic fields. It is an easy matter to show from the above discussion, for example, that different components of either the electric or magnetic field evaluated at the same time but at different space points can be measured simultaneously, as can a component of **E** and a component of **B** at the same space point but at different times. However, the general situation involving field components in different regions of space–time connected by light rays is more difficult. Bohr and Rosenfeld[4] first considered this problem in detail, and the reader is referred to that paper for further study. The conclusion is that, in general, a single field strength can be measured with unlimited accuracy, but the exact, simultaneous measurement of *two* field strengths is prevented by the quantum mechanical properties of a test charge needed to perform such measurements.

Exercise. Show that the field strength **E** does not commute with the number operator $\mathsf{N}_\mathbf{k}$ for a particular state. More precisely, demonstrate that the expectation value of the commutator $[\mathbf{E}, \mathsf{N}_\mathbf{k}]$ between two arbitrary number states is generally nonzero.

B. PHOTONS

In the previous section we have seen that the free, or transverse, electromagnetic field can be represented by a countably infinite set of quantized harmonic oscillators, each of which can have an energy $n_i\hbar\omega_i$, within a factor of $\frac{1}{2}$. In terms of the wave vector **k**, the energy of the quantized wave is

$$\hbar\omega_\mathbf{k} = \hbar ck . \tag{9-33}$$

It will be demonstrated in a moment that this last equation can be interpreted more physically by assigning to each wave a momentum $\mathbf{p} = \hbar\mathbf{k}$. As we have seen in Chapter IV, this quantized wave vector has Lorentz transformation properties identical to those of a particle with the same energy and momentum, but the invariant associated with the energy-momentum, or propagation, 4-vector is zero [see Eq. (4-30)]. Thus, we are led to the conclusion that the electromagnetic field consists of light particles, or quanta, having zero mass, which we call *photons*. A field oscillator can emit from zero to a countably infinite number of photons at that oscillator frequency, and the total field is represented by an infinite number of discrete frequencies. In a macroscopic volume the frequency spectrum is essentially continuous, the classical limit being the usual one of large quantum numbers. This quantum view of the electromagnetic field is originally due to Fermi.[5]

The lowest energy level of the oscillator occurs for $n_i = 0$ and, if no photons are excited in V, we describe this state as the *vacuum state*. One observes from Eq. (9-5b), however, that we are then forced to ascribe to the vacuum an infinite *zero-point energy*

$$E_0 = \sum_i \tfrac{1}{2}\hbar\omega_i \,. \tag{9-34}$$

The significance of this quantity is discussed in detail below.

The dual nature of light—waves and quanta—is complementary in the sense of the Copenhagen interpretation of quantum theory emphasized by Bohr and others. The experimental evidence for the existence of photons (the photoelectric effect and blackbody radiation spectrum, for example) is as ample as that for wave phenomena, so that it is necessary to live with both aspects. The wave nature of light has been studied extensively in classical electromagnetic theory, so that the emphasis here will be on the particle behavior.

In order to exhibit explicitly the relation of the wave vector $\mathbf{k}$ to the photon momentum, recall the classical definition of field momentum, Eq. (4-45).* Because the field operators do not necessarily commute, the quantum mechanical definition must be written in symmetric form:

$$\mathbf{p} = (1/8\pi c)\int (\mathbf{E}^\dagger \times \mathbf{B} - \mathbf{B}^\dagger \times \mathbf{E})\, d^3r \,. \tag{9-35}$$

(Note that $\mathbf{E}$ and $\mathbf{B}$ are actually Hermitian.) One now substitutes the expansions (9-22) and (9-23) into this definition, incorporating the transversality condition (8-12). As a result of the ensuing algebra, which is

* The difficulties in applying this definition to the electromagnetic momentum of the electron are irrelevant here, because the present discussion involves only the free field.

left to the reader as an exercise, we find that

$$\mathbf{p} = \hbar \sum_{\mathbf{k}} \mathbf{k}(a_{\mathbf{k}}^{\dagger} a_{\mathbf{k}} + \tfrac{1}{2}) . \tag{9-36}$$

Thus, up to a factor of $\frac{1}{2}$, each photon has a linear momentum $\hbar\mathbf{k}$.

It is well known that electromagnetic waves carry angular momentum, so that one might expect the photon to have an intrinsic angular momentum, or spin. To examine this point further, let us first note that the particle must be a boson. In any collection of identical particles, more than one particle can occupy the same state only if the assembly obeys Bose–Einstein statistics. The Pauli principle forbids such a macroscopic occupation for particles obeying Fermi–Dirac statistics. Thus, an electromagnetic wave could never achieve a significant intensity at a given frequency, such as radio waves, unless photons are indeed bosons. Moreover, the success of Planck's theory of blackbody radiation is directly dependent on the boson nature of light particles.[6] The spin-statistics theorem of Pauli[7] then ensures us that the photon has a spin quantum number equal to an *integer* multiple of $\hbar$.

Before deciding upon the value of the integer to be assigned to the spin of the photon, it is instructive to outline qualitatively a group-theoretic description of such a particle. In the nonrelativistic discussion undertaken here we shall describe photons by the quantum numbers $\mathbf{k}$ and the two linearly independent directions of polarization, since spin-dependent interactions will never be considered. However, to treat the spin correctly, it is necessary to use a relativistic treatment and acknowledge the fact that photons are not really transverse. From the relativistic energy–momentum relation

$$E = [c^2\mathbf{p}^2 + m_0^2c^4]^{1/2} , \tag{9-37}$$

it should by now be clear that the photon must have zero rest mass. In fact, quantum electrodynamics could not be maintained gauge invariant were the photon to have a finite mass.* Now every fundamental physical system can be described by a symmetry group, and for QED this is the group of inhomogeneous Lorentz transformations.‡ For particles with zero rest mass there are two *types* of irreducible representation of the Lorentz group, only one of which seems to be realized in nature (the other corresponding to particles with a continuous spin variable). Wigner[9] has analyzed the physical type of representation quite fully and shown that only two irreducible representations exist for the mass-zero particles, no

* See, however, recent discussions of this point in the literature.[8]

‡ Actually, Maxwell's equations are invariant under the more general group of conformal transformations, but it is not clear at this time exactly what symmetries are introduced into the physical picture by this larger group.

matter how large their spin. Thus, a photon has only two possible spin states, rather than the $(2S+1)$ usually obtained. These two spin states can be taken parallel and antiparallel to the momentum vector $\mathbf{k}$, and this is a Lorentz-invariant statement for mass-zero particles. It is customary to label these two states as those of *positive* and *negative helicity*, and they correspond respectively to left and right circular polarization. Moreover, they can be combined linearly to give plane polarized states, which we shall use almost exclusively in the remainder of the book.

The preceding discussion of the photon spin can be put on a more quantitative basis by considering the classical definition of field angular momentum, in a manner analogous to Eq. (9-35). Thus, the total angular momentum of the field can be written

$$\begin{aligned}\mathbf{J} &= (1/4\pi c)\int \mathbf{r}\times(\mathbf{E}\times\mathbf{B})\,d^3r \\ &= \mathbf{L} + M\,,\end{aligned} \tag{9-38}$$

where

$$\mathbf{M} = (1/4\pi c)\int \mathbf{E}\times\mathbf{A}\,d^3r\,, \tag{9-39a}$$

$$\mathbf{L} = (1/4\pi c)\int (\mathbf{r}\times\nabla A_i)E^i\,d^3r\,. \tag{9-39b}$$

In the same way as above, the operator form must be written symmetrically to account for the noncommutativity, and we have the quantum mechanical definitions

$$\mathbf{M} = (1/8\pi c)\int (\mathbf{E}\times\mathbf{A} - \mathbf{A}\times\mathbf{E})\,d^3r\,, \tag{9-40a}$$

$$\mathbf{L} = (1/8\pi c)\int [E^i(\mathbf{r}\times\nabla A_i) + (\mathbf{r}\times\nabla A_i)E^i]\,d^3r\,. \tag{9-40b}$$

Exercise. Verify the algebra leading from Eq. (9-38) to Eqs. (9-40), remembering that $\nabla\cdot\mathbf{E} = 0$, and that the transverse nature of $\mathbf{A}$ renders the splitting of $\mathbf{J}$ gauge independent.

It is seen that $\mathbf{L}$ contains the coordinate $\mathbf{r}$ explicitly, while $\mathbf{M}$ does not, so that it is natural to identify these operators as orbital and spin angular momentum, respectively. Clearly, $\mathbf{M}$ does not depend on the choice of origin. In the following it is momentarily convenient to suppress the polarization index by replacing the creation and annihilation operators by the vector operators*

$$a_{\mathbf{k}} = \sum_i \epsilon_{\mathbf{k}i}a_{\mathbf{k}i}\,, \qquad \epsilon_{\mathbf{k}1}\times\epsilon_{\mathbf{k}2} = \hat{\mathbf{k}}\,. \tag{9-41}$$

* The caret, as in $\hat{\mathbf{k}}$, denotes a unit vector.

Then, substituting Eqs. (9-21) and (9-22) into Eq. (9-40a), we obtain

$$\mathbf{M} = -i\hbar \sum_{\mathbf{k}} (a_{\mathbf{k}}^{\dagger} \times a_{\mathbf{k}}) . \tag{9-42}$$

The operator $\mathbf{M}$ can be brought into a form which transparently commutes with both H and $\mathbf{p}$ by changing from linear to circular polarization. Thus, circular polarization unit vectors are defined as

$$\mathbf{e}_{\mathbf{k}+} = -(1/\sqrt{2})(\epsilon_{\mathbf{k}1} + i\epsilon_{\mathbf{k}2}) , \tag{9-43a}$$

$$\mathbf{e}_{\mathbf{k}-} = (1/\sqrt{2})(\epsilon_{\mathbf{k}1} - i\epsilon_{\mathbf{k}2}) . \tag{9-43b}$$

With the notation $\lambda = \pm 1$, these vectors satisfy the relations

$$\mathbf{e}_{\mathbf{k}\lambda}^{*} \cdot \mathbf{e}_{\mathbf{k}\lambda'} = \delta_{\lambda\lambda'} , \tag{9-44a}$$

$$\mathbf{e}_{\mathbf{k}\lambda}^{*} \times \mathbf{e}_{\mathbf{k}\lambda'} = i\lambda\hat{\mathbf{k}}\delta_{\lambda\lambda'} . \tag{9-44b}$$

The creation and annihilation operators must have the property

$$a_{\mathbf{k}} = \sum_{i} \epsilon_{\mathbf{k}i} a_{\mathbf{k}i} = \sum_{\lambda} \mathbf{e}_{\mathbf{k}\lambda} a_{\mathbf{k}\lambda} , \tag{9-45}$$

so that

$$a_{\mathbf{k}+} = -(1/\sqrt{2})(a_{\mathbf{k}1} - ia_{\mathbf{k}2}) , \tag{9-46a}$$

$$a_{\mathbf{k}-} = (1/\sqrt{2})(a_{\mathbf{k}1} + ia_{\mathbf{k}2}) . \tag{9-46b}$$

Exercise. Prove that this transformation leaves Eqs. (9-17), (9-19), and (9-36) invariant.

Substitution of Eqs. (9-44b) and (9-45) into Eq. (9-42) now yields

$$\mathbf{M} = \hbar \sum_{\mathbf{k}\lambda} \lambda\hat{\mathbf{k}}(a_{\mathbf{k}\lambda}^{\dagger} a_{\mathbf{k}\lambda} + \tfrac{1}{2}) , \tag{9-47}$$

so that H, $\mathbf{p}$, and $\mathbf{M}$ are simultaneously observable in any number state.

Consider now any one-photon state labeled by a momentum vector $\mathbf{k}$ and polarization index λ in the circular representation. In Problem 9-2 the reader is asked to show that $(\mathbf{L} \cdot \hat{\mathbf{k}})$ operating on this state is zero. Therefore, since $\mathbf{M} \cdot \hat{\mathbf{k}}$ is essentially the number operator, it is clear that the one-photon state $|\mathbf{k}\lambda\rangle$ is an eigenstate of $(\mathbf{J} \cdot \hat{\mathbf{k}})$, the component of total angular momentum along $\mathbf{k}$:

$$(\mathbf{J} \cdot \hat{\mathbf{k}}) \, |\mathbf{k}\lambda\rangle = \hbar\lambda \, |\mathbf{k}\lambda\rangle . \tag{9-48}$$

However, these are just the helicity states of the photon (± 1), and we conclude that the photon is a particle of spin one.

As a final matter for this section it is necessary to discuss the infinite zero-point energy, Eq. (9-34), which is a purely quantum mechanical entity.

There are two questions of fundamental significance which must be asked regarding this infinite energy of the vacuum state: is it really infinite, and does it have any physical significance?

The sum in Eq. (9-34) is, of course, infinite. This is verified by considering it to be a sum over photon momentum states and using the prescription (9-26) to convert the sum to an integral, as follows:

$$\tfrac{1}{2}\sum_{\mathbf{k}} \hbar ck \rightarrow (\hbar c V/4\pi^2)\int_0^\infty k^3\,dk\,. \tag{9-49}$$

However, physical considerations would seem to imply a cutoff at the upper end of the integration interval because it is impossible to contain the higher frequencies in a finite volume V with conducting walls. As in the case of the classical cavity resonator with nonperfectly conducting walls, these frequencies are not observable as excitations of the system. There are other approaches to discussing this divergence, but all appear to imply the necessary existence of a cutoff. One might at this point recall other harmonically vibrating systems, such as the Debye theory of lattice vibrations.[10] Physical considerations again imply a cutoff of the vibrational energies at the Debye frequency.

Suppose, on the other hand, that the zero-point energy were infinite. Even if this is the case, it is possible to formulate the theory so that this ubiquitous quantity does not appear, as it does in Eqs. (9-5b), (9-36), and (9-47), since the classical Hamiltonian, Eq. (9-2), can also be written in terms of q_i and $q_i{}^*$:

$$H_i = \omega_i{}^2(q_i q_i{}^* + q_i{}^* q_i)\,. \tag{9-50}$$

The quantization of this quantity is *not* unique, since classically it can equally well be written as

$$H_i = 2\omega_i{}^2 q_i{}^* q_i\,. \tag{9-51}$$

If one *now* introduces P_i and Q_i, the resulting quantized form is

$$\mathsf{H}_i = \tfrac{1}{2}(P_i{}^2 + \omega_i{}^2 Q_i{}^2) - \tfrac{1}{2}\hbar\omega_i\,,$$

and in this form the energy levels are

$$E_i = n_i\hbar\omega_i\,. \tag{9-52}$$

Thus, the zero-point energy has disappeared and the total Hamiltonian becomes

$$\mathsf{H} = \sum_{\mathbf{k}} (\hbar ck) a_{\mathbf{k}}{}^\dagger a_{\mathbf{k}}\,. \tag{9-53}$$

This procedure can be formalized by adopting the view that the Hamiltonian must always be written in this form in order for the theory

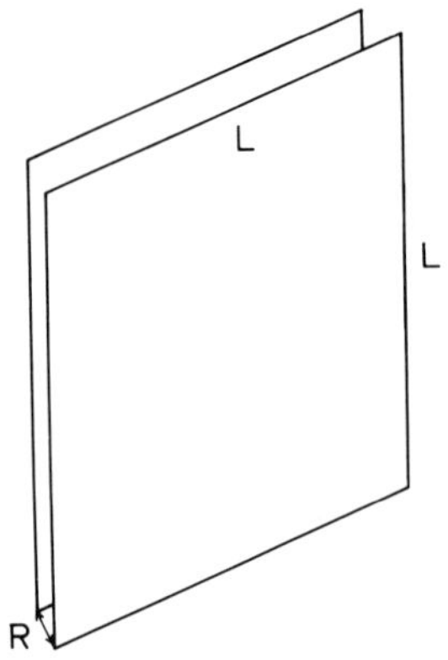

Fig. 24. Two flat conducting plates of area L^2 and a distance $R \ll L$ apart. The plates are attracted towards one another by the zero-point force given by Eq. (9-54).

to make sense. That is, in quantizing the canonical equations, one sometimes obtains creation operators standing to the right of annihilation operators, and, acting on the vacuum state, such an arrangement would constantly produce infinite energies. Therefore, we shall assume from now on that the fundamental operators of the theory are to be written in *normal form*, with creation operators to the *left* of annihilation operators. The terms of $\frac{1}{2}$ in Eqs. (9-36) and (9-47) are therefore to be discarded as having no physical meaning in the theory.

It would seem now that the zero-point energy is merely a formal difficulty of the theory and should be consistently ignored, as will be done in the following chapters. On the other hand, the second question raised above appears to have an affirmative answer: there *is* a physical significance to be ascribed to the energy of the vacuum. For instance, it is exactly the zero-point vibrational energy that prevents He from solidifying under its own vapor pressure, as is discussed in detail by London.[11] More in keeping with the physical system under study here, consider a volume V consisting of two flat conducting plates of macroscopic area $A = L^2$, which are parallel and a distance $R \ll L$ apart (Fig. 24). Casimir[12] has calculated the zero-point energy in V by using a physically reasonable cutoff of about $3R$, and predicted a force of attraction per unit area between the two plates of magnitude

$$F = -\frac{\pi}{480}\frac{hc}{R^4}. \tag{9-54}$$

This force between the plates in no way depends on e, since no charges are present, and it is purely quantum mechanical. The interesting point to be made here is that this force has actually been observed,[13] so that it is clear that an effect exists. Whether or not it is actually due to the zero-point energy, however, is not so clear, although one can readily attribute the effect to the quantum nature of the field, as follows.

We have seen in the exercise at the end of the last section that the number operator N_i and the field strength **E** do not commute. Hence, according to Eqs. (9-31), it is not possible to know simultaneously the strength of the field and the number of photons present. In particular, in the vacuum it is known with certainty that no quanta are present, so that the field intensity is very uncertain. Reference to Eqs. (9-20) and (9-22) shows that the vacuum expectation value of **E** is indeed zero, but the expectation value of $\mathbf{E}^2$ is not. Thus, according to the definition (9-32), there will be significant fluctuations of **E** about its average value of zero. One can therefore interpret these *vacuum fluctuations* as being responsible for the observed effect predicted by Casimir, and we shall see later that they are, in fact, the cause of other well-observed phenomena.

Exercise. Evaluate $\langle 0 | \mathbf{E}^2 | 0 \rangle$ explicitly, and interpret the result.

The obvious question, then, is whether the zero-point energy and the vacuum fluctuations are one and the same thing. If they are, why is it that the former can be eliminated from the theory? The answer is not yet clear, and a deeper significance has yet to be discovered. Therefore, we will adopt the view that the zero-point energies are to be formally removed from the theory by defining operators in normal form, and all physical effects of the type just discussed are to be ascribed to the quantum fluctuations of the vacuum. Moreover, in the next chapter we shall develop a calculus of creation and annihilation operators with which to describe quantum electrodynamics, and the zero-point energy never appears in this formulation.

C. COHERENT PHOTON STATES*

It is interesting to make a small digression from the formal development at this point in order to demonstrate the efficacy of the preceding formulation in terms of creation and annihilation operators. If the radiation field can be described by a particular frequency ω, then the description of Section A can be used without the need for retaining subscripts. Consider the eigenvalue problem for the annihilation operator:

$$a | x \rangle = x | x \rangle . \tag{9-55}$$

Since a is not a Hermitian operator, the eigenvalue x may be complex, and we write it as

$$x = | x | e^{i\phi} . \tag{9-56}$$

The physical meaning of the modulus $|x|$ is easily discerned by

* The discussion of coherent states presented here was first formulated by Glauber.[14]

noting that the average value of the excitation of the oscillator in these states is

$$N \equiv (\langle x \,|\, a^\dagger a \,|\, x\rangle / \langle x \,|\, x\rangle) = |\, x\,|^2 . \tag{9-57}$$

Thus, Eq. (9-56) can be rewritten

$$x = N^{1/2} e^{i\phi} . \tag{9-58}$$

The states $|\, x\rangle$ can be determined by expanding them in terms of number states,

$$|\, x\rangle = \sum_{n=0}^{\infty} |\, n\rangle \langle n \,|\, x\rangle , \tag{9-59}$$

and applying the prescripion (9-20a). Thus, with the aid of Eq. (9-55),

$$\langle n \,|\, x\rangle = (x/\sqrt{n})\langle (n-1) \,|\, x\rangle . \tag{9-60}$$

This procedure can be carried out n times to obtain the more useful form

$$\langle n \,|\, x\rangle = [x^n/(n!)^{1/2}]\, \langle 0 \,|\, x\rangle . \tag{9-61}$$

One can then normalize the states $|\, x\rangle$ to unity in order to determine $\langle 0 \,|\, x\rangle$ up to an arbitrary phase factor. Thus,

$$\begin{aligned} \langle x \,|\, x\rangle = 1 &= |\, \langle 0 \,|\, x\rangle\,|^2 \sum_{n=0}^{\infty} \{|\, x\,|^{2n}/n!\} \\ &= (\exp |\, x\,|^2)\, |\, \langle 0 \,|\, x\rangle\,|^2 , \end{aligned} \tag{9-62}$$

so that

$$|\, x\rangle = [\exp(-\tfrac{1}{2}\,|\, x\,|^2)] \sum_{n=0}^{\infty} [x^n/(n!)^{1/2}]\, |\, n\rangle . \tag{9-63}$$

Now, the nth number state can be generated from the vacuum by applying the creation operator, $a^\dagger$, n times and introducing a suitable normalization factor. Hence,

$$|\, n\rangle = [(a^\dagger)^n/(n!)^{1/2}]\, |\, 0\rangle . \tag{9-64}$$

Exercise. Verify explicitly the equality of both sides of this last equation.

Substituting Eq. (9-64) into (9-63), we can perform the indicated summation and now write

$$|\, x\rangle = \exp(x a^\dagger - \tfrac{1}{2}\,|\, x\,|^2)\, |\, 0\rangle . \tag{9-65}$$

There are several important observations which can be made about the expression (9-65), the foremost of which is that the state $|\, x\rangle$ represents the minimum-uncertainty wave packet describing a displaced harmonic-oscillator ground state. As is well known,[15] these states oscillate back and forth at frequency ω without any change of shape or spreading. The eigenstates $|x\rangle$ of the annihilation operator are, therefore, *coherent states* and are suitable for describing laser action.

It is possible to define a single operator accomplishing the same objectives as that of Eq. (9-65) by writing

$$\mathsf{A}^\dagger(x) \equiv \exp(xa^\dagger - x^*a)\,, \tag{9-66}$$

so that

$$|x\rangle = \mathsf{A}^\dagger(x)\,|0\rangle\,. \tag{9-67}$$

The equivalence of Eqs. (9-67) and (9-65) is most easily demonstrated by applying the Baker–Hausdorff theorem[16] to the product

$$\exp(xa^\dagger)\exp(-x^*a) = \exp(xa^\dagger - x^*a)\exp(\tfrac{1}{2}\,|x|^2)\,,$$

and referring to the footnote following Eqs. (9-20).

Exercise. Verify the stated equivalence of Eqs. (9-65) and (9-67).

The operator $\mathsf{A}^\dagger(x)$ thus generates a coherent state from the vacuum, and in this sense it can be considered a creation operator for these states. It is clear that

$$\mathsf{A}(x) = \mathsf{A}^\dagger(-x)\,, \tag{9-68}$$

which shows that the operators are unitary,

$$\mathsf{A}^\dagger(x)\mathsf{A}(x) = \mathsf{A}(x)\mathsf{A}^\dagger(x) = 1\,,$$

and $\mathsf{A}(x)$ is an annihilation operator for coherent states:

$$\mathsf{A}(x)\,|x\rangle = |0\rangle\,. \tag{9-69}$$

Exercise. Show that A and $\mathsf{A}^\dagger$ satisfy anticommutation relations,* and that these operators are displacement operators for a and $a^\dagger$. That is,

$$\mathsf{A}(x)a\mathsf{A}^\dagger(x) = a + x\,, \tag{9-70a}$$

$$\mathsf{A}(x)a^\dagger\mathsf{A}^\dagger(x) = a^\dagger + x^*\,. \tag{9-70b}$$

Thus, it is possible to discuss the entire radiation field in terms of coherent states, and this formulation has been found quite useful in studying photon statistics. For example, Sudarshan[17] and Glauber[18] have shown the ease with which one can construct density matrices applicable to such

* The anticommutator of two operators is defined as $[\mathsf{A}, \mathsf{B}]_+ = \mathsf{AB} + \mathsf{BA}$. These relations are relevant to fermions, as will be seen in the following chapter.

problems as blackbody radiation, and Carruthers and Nieto[19] have applied these states to a study of the number–phase uncertainty relation for the electromagnetic field. The reader is referred to these articles for further details.

PROBLEMS

9-1. For a given propagation vector **k**, show that for two components of the associated polarization vector

$$\sum_{m=1}^{2} \epsilon_i^{(m)}(\mathbf{k})\epsilon_j^{(m)}(\mathbf{k}) = \delta_{ij} - \hat{k}_i\hat{k}_j \underset{i=j}{\longrightarrow} \sin^2\theta\,,$$

where θ is the angle between **k** and the x_i axis.

9-2. Consider a one-photon state $|\mathbf{k}\rangle$ of arbitrary polarization, and the operator **L**, Eq. (9-40b). Demonstrate that $\langle\mathbf{k}|\,\hat{\mathbf{k}}\cdot\mathbf{L}\,|\mathbf{k}\rangle = 0$, under the restriction that the vacuum expectation value vanishes.

9-3. Show that the coherent states $|x\rangle$, Eqs. (9-65) and (9-67), are *not* orthogonal, but are indeed complete in that the completeness relation

$$\frac{1}{\pi}\int |x\rangle\,\langle x|\,d^2x = 1$$

is valid. The integral is over the entire complex plane.

9-4. Assume the photon to have a nonzero mass m_γ and demonstrate that it is then impossible to maintain gauge invariance in the theory of QED as developed thus far.

9-5. Using relativistic considerations, show that a free electron *cannot* absorb a photon. In the atomic photoelectric effect, which of the bound electrons would you suspect has the greatest probability for absorbing an incident photon?

9-6. Derive an expression for the distribution of photons in a classical wave of given frequency.

REFERENCES

1. M. Planck, *Ann. Physik* **1,** 69 (1900).
2. K. Huang, "Statistical Mechanics," pp. 193, 445. Wiley, New York, 1963.
3. P. Jordan and W. Pauli, *Z. Physik* **47,** 151 (1928).
4. N. Bohr and L. Rosenfeld, *Kgl. Danske Videnskab. Selskab.* **12,** 8 (1933).
5. E. Fermi, *Rev. Mod. Phys.* **4,** 87 (1932).
6. S. N. Bose, *Z. Physik* **26,** 178 (1924); **27,** 384 (1924).
7. W. Pauli, *Phys. Rev.* **58,** 716 (1940).
8. J. Schwinger, *Phys. Rev.* **125,** 397 (1962); V. I. Ogievetskij and I. V. Polubarinov, *Ann. Intern. Conf. High Energy Phys.* **11,** *CERN, 1962,* p. 666. CERN, Geneva, 1962; P. Bandyopadhyay, *Nuovo Cimento* **55A,** 367 (1968).
9. E. P. Wigner, *Nuovo Cimento* **3,** 517 (1956); *Rev. Mod. Phys.* **29,** 255 (1957).

10. J. E. Mayer and M. G. Mayer, "Statistical Mechanics," Chapter 11. Wiley, New York, 1940.
11. F. London, "Superfluids," Vol. II, Section 4. Wiley, New York, 1954.
12. H. B. G. Casimir, *Koninkl. Ned. Akad. Wetenschap. Proc.* **60,** 793 (1948); H. B. G. Casimir and D. Polder, *Phys. Rev.* **73,** 360 (1948). See also: T. H. Boyer, *ibid.* **174,** 1631 (1968), who makes a similar calculation for a spherical, uncharged conducting shell. The zero-point energy ($\cong 0.09\hbar c/2r$) tends to expand the shell.
13. B. V. Deryagin and I. I. Abrikosava, *Soviet Phys. JETP (English Transl.)* **30,** 993 (1956); B. V. Deryagin, I. I. Abrikosava, and E. M. Lifshitz, *Quart. Rev. (London)* **10,** 295 (1956); J. A. Kitchener and A. P. Prosser, *Proc. Roy. Soc. (London)* **A242,** 403 (1957); M. J. Sparnaay, *Physica* **24,** 751 (1958); A. Van Silfhout, "Dispersion Forces between Macroscopic Objects." Drukkerij Holland, N. V., Amsterdam, 1966.
14. R. J. Glauber, *Phys. Rev.* **131,** 2766 (1963).
15. E. Merzbacher, "Quantum Mechanics," p. 160. Wiley, New York, 1961.
16. E. Merzbacher, "Quantum Mechanics," p. 162, Equation (8.106). Wiley, New York, 1961.
17. E. C. G. Sudarshan, *Phys. Rev. Letters.* **10,** 277 (1963).
18. R. J. Glauber, *Phys. Rev.* **130,** 2529 (1963).
19. P. Carruthers and M. M. Nieto, *Phys. Rev. Letters.* **14,** 387 (1965).

X THE QUANTUM MECHANICS OF N-PARTICLE SYSTEMS

In the last chapter the quantization of the free electromagnetic field was carried out, and with it was achieved a partial understanding of the dual nature of light. Particular emphasis is directed toward the fact that light particles, or photons, can be created and destroyed in the field. This concept is foreign to both classical and elementary quantum mechanics, and is not completely appreciated in the context of the direct, heuristic quantization program pursued in Chapter IX.

Thus, in the present chapter a calculus for systems of N particles is developed from the fundamental principles of quantum mechanics, in which particles can be created and destroyed. It is clear from the previous development that the particle characterization of the electromagnetic field necessitates its interpretation as a many-body system, and therefore the study of the field should be considered an N-body problem. Since the photon is a boson, the problem is then one of studying systems containing a large number of identical particles satisfying Bose–Einstein statistics.

Although it appears only necessary to examine assemblies of bosons when considering the radiation field and nonrelativistic sources, it is quite easy to develop a unified theory encompassing both types of particle found in nature, and so we shall make room in the formalism for both fermions and bosons. For the latter, the results obtained closely parallel the description of the radiation field by harmonic oscillators developed previously. In the case of fermions, however, there is no classical field limit and the resulting physical interpretation is purely quantum mechanical. The inclusion of fermions is not merely an academic exercise, though, because later on in QED this formalism for electrons becomes both necessary and useful.

A. IDENTICAL PARTICLES

The initial assumption to be made in the program outlined above is that a collection of N (possibly interacting) particles can be described by means of a generalization of the single-particle Schroedinger equation. The N-body state vector characterizing a state of the system then satisfies the equation

$$i\hbar\partial_t \langle \mathbf{x}_1 \cdots \mathbf{x}_N, t \,|\, \alpha \rangle = \mathsf{H}_N \langle \mathbf{x}_1 \cdots \mathbf{x}_N, t \,|\, \alpha \rangle , \tag{10-1}$$

where H_N is the N-particle Hamiltonian operating on a function of all the particle coordinates, the time, and other quantum numbers represented collectively by α. The quantity $\langle \mathbf{x}_1 \cdots \mathbf{x}_N, t \,|\, \alpha \rangle$ is sometimes called the wave function.

By definition, the N particles are said to be identical if H_N is unchanged by a permutation of the particle coordinates. As is well known, this concept of identical particles introduces nonclassical *exchange effects* into the theory, and so it is important to clearly establish at the outset a notation for the algebra of permutations.

A permutation $\mathscr{P}$ on N objects is denoted by the following series of replacements:

$$\mathscr{P}: \quad \begin{pmatrix} a_1 a_2 \cdots a_N \\ b_1 b_2 \cdots b_N \end{pmatrix} , \tag{10-2}$$

where the order of the columns is irrelevant. A *transposition* is a permutation which interchanges only two elements, and every permutation can be written as a product of transpositions. Clearly, every permutation has an inverse; in fact, the set of all $N!$ permutations on N objects forms the *symmetric group*, in which the permutation leaving the positions of the objects unchanged is the identity permutation. The *signature*, or sign of the permutation, is $(-1)^P$, where the superscript P is the number of transpositions into which the permutation is factored. Thus, the permutation is *odd* if the signature is (-1), and *even* if it is $(+1)$.

If the Hamiltonian for N identical particles possesses symmetry properties, then one concludes from Eq. (10-1) that the wave functions must also be examined for such properties. Denoting by $\mathbf{x}$ the set of N particle coordinates, we see that the effect of a permutation on the wave function is

$$\psi(\mathbf{x}) = \psi'(\mathbf{x}) = \psi(\mathscr{P}\mathbf{x}) . \tag{10-3}$$

Mathematically, this transformation is induced by a linear operator in the Hilbert space of state vectors for the system. These linear operators form a representation of the symmetric group and commute with H_N.

Now, for $N > 2$, there are numerous alternatives as to the kind of symmetry properties one can ascribe to a wave function. However, it is found in nature that only two kinds of particle appear: those with completely symmetric wave functions (bosons), and those with completely antisymmetric wave functions (fermions). There exist formal arguments for this state of affairs,[1] but we shall here agree to accept this (experimentally supported) statement as a *symmetrization postulate*, and discuss its plausibility further below. Note well that this postulate is distinct from a statement regarding the amount of spin a particle can have,* a question resolved previously by Pauli.[2]

As is amply clear even in classical mechanics, the dynamic problem of N bodies is virtually intractable, at least in closed form. Consequently, our objective here must be to express the N-body state vectors in terms of the well-understood quantities for the one- or two-body problem, in order that the theory can be carried as far as possible before the inherent difficulties become overwhelming.

The easily understood quantities, of course, are the eigenfunctions of an observable relative to a single particle. An example is the complete set of orthonormal wave functions satisfying the single-particle, time-independent Schroedinger equation:

$$\mathsf{H}\phi_\lambda = \lambda\phi_\lambda . \tag{10-4}$$

For the purposes of this exposition it is not necessary that the operator and its eigenvalues λ be related to the energy, as long as the eigenvalues are discrete and are quantum numbers characterizing a single-particle state. Then it is useful to introduce functions which are products of the single-particle functions belonging to the complete set $\{\phi_\lambda\}$ such that the product contains one function for each particle of the N-body system. Thus, we define

$$G_{\lambda_1\cdots\lambda_N}(\mathbf{x}_1 \cdots \mathbf{x}_N) \equiv \phi_{\lambda_1}(\mathbf{x}_1)\phi_{\lambda_2}(\mathbf{x}_2) \cdots \phi_{\lambda_N}(\mathbf{x}_N) , \tag{10-5}$$

which, it is to be emphasized, is *not* an N-particle wave function, but only the indicated product. Each subscript λ_i runs over the single-particle spectrum for the ith particle, as indicated in Eq. (10-4). The entire set $\{G_\lambda(\mathbf{x})\}$ is assumed to be complete, which follows from the completeness of the set $\{\phi_\lambda(\mathbf{x})\}$. Physically, this amounts to saying that the N-body system retains to a large extent the properties of the individual particles. The *true* wave function of the system can then be expanded as

$$\psi(\mathbf{x}_1 \cdots \mathbf{x}_N) = \sum_{\{\lambda_i\}} g(\lambda_i) G_{\lambda_1\cdots\lambda_N}(\mathbf{x}_1 \cdots \mathbf{x}_N) , \tag{10-6}$$

an expansion which we will not find explicit enough to be useful.

In order to construct wave functions, or state vectors, with the appropriate symmetry properties, recall that it was observed above that these quan-

* Recall the discussion following Eq. (9-36).

tities must be either completely symmetric or completely antisymmetric. It is apparent that a symmetric state vector for the N-body system can be obtained by merely adding together all possible functions obtained by permuting the particle coordinates in Eq. (10-5). Clearly, any interchange of particle coordinates will leave this sum unchanged. Likewise, an antisymmetric state vector can be constructed by adding together all those functions obtained by even permutations of coordinates in Eq. (10-5) and subtracting from this sum all the odd permutations. An interchange of coordinates for any two particles will yield the same function with a minus sign. Thus, we can define a set of *symmetrized state vectors* by writing

$$\langle \mathbf{x}_1 \cdots \mathbf{x}_N \mid \lambda_1 \cdots \lambda_N \rangle \equiv n^{(s)} \sum_{\mathscr{P}} \varepsilon^P \mathscr{P} G_{\lambda_1 \cdots \lambda_N}(\mathbf{x}_1 \cdots \mathbf{x}_N) , \qquad (10\text{-}7)$$

where $n^{(s)}$ is a symmetrized normalization factor,

$$\varepsilon = \begin{cases} +1, & \text{for bosons} \\ -1, & \text{for fermions,} \end{cases} \qquad (10\text{-}8)$$

ε^P is the signature of the permutations, and the sum is over all $N!$ permutations of particle coordinates in the function of Eq. (10-5). Note that we could equally well permute the state labels λ_i, and the choice of using the coordinates is merely a convenient convention. It is also useful to adopt the convention that the identity permutation corresponds to that value of G_λ in which the state labels are ordered with $\lambda_1 \leqslant \lambda_2 \leqslant \cdots \leqslant \lambda_N$. This set of symmetrized vectors spans the Hilbert space of the N-particle system and clearly represents a suitable basis for describing a collection of bosons or fermions.

As an example of the scheme adopted here, consider the case $N = 2$. The only two possible symmetrized state vectors that can be constructed are

$$\phi_{\lambda_1}(\mathbf{x}_1)\phi_{\lambda_2}(\mathbf{x}_2) \pm \phi_{\lambda_1}(\mathbf{x}_2)\phi_{\lambda_2}(\mathbf{x}_1) . \qquad (10\text{-}9)$$

The upper sign gives the symmetric state and the lower sign the antisymmetric state.

For $N = 3$, there are $3\,! = 6$ possible combinations of single-particle product wave functions. The completely symmetric and completely antisymmetric combinations are

$$\begin{aligned} &\phi_{\lambda_1}(\mathbf{x}_1)\phi_{\lambda_2}(\mathbf{x}_2)\phi_{\lambda_3}(\mathbf{x}_3) \pm \phi_{\lambda_1}(\mathbf{x}_1)\phi_{\lambda_2}(\mathbf{x}_3)\phi_{\lambda_3}(\mathbf{x}_2) \\ &+ \phi_{\lambda_1}(\mathbf{x}_2)\phi_{\lambda_2}(\mathbf{x}_3)\phi_{\lambda_3}(\mathbf{x}_1) \pm \phi_{\lambda_1}(\mathbf{x}_2)\phi_{\lambda_2}(\mathbf{x}_1)\phi_{\lambda_3}(\mathbf{x}_3) \\ &+ \phi_{\lambda_1}(\mathbf{x}_3)\phi_{\lambda_2}(\mathbf{x}_1)\phi_{\lambda_3}(\mathbf{x}_2) \pm \phi_{\lambda_1}(\mathbf{x}_3)\phi_{\lambda_2}(\mathbf{x}_2)\phi_{\lambda_3}(\mathbf{x}_1) \end{aligned} \qquad (10\text{-}10)$$

where the choice of all upper or all lower signs is to be made. There are, of course, four other linearly independent combinations which can be formed, all of which possess quite different symmetry characteristics. These four anomalous combinations, along with similar types obtained for higher values of N, have *no common* symmetry properties. This means that, were they physically realizable, the qualitative description of a collection of N identical particles would depend on N, thereby violating the very concept of indistinguishable particles. Moreover, it is easily shown that for all combinations, except those which are completely symmetric or completely antisymmetric, the product wave functions do not retain their symmetry under an arbitrary unitary transformation, nor are they simultaneous eigenfunctions of *all* the permutations.

Exercise. Verify these last statements explicitly for the case $N = 3$.

In group-theoretic language, only the symmetric and antisymmetric combinations can form a basis for the one-dimensional representations of the symmetric group. With all of the foregoing theoretical observations, and supported by the overwhelming experimental evidence, we are forced to conclude that the basis vectors of Eq. (10-7) are the correct ones for describing the state space of N identical particles, either fermions or bosons.

In order to completely specify the state vectors of Eq. (10-7) it is necessary to determine the normalization constants $n^{(s)}$, according to the condition

$$\begin{aligned} 1 &= \int \langle \lambda_1 \cdots \lambda_N \mid \mathbf{x}_1 \cdots \mathbf{x}_N \rangle \, d^3x_1 \cdots d^3x_N \, \langle \mathbf{x}_1 \cdots \mathbf{x}_N \mid \lambda_1 \cdots \lambda_N \rangle \\ &= [n^{(s)}]^2 \int d^3x_1 \cdots d^3x_N \Big[\sum_{\mathscr{P}} \varepsilon^P \mathscr{P} \phi_{\lambda_1}(\mathbf{x}_1) \cdots \phi_{\lambda_N}(\mathbf{x}_N) \Big]^* \\ &\quad \times \Big[\sum_{\mathscr{P}'} \varepsilon^{P'} \mathscr{P}' \phi_{\lambda_1}(\mathbf{x}_1{}') \cdots \phi_{\lambda_N}(\mathbf{x}_N{}') \Big]_{\mathbf{x}_i{}' = \mathbf{x}_i} . \end{aligned} \tag{10-11}$$

Since the permutation operators form a group, the product of any two permutations is again a permutation in the group. Because we are summing over all permutations, it is clear that Eq. (10-11) can be rewritten

$$\begin{aligned} 1 = [n^{(s)}]^2 \int d^3x_1 \cdots d^3x_N \Big[\sum_{\mathscr{P}} \varepsilon^P \mathscr{P} \phi_{\lambda_1}(\mathbf{x}_1) \cdots \phi_{\lambda_N}(\mathbf{x}_N) \Big]^* \\ \times \sum_{\mathscr{P}\mathscr{P}'} \varepsilon^{P+P'} \mathscr{P}' \mathscr{P} \phi_{\lambda_1}(\mathbf{x}_1) \cdots \phi_{\lambda_N}(\mathbf{x}_N) . \end{aligned} \tag{10-12}$$

Consider first the problem for fermions, in which case the Pauli principle requires all particles to be in different states. Clearly, the orthonormality of the set $\{\phi_{\lambda_i}\}$ requires that the only nonzero products in the integrand be those corresponding to the same $\mathscr{P}$, which introduces $N!$

identical terms. The same argument shows that in the sum over $\mathscr{P}'$ only the identity permutation will yield a nonzero result. Thus, in this case,

$$\text{Fermions:} \qquad 1 = [n^{(s)}]^2 N! \,. \tag{10-13a}$$

For bosons, a single state may be occupied by more than one particle, so that if particles 1 and 2 are both in the state λ_1 a permutation interchanging $\mathbf{x}_1$ and $\mathbf{x}_2$ will lead to two indentical terms in the sum over $\mathscr{P}'$ in Eq. (10-12), thereby introducing a factor of 2. In general, if there are n_i bosons occupying the state λ_i, there will arise an additional factor of $(n_i!)$. Carrying out the sums over $\mathscr{P}'$ and $\mathscr{P}$ and performing the integrals, we find that

$$\text{Bosons:} \qquad 1 = [n^{(s)}]^2 (N)! \prod_{i=1}^{\infty} (n_i!) \,, \tag{10-13b}$$

where the product ranges over the countably infinite set of discrete values assumed by the quantum number λ_i.

One now observes that the *occupation numbers* n_i can take only the values 0 or 1 in the case of fermions. Thus, Eq. (10-13b) is really valid for both types of particle, so that, quite generally,

$$n^{(s)} = \left[(N!) \prod_i (n_i!)\right]^{-1/2} . \tag{10-14}$$

Exercise. Verify *explicitly* for $N = 3$ the result of Eq. (10-14) and the arguments leading to it.

B. THE NUMBER REPRESENTATION

While the basis vectors of Eq. (10-7) provide an adequate scheme for describing the state of an N-particle assembly of identical particles, they are still not sufficiently flexible for achieving the objective of this chapter; namely, the construction of a calculus incorporating the possible creation and annihilation of nonconserved particles, such as photons. As an immediate step toward this goal, it is useful to change representations from that labeled by the single-particle eigenvalues to one labeled by the occupation numbers. Thus, in place of Eq. (10-5) we can write

$$G_{n_1 \cdots n_i \cdots}(\mathbf{x}_1 \cdots \mathbf{x}_N) \equiv G_{\lambda_1 \cdots \lambda_N}(\mathbf{x}_1 \cdots \mathbf{x}_N) \,, \tag{10-15}$$

which is only a change of notation. If we notice that

$$\sum_{i=1}^{\infty} n_i = N \,, \tag{10-16}$$

then a little thought shows that specification of which and how many particles are in each state is completely equivalent to specifying the state of each particle. The reader should note well that the subscript i in n_i, labeling the ith state, refers to the type of single-particle states used as a basis in Eq. (10-5). Thus, i may refer to a position state, momentum state, energy state, or any other suitably complete set of quantum numbers. It is obvious that the functions of Eq. (10-15) are orthonormal, since the set $\{\phi_\lambda\}$ possesses this property. In the new notation this is expressed by

$$\int G^*_{n_1\cdots n_i\cdots}(\mathbf{x}_1 \cdots \mathbf{x}_N) G_{m_1\cdots m_i\cdots}(\mathbf{x}_1 \cdots \mathbf{x}_N)\, d^3x_1 \cdots d^3x_N$$
$$= \delta_{n_1 m_1} \cdots \delta_{n_i m_i} \cdots . \tag{10-17}$$

Exercise. Verify this orthonormality relation explicitly for three particles which have only four states available to them.*

We now write for the basis vectors spanning the Hilbert space of N indentical particles

$$\langle \mathbf{x}_1 \cdots \mathbf{x}_N \,|\, n_1 \cdots n_i \cdots \rangle = n^{(s)} \sum_{\mathscr{P}} \varepsilon^P \mathscr{P} G_{n_1\cdots n_i\cdots}(\mathbf{x}_1 \cdots\cdot \mathbf{x}_N)\,, \tag{10-18}$$

which is completely equivalent to the basis of Eq. (10-7). The factor $n^{(s)}$ is still given by Eq. (10-14). It should be noted that the identity permutation is taken to be that in which the occupation numbers n_i are ordered in i with respect to increasing values of λ_i, but not necessarily with respect to the numerical value of the n_i themselves. In analogy with Eq. (10-6), a many-body wave function describing a state of the system has the expansion

$$\psi(\mathbf{x}_1 \cdots \mathbf{x}_N) = \sum_{\{n_i\}} C(n_i) \langle \mathbf{x}_1 \cdots \mathbf{x}_N \,|\, n_1 \cdots n_i \cdots \rangle\,, \tag{10-19}$$

where the sum is over all possible *sets* of occupation numbers consistent with the statistics of the particles involved. The number $C(n_i)$ can be obtained in the usual manner for Fourier coefficients.

The functions $\langle \mathbf{x}_1 \cdots \mathbf{x}_N \,|\, n_1 \cdots n_i \cdots \rangle$ have the appearance of matrix elements defining a unitary transformation from the position to the *number representation*, within the same Hilbert space. Although this nomenclature is common, it is somewhat misleading because the basis labeled by the occupation numbers is not really a different representation in the quantum mechanical sense. If the subscripts on the occupation numbers refer to momentum states, for example, the functions (10-18) still connect the position and momentum representations, but with a different labeling for momentum states. Nevertheless, this seemingly trivial change of notation

* The general verification of Eq. (10-17) is representative of many similar theorems in many-body theory, which are verified through "proof by meditation"!

will presently emerge as a tremendous step toward reaching the goals of this chapter.

As a final matter for this section, let us deduce an explicit expression of the Pauli principle. From the prescription for evaluating an $N \times N$ determinant, it is easy to show that Eq. (10-18) for fermions can be written

$$\langle \mathbf{x}_1 \cdots \mathbf{x}_N \,|\, n_1 \cdots n_i \cdots \rangle = (N!)^{-1/2} \begin{vmatrix} \phi_{\lambda_1}(\mathbf{x}_1) & \cdots & \phi_{\lambda_N}(\mathbf{x}_1) \\ \phi_{\lambda_1}(\mathbf{x}_2) & \cdots & \phi_{\lambda_N}(\mathbf{x}_2) \\ \vdots & & \vdots \\ \phi_{\lambda_1}(\mathbf{x}_N) & \cdots & \phi_{\lambda_N}(\mathbf{x}_N) \end{vmatrix}, \quad (10\text{-}20)$$

called the *Slater–Fock determinant*.[3] In this form it is evident that no two particles can occupy the same state, for this would render two rows of the determinant indentically equal, and consequently it would vanish.

C. FOCK SPACE

It is well to reiterate at this point the goal of the present chapter, which is to construct a calculus for the N-body problem including in a natural way the concept of creation and annihilation operators. This is very desirable because, as we have seen, physical processes exist in real systems which create and destroy particles, and so one must work with a theory capable of accommodating such interaction terms in the Hamiltonian. Thus, we will now generalize the preceding formalism by introducing the notion of *Fock space*,[4] in which the totality of all N-body spaces is considered, for $N = 0, 1, 2, \ldots$. In this space all functions of the type (10-18) appear, for *all N*.

From the last paragraph it should be clear that Fock space is merely the direct sum of all the Hilbert spaces for N particles, $N = 0, 1, 2, \ldots,$ and can be written symbolically as

$$\mathscr{H}_F = \mathscr{H}_0 \oplus \mathscr{H}_1 \oplus \mathscr{H}_2 \oplus \cdots . \quad (10\text{-}21)$$

It follows that one can then define basis vectors in this space in the following manner:

$$|n_1 \cdots n_i \cdots \rangle_N \equiv \langle \mathbf{x}_1 \cdots \mathbf{x}_N \,|\, n_1 \cdots n_i \cdots \rangle \begin{pmatrix} 0 \\ 0 \\ \vdots \\ 0 \\ 1 \\ 0 \\ \vdots \end{pmatrix}, \quad (10\text{-}22)$$

where the first position in the column matrix represents the vacuum (a realizable state of the system), and the "1" appears in the $(N+1)$th position. The subscript N on the *Fock vector* indicates the number of particles in this Fock state, but, because there exists rare cause for confusion, it will generally be omitted. It is to be noted that this generalized space is infinite dimensional; that is, it contains infinitely many N-particle subspaces, to allow for creation of an arbitrary number of nonconserved particles. The normalization of the basis vectors is guaranteed by

$$1 = \int |\langle \mathbf{x}_1 \cdots \mathbf{x}_N \,|\, n_1 \cdots n_i \cdots \rangle|^2 \, d^3x_1 \cdots d^3x_N , \qquad (10\text{-}23)$$

where it is understood that the normalization of the vacuum is

$$\langle 0 \,|\, 0 \rangle = 1 . \qquad (10\text{-}24)$$

In Fock space an arbitrary state vector is written

$$f = \begin{pmatrix} f_0 \\ f_1(\mathbf{x}_1) \\ f_2(\mathbf{x}_1, \mathbf{x}_2) \\ \vdots \\ f_N(\mathbf{x}_1, \ldots, \mathbf{x}_N) \\ \vdots \end{pmatrix} \qquad (10\text{-}25)$$

Each of the f_i can be expanded in the N-particle subspaces in terms of the basis vectors (10-18), as demonstrated in Eq. (10-19). Equations (10-22), (10-23), and (10-25) constitute a precise definition of Fock space.

It is at this point that the concept of *annihilation operator* can be introduced into Fock space, by means of the defining equation*

$$a_i \,|\, n_1 \cdots n_i \cdots \rangle = \varepsilon^{s_i} \sqrt{n_i} \,|\, n_1 \cdots (n_i - 1) \cdots \rangle , \qquad (10\text{-}26)$$

where

$$s_i = \sum_{j=1}^{i-1} n_j . \qquad (10\text{-}27)$$

The quantity s_i is the number of occupied states *up to* the ith, and is a necessary element for characterizing fermion operators (and irrelevant for bosons, since in that case $\varepsilon = +1$). The reader will recall that we had agreed to maintain a definite ordering of states for the identity permutation, so that the sign factor is inserted in Eq. (10-26) in order to unambiguously identify the identity in the antisymmetric case.

* Note that this definition implicitly includes the convention that $a_i | n_1 \cdots n_i \cdots \rangle = 0$ if $n_i = 0$.

In a similar manner, a *creation operator* in Fock space is defined by

$$a_i^\dagger \,|\, n_1 \cdots n_i \cdots \rangle = \varepsilon^{s_i}(1 + \varepsilon n_i)^{1/2} \,|\, n_1 \cdots (n_i + 1) \cdots \rangle\,, \qquad (10\text{-}28)$$

where the factor of ε in the square root ensures that no fermion can be created in an already occupied state.*

Exercise. Show that $a_i^\dagger$ is indeed the Hermitian conjugate of a_i.

One aim of the present development is to include both bosons and fermions in the formalism, so that it is convenient at this time to generalize the definition of commutator in Fock space in the following manner:

$$[\mathsf{A}, \mathsf{B}] = \mathsf{AB} - \varepsilon \mathsf{BA}\,, \qquad (10\text{-}29)$$

because we shall now show that the operators of Eqs. (10-27) and (10-28) behave in exactly this manner. For instance,

$$a_i a_j \,|\, n_1 \cdots n_i \cdots n_j \cdots \rangle$$
$$= \sqrt{n_i}\sqrt{n_j}\,\varepsilon^{s_i + s_j} \,|\, n_1 \cdots (n_i - 1) \cdots (n_j - 1) \cdots \rangle\,,$$

so that

$$[a_i, a_j] = [a_i^\dagger, a_j^\dagger] = 0\,. \qquad (10\text{-}30a)$$

On the other hand, if $i \neq j$, the definition (10-27) leads to

$$(a_i a_j^\dagger - \varepsilon a_j^\dagger a_i) \,|\, n_1 \cdots n_i \cdots n_j \cdots \rangle = 0\,,$$

while, for $i = j$,

$$(a_i a_i^\dagger - \varepsilon a_i^\dagger a_i) \,|\, n_1 \cdots n_i \cdots \rangle$$
$$= \{(1 + n_i)^{1/2}(1 + \varepsilon n_i)^{1/2} - \varepsilon\sqrt{n_i}\,[1 + \varepsilon(n_i - 1)]^{1/2}\} \,|\, n_1 \cdots n_i \cdots \rangle$$
$$= |\, n_1 \cdots n_i \cdots \rangle\,,$$

for either fermions (in which case n_i is 0 or 1) or bosons. Hence,

$$[a_i, a_j^\dagger] = \delta_{ij}\,. \qquad (10\text{-}30b)$$

One should notice that, in making the definition (10-29), it was assumed that the fermion operators would *anticommute*, and this indeed turned out to be the case. Jordan and Wigner[5] first showed this to be true for any operators describing fermions.

Let us now define an operator

$$\mathcal{N}_i \equiv a_i^\dagger a_i\,, \qquad (10\text{-}31)$$

* In this definition the utility of the phase factor s_i is more transparent, because in creating a particle in a certain state it is necessary to include an additional plane-wave function in Eq. (10-5), and this must be inserted in the proper location in order to maintain the ordering in $\{\lambda_i\}$. Our convention clearly accomplishes this.

which has as eigenfunctions the state vectors (10-22):

$$\mathcal{N}_i | n_1 \cdots n_i \cdots \rangle = a_i^\dagger \varepsilon^{s_i} \sqrt{n_i} \, | n_1 \cdots (n_i - 1) \cdots \rangle$$
$$= n_i | n_1 \cdots n_i \cdots \rangle . \tag{10-32}$$

The eigenvalues of $\mathcal{N}_i$ are just the number of particles in the ith state, so that Eq. (10-31) defines the *number operator* for that state. The total-number operator for the system is

$$\mathcal{N} = \sum_{i=1}^{\infty} \mathcal{N}_i = \sum_{i=1}^{\infty} a_i^\dagger a_i , \tag{10-33}$$

and

$$\mathcal{N} | n_1 \cdots n_i \cdots \rangle = N | n_1 \cdots n_i \cdots \rangle . \tag{10-34}$$

Exercise. Show that

$$[a_i, \mathcal{N}_i] = a_i , \qquad [a_i^\dagger, \mathcal{N}_i] = -a_i^\dagger . \tag{10-35}$$

An important feature of the operators a_i and $a_i^\dagger$ is that the basis vectors of Eq. (10-22) can now be written in the transparent form

$$| n_1 \cdots n_i \cdots \rangle = \prod_i (n_i!)^{-1/2} (a_i^\dagger)^{n_i} | 0 \rangle , \tag{10-36}$$

which will be very useful in the next chapter.

Exercise. Demonstrate explicitly, by use of the commutation relations, that these state vectors are orthonormal and satisfy Eq. (10-32).

In order to complete the description of an N-particle system in Fock space, it is necessary to obtain expressions for the operators of the theory, particularly the Hamiltonian. To do this, however, one must first decide on a general form for H_N in order to calculate the matrix elements, although it is unnecessary to specify an exact functional form for the interaction potentials.

Any description of an N-particle system must, in principle, allow for the possibility of N-body interactions. But, as is well known, not even the three-body problem can be solved in closed form, classically or quantum mechanically, let alone the many-body situation. A major difficulty in this respect is that one does not even know how to write a functional expression for such an interaction. Thus, it has become necessary in theories such as those of statistical mechanics to assume that N-body forces are negligible in actual systems, and that the real interaction among particles is well represented by the sum of all possible two-body interactions among the N particles. The negligible effect of many-body forces has been quite amply demonstrated experimentally, except possibly at very low temperatures. In the latter case, the difficulty has been resolved by describing such systems in terms of quasi-particles and other excitations,* and then neglecting "many-quasi-particle interactions."

Therefore, in accordance with both practice and necessity, we will assume that only one- and two-body interactions are to be included in the Hamiltonian. This assumption does not, of course, refer to the emission and absorption of photons or other quanta by particles with mass. Thus, an important feature of the electromagnetic interaction is that it is not really affected by this assumption [recall the *exact* Hamiltonian of Eq. (8-47)]. An additional bonus given by the present treatment is that the nonrelativistic nature of the Hamiltonian limits the theory to one- and two-photon processes, as will become clear below. With all these points in mind, we write for the Hamiltonian

$$\mathsf{H}_N = \mathsf{H}^{(0)} + \mathsf{V}_1 + \mathsf{V}_2 = \sum_{k=1}^{N} \mathsf{H}_k + \tfrac{1}{2} \sum_{i \neq j}^{N} \mathsf{V}_{ij}\,, \tag{10-37}$$

where, for the kth particle,

$$\mathsf{H}_k = \mathsf{H}_0(k) + \mathsf{V}_1(k) \tag{10-38}$$

is the free-particle Hamiltonian and any one-particle interactions, such as those with an external field. The V_{ij} are two-particle interactions between particles i and j, and the factor of $\frac{1}{2}$ arises from the assumed validity of Newton's third law.

The theory is general enough to include *nonlocal interactions*, in which case the operators are written

$$\mathsf{H}_0\psi(\mathbf{x}) = -(\hbar^2/2M)\,\nabla^2\psi(\mathbf{x})\,, \tag{10-39a}$$

$$\mathsf{V}_1\psi(\mathbf{x}) = \int \mathsf{V}_1(\mathbf{x}, \mathbf{x}')\psi(\mathbf{x}')\,d^3x'\,, \tag{10-39b}$$

$$\mathsf{V}_2\psi(\mathbf{x}_1, \mathbf{x}_2) = \int \mathsf{V}_2(\mathbf{x}_1, \mathbf{x}_2; \mathbf{x}_1', \mathbf{x}_2')\psi(\mathbf{x}_1', \mathbf{x}_2')\,d^3x_1'\,d^3x_2'\,. \tag{10-39c}$$

For point, or local interactions these expressions are reduced by the definitions

$$\mathsf{V}_1(\mathbf{x}, \mathbf{x}') = \mathsf{V}_1(\mathbf{x})\,\delta(\mathbf{x} - \mathbf{x}')\,, \tag{10-40a}$$

$$\mathsf{V}_2(\mathbf{x}_1, \mathbf{x}_2; \mathbf{x}_1', \mathbf{x}_2') = \mathsf{V}_2(\mathbf{x}_1, \mathbf{x}_2)\,\delta(\mathbf{x}_1 - \mathbf{x}_1')\,\delta(\mathbf{x}_2 - \mathbf{x}_2')\,. \tag{10-40b}$$

It is clear from the way in which we have constructed Fock space that H must be represented by an infinite matrix with the H_N corresponding to each N-particle subspace on the main diagonal. For $N > 1$,

* See, for example, the first chapter of the book by Pines.[6]

the *form* of each H_N will, of course, be the same, since many-body interactions have been eliminated. Referring to Eq. (10-22), we see that it is necessary to examine the operation

$$\mathsf{H}_N \langle \mathbf{x}_1 \cdots \mathbf{x}_N \,|\, n_1 \cdots n_i \cdots \rangle .$$

Let us first study the part $\mathsf{H}^{(0)} + \mathsf{V}_1$, which, from Eqs. (10-18), (10-15), and (10-5), has the form

$$\Big(\sum_{k=1}^{N} \mathsf{H}_k\Big) \langle \mathbf{x}_1 \cdots \mathbf{x}_N \,|\, n_1 \cdots n_i \cdots \rangle$$
$$= \Big[(N!)^{-1/2} \prod_i (n_i!)^{-1/2} \sum_{\mathscr{P}} \varepsilon^P \mathscr{P}\Big] \Big(\sum_{k=1}^{N} \mathsf{H}_k\Big) [\phi_{\lambda_1}(\mathbf{x}_1) \cdots \phi_{\lambda_N}(\mathbf{x}_N)] , \tag{10-41}$$

where it must be remembered that the set $\{\lambda_i\}$ is ordered.

Recall that the effect of a single-particle operator acting on a single-particle wave function can be written

$$\mathsf{H}_k \phi_k(\mathbf{x}_k) = \sum_i \langle i \,|\, \mathsf{H}_k \,|\, k\rangle\, \phi_i(\mathbf{x}_k) , \tag{10-42}$$

which can be proven by taking the scalar product of both sides with $\phi_j(\mathbf{x}_k)$. The right-hand side of Eq. (10-41) can therefore be rewritten as

$$\Big[(N!)^{-1/2} \prod_i (n_i!)^{-1/2} \sum_{\mathscr{P}} \varepsilon^P \mathscr{P}\Big] \sum_{k=1}^{N} \sum_{\lambda_i} \langle \lambda_i \,|\, \mathsf{H}_k \,|\, \lambda_k\rangle$$
$$\times [\phi_{\lambda_1}(\mathbf{x}_1) \cdots \phi_{\lambda_i}(\mathbf{x}_k) \cdots \phi_{\lambda_N}(\mathbf{x}_N)]$$
$$= \sum_{k=1}^{N} \sum_{\lambda_i} \langle \lambda_i \,|\, \mathsf{H}_k \,|\, \lambda_k\rangle \Big[(N!)^{-1/2} \prod_i (n_i!)^{-1/2}\Big]$$
$$\times \sum_{\mathscr{P}} \varepsilon^P \mathscr{P} [\phi_{\lambda_1}(\mathbf{x}_1) \cdots \phi_{\lambda_i}(\mathbf{x}_k) \cdots \phi_{\lambda_N}(\mathbf{x}_N)] . \tag{10-43}$$

Due to the sum over λ_i, the set $\{\lambda_j\}$ is no longer ordered, and one must also observe that the effect of this procedure (for a given λ_i in the sum) has been to add a particle to the state λ_i and remove one from the state λ_k. To adjust for the unordering of the λ_i, one must count the number of positions λ_i has been removed from its proper position and into that of λ_k, which clearly corresponds to a sign factor $\varepsilon^{s_i + s_k}$. Since the sum over $\mathscr{P}$ contains all permutations, the product in Eq. (10-43) can be returned to its regularly ordered form by introducing this sign factor.

We now wish to identify the sum over $\mathscr{P}$ with the basis vectors $\langle \mathbf{x}_1 \cdots \mathbf{x}_N \,|\, n_1 \cdots n_i \cdots \rangle$, Eq. (10-18), but the normalization factors are incorrect because the numbers of particles in the states λ_i and λ_k have been changed. Multiplying and dividing by the proper factors of $(n_i + 1)!$ and $(n_k - 1)!$ in the product over occupation numbers, extracting the factors $(n_i!)$ and $(n_k!)$, and gathering together the extraneous numbers, we can then make the desired identification. The final step in this procedure is the observation that the replacement

$$\sum_{k=1}^{N} \to \sum_{\lambda_k} n_k \tag{10-44}$$

can be made. The right-hand side of Eq. (10-43) now becomes

$$\left[\sum_{\lambda_k} \sum_{\lambda_i} \langle \lambda_i | \mathsf{H}_k | \lambda_k \rangle \right] \varepsilon^{s_i + s_k} [n_k(1 + \varepsilon n_i)]^{1/2}$$
$$\times \langle \mathbf{x}_1 \cdots \mathbf{x}_N | n_1 \cdots (n_i + 1) \cdots (n_k - 1) \cdots \rangle . \tag{10-45}$$

The expression (10-45) can now be combined with the definitions of a_i and $a_i^\dagger$, Eqs. (10-26) and (10-28), to write in place of Eq. (10-41)

$$\left(\sum_{k=1}^{N} \mathsf{H}_k \right) \langle \mathbf{x}_1 \cdots \mathbf{x}_N | n_1 \cdots n_i \cdots \rangle$$
$$= \left[\sum_{\lambda_i \lambda_k} a_i^\dagger a_k \langle \lambda_i | \mathsf{H}_k | \lambda_k \rangle \right] \langle \mathbf{x}_1 \cdots \mathbf{x}_N | n_1 \cdots n_i \cdots \rangle . \tag{10-46}$$

Thus, insofar as its effect on the basis vectors is concerned, the one-particle portion of the Hamiltonian is

$$\mathsf{H}^{(0)} + \mathsf{V}_1 = \sum_{k=1}^{N} \mathsf{H}_k$$
$$= \sum_{\lambda_i \lambda_k} a_i^\dagger a_k \langle \lambda_i | \mathsf{H}_0 + \mathsf{V}_1 | \lambda_k \rangle , \tag{10-47}$$

and if the states $|\lambda_k\rangle$ are eigenstates of H_0, as is usually desired, then

$$\langle \lambda_i | \mathsf{H}_0 | \lambda_k \rangle = \omega_i \delta_{ik} , \tag{10-48}$$

where the ω_i are the free-particle energy states.

In a similar manner, the representation of V_2 in Fock space can be calculated. A straightforward, but algebraically more tedious, calculation yields

$$\mathsf{V}_2 = \frac{1}{2} \sum_{\substack{\lambda_i \lambda_j \\ \lambda_k \lambda_l}} a_k^\dagger a_l^\dagger \langle \lambda_k \lambda_l | \mathsf{V}_2 | \lambda_i \lambda_j \rangle a_j a_i . \tag{10-49}$$

Note that for fermions it is important to observe the ordering of the labels i and j.

Exercise. Derive Eq. (10-49) with the aid of the valid replacement

$$\sum_{i \neq j} \to \sum_{\lambda_i \lambda_j} n_{ij}\,, \tag{10-50}$$

where $n_{ij} = n_i n_j$ for $i \neq j$, and $n_{ij} = n_i(n_j - 1)$ for $i = j$.

It will be useful in what follows to have explicit expressions for the matrix elements of V_1 and V_2 in the case of local (or point) interactions. From Eqs. (10-39), (10-40), (10-47), and (10-49), we have

$$\langle \lambda_i \,|\, \mathsf{V}_1 \,|\, \lambda_k \rangle = \int \phi^*_{\lambda_i}(\mathbf{x}) \mathsf{V}_1(\mathbf{x}) \phi_{\lambda_k}(\mathbf{x})\, d^3x\,, \tag{10-51}$$

$$\langle \lambda_k \lambda_l \,|\, \mathsf{V}_2 \,|\, \lambda_i \lambda_j \rangle = \int \phi^*_{\lambda_k}(\mathbf{x}_1) \phi^*_{\lambda_l}(\mathbf{x}_2) \mathsf{V}_2(\mathbf{x}_1, \mathbf{x}_2) \phi_{\lambda_i}(\mathbf{x}_1) \phi_{\lambda_j}(\mathbf{x}_2)\, d^3x_1\, d^3x_2\,. \tag{10-52}$$

In the following chapters we shall want to apply the formalism developed here to an understanding of the radiation processes of QED. Although this program is the major concern of this book, it is to be emphasized that the preceding theory is very general and applies to any collection of nonrelativistic particles. In order to exhibit the power of Fock-space techniques, an application to the calculation of the ground-state energy for a system of electrons is presented in Appendix B.

PROBLEMS

10-1. Consider the creation and annihilation operators a_i and $a_i^\dagger$, for either bosons or fermions, and define new operators by the linear transformation

$$b_i = u_i a_i + v_i a_i^\dagger\,, \qquad b_i^\dagger = u_i a_i^\dagger + v_i a_i\,.$$

Find the conditions which must be imposed on the coefficients u_i and v_i in order that the transformation be canonical. Find the inverse transformation.

10-2. Prove the following relations for boson operators:

$$[a^\dagger a, a^{\dagger m}] = m a^{\dagger m}\,, \qquad [a^\dagger a, a^m] = -m a^m\,.$$

REFERENCES

1. A. Messiah and O. W. Greenberg, *Phys. Rev.* **136,** B248 (1964); M. Flicker and H. S. Leff, *ibid.* **163,** 1353 (1967); M. D. Girardeau, *ibid.* **139,** B500 (1965); *J. Math. Phys.* **10,** 1302 (1969).
2. W. Pauli, *Phys. Rev.* **58,** 716 (1940).
3. J. C. Slater, *Phys. Rev.* **35,** 210 (1930); V. Fock, *Z. Physik* **61,** 126 (1930).
4. V. Fock, *Z. Physik* **75,** 622 (1932).
5. P. Jordan and E. P. Wigner, *Z. Physik* **47,** 631 (1928).
6. D. Pines, "The Many-Body Problem." Benjamin, New York, 1961.

XI METHODS OF CALCULATION

The theoretical developments of the last two chapters have provided the basis for a completely general, quantum mechanical description of the coupled system of charged particles and radiation field. We now wish to exhibit these results more explicitly, so as to study several actual physical processes. Although only the quantization of the transverse radiation field has been studied, the remaining terms in the classical Hamiltonian, Eq. (8-47), are easily quantized by requiring the canonical variables to be linear operators satisfying the commutation relations of elementary quantum mechanics. As long as one works in the Coulomb gauge, there are no longitudinal waves in the nonrelativistic theory, and the charged particles interact via the (quantum mechanical) static Coulomb interaction. The quantization procedure is then to pay due attention to the commutation rules for the operators and follow the classical development of Chapter VIII.

A. COUPLED SYSTEM OF RADIATION AND CHARGED PARTICLES

We wish to write the Fock-space Hamiltonian for the system in the form indicated by Eqs. (10-37) and (10-38). This can be done by referring to Eq. (8-48) and making the expansion

$$\frac{1}{2m}\left(\mathbf{p}-\frac{e}{c}\mathbf{A}\right)^2=\frac{p^2}{2m}-\frac{e}{2mc}\mathbf{p}\cdot\mathbf{A}-\frac{e}{2mc}\mathbf{A}\cdot\mathbf{p}+\frac{e^2}{2mc^2}\mathbf{A}^2\,, \tag{11-1}$$

where one must remember that $\mathbf{p}$ and $\mathbf{A}$ are quantum mechanical operators. The identification with Eq. (10-38) is now made by equating the first term on the right-hand side of Eq. (11-1) with H_0, and the remaining three terms with the one-particle interaction V_1. It will be seen, however, that in this case the matrix element (10-51) for V_1 will have to be modified slightly to accommodate the extra quantum numbers due to the photons.

A great advantage in using the Coulomb gauge manifests itself at this point, because the commutator of **p** and **A** is proportional to $\nabla \cdot \mathbf{A}$, and the two operators therefore commute in this gauge.

Exercise. Show that in the Coulomb gauge

$$[\mathbf{p}, \mathbf{A}] = -i\hbar\, \nabla \cdot \mathbf{A} = 0\,.$$

Thus, referring to Eqs. (10-37) and (10-38), we can write the Fock-space Hamiltonian of Eq. (8-47) for the system of charged particles and radiation as

$$\mathsf{H}_N = \mathsf{H}^{(0)} + \mathsf{V}_1 + \mathsf{V}_2 = \mathsf{H}^{(0)} + \mathsf{V}_\gamma + \mathsf{V}_\mathrm{C}\,, \tag{11-2}$$

where $\mathsf{H}^{(0)}$ contains the first term on the right-hand side of Eq. (11-1) *and* the Hamiltonian for the free radiation field, V_C is the two-body Coulomb interaction between charged particles, and V_γ is the one-body interaction

$$\mathsf{V}_\gamma = -\frac{e}{mc}\,\mathbf{p}\cdot\mathbf{A} + \frac{e^2}{2mc^2}\,\mathbf{A}^2\,. \tag{11-3}$$

According to Eqs. (10-47) and (10-49), the Hamiltonian (11-2) in Fock space can now be written

$$\begin{aligned}\mathsf{H} = \sum_{\lambda_i\lambda_k} a_i^\dagger a_k \langle\lambda_i\,|\,\mathsf{H}_0\,|\,\lambda_k\rangle + \sum_{\lambda_i\lambda_k} a_i^\dagger a_k \langle\lambda_i\,|\,\mathsf{V}_\gamma\,|\,\lambda_k\rangle \\ + \tfrac{1}{2}\sum_{\lambda_i\lambda_j,\lambda_k\lambda_l} a_k^\dagger a_l^\dagger \langle\lambda_k\lambda_l\,|\,\mathsf{V}_\mathrm{C}\,|\,\lambda_i\lambda_j\rangle\, a_j a_i\,,\end{aligned} \tag{11-4}$$

where the appropriate matrix elements are given by Eqs. (10-51) and (10-52).

In order to provide an example for calculating the matrix elements appearing in Eq. (11-4), and to obtain some expressions which will be quite useful later, let us choose to work for the moment in the momentum representation. Then the single-particle basis functions of Fock space are taken to be box-normalized plane waves:

$$\phi_{\lambda_i}(\mathbf{x}) = \frac{1}{V^{1/2}} \exp(i\mathbf{k}_i\cdot\mathbf{x})\,. \tag{11-5}$$

Clearly, for the charged particles

$$\begin{aligned}\langle\mathbf{k}_1\,|\,\mathsf{H}_0\,|\,\mathbf{k}_2\rangle &= \frac{1}{V}\int [\exp(-i\mathbf{k}_1\cdot\mathbf{x})]\left(-\frac{\hbar^2}{2m}\nabla^2\right)\exp(i\mathbf{k}_2\cdot\mathbf{x})\,d^3x \\ &= \frac{\hbar^2 k_1^{\,2}}{2m}\,\delta_{\mathbf{k}_1,\mathbf{k}_2}\,.\end{aligned} \tag{11-6}$$

Evaluation of the matrix element in the second line of Eq. (11-4) requires consideration of the two terms of Eq. (11-3). Let us refer to these as $\mathsf{V}_{1\gamma}$ and $\mathsf{V}_{2\gamma}$, respectively, and substitute the expansion (9-21) for the vector potential **A**. The matrix element of $\mathsf{V}_{1\gamma}$, or the $(\mathbf{p}\cdot\mathbf{A})$ interaction, then depends on the integral

$$\frac{1}{V}\int[\exp(-i\mathbf{k}_1\cdot\mathbf{x})](\mathbf{p}\cdot\mathbf{A})\exp(i\mathbf{k}_2\cdot\mathbf{x})\,d^3x$$

$$=\left(\frac{2\pi\hbar c}{V}\right)^{1/2}\sum_{\mathbf{k}_3}\frac{\hbar\mathbf{k}_2\cdot\boldsymbol{\epsilon}_3}{Vk_3^{1/2}}\left[a_{\mathbf{k}_3}\int\exp[i(\mathbf{k}_2+\mathbf{k}_3-\mathbf{k}_1)\cdot\mathbf{x}]\,d^3x\right.$$

$$\left.+\,a_{\mathbf{k}_3}^{\dagger}\int\exp[-i(\mathbf{k}_2+\mathbf{k}_3-\mathbf{k}_1)\cdot\mathbf{x}]\,d^3x\right] \tag{11-7}$$

and we note the appearance of an additional quantum number due to the photon, as observed above. The second term in Eq. (11-7) is just the Hermitian conjugate of the first,* so that the first contribution to V_γ really consists of a sum of two terms: $\mathsf{V}_{1\gamma}+\mathsf{V}_{1\gamma}^{\dagger}$, where

$$\mathsf{V}_{1\gamma}=\sum_{\mathbf{k}_1\mathbf{k}_2\mathbf{k}_3}a_{\mathbf{k}_1}^{\dagger}\langle\mathbf{k}_1\,|\,\mathsf{V}_{1\gamma}\,|\,\mathbf{k}_2\mathbf{k}_3\rangle\,a_{\mathbf{k}_2}a_{\mathbf{k}_3}^{(\gamma)}\,, \tag{11-8}$$

and

$$\langle\mathbf{k}_1\,|\,\mathsf{V}_{1\gamma}\,|\,\mathbf{k}_2\mathbf{k}_3\rangle=-Z\left(\frac{2\pi\alpha}{Vk_3}\right)^{1/2}\frac{\hbar^2}{m}(\mathbf{k}_1\cdot\boldsymbol{\epsilon}_3)\,\delta_{\mathbf{k}_1,\mathbf{k}_2+\mathbf{k}_3}\,\delta_{n_1n_2}\,, \tag{11-9}$$

which is real. In this last expression Z is the charge number of the charged particle, $\alpha=e^2/\hbar c$ is the fine-structure constant, and a factor conserving spin-projection quantum numbers n_1 and n_2 has been explicitly included because we will not consider spin-dependent interactions. It is to be emphasized again that the momentum sums implicitly include sums over spin states and photon polarization indices.‡ Equation (11-8) represents the annihilation of one photon, while its Hermitian conjugate, $\mathsf{V}_{1\gamma}^{\dagger}$, represents the creation of a single photon.

In like manner, the second, or $\mathbf{A}^2$-interaction contribution to the matrix element of V_γ consists of a sum of two terms: $\mathsf{V}_{2\gamma}+\mathsf{V}_{2\gamma}^{\dagger}$, where

$$\mathsf{V}_{2\gamma}=\sum_{\mathbf{k}_1\mathbf{k}_2\mathbf{k}_3\mathbf{k}_4}a_{\mathbf{k}_1}^{\dagger}\langle\mathbf{k}_1\,|\,\mathsf{V}_{2\gamma}\,|\,\mathbf{k}_2\mathbf{k}_3\mathbf{k}_4\rangle\,a_{\mathbf{k}_2}a_{\mathbf{k}_3}^{(\gamma)}a_{\mathbf{k}_4}^{(\gamma)}$$

$$+\sum_{\mathbf{k}_1\mathbf{k}_2\mathbf{k}_3\mathbf{k}_4}a_{\mathbf{k}_1}^{\dagger}a_{\mathbf{k}_2}^{(\gamma)\dagger}\langle\mathbf{k}_1\mathbf{k}_2\,|\,\mathsf{V}_{2\gamma}\,|\,\mathbf{k}_3\mathbf{k}_4\rangle\,a_{\mathbf{k}_3}a_{\mathbf{k}_4}^{(\gamma)}\,, \tag{11-10}$$

* Since they appear in a sum over all values in Eq. (11-4), we have relabeled $\mathbf{k}_1$ and $\mathbf{k}_2$ in the second term of Eq. (11-7) in order to identify the Hermitian conjugate.

‡ Since the photon cannot be transformed to a rest frame, one can actually omit the value $\mathbf{k}=\mathbf{0}$ from photon momentum sums.

and

$$\langle \mathbf{k}_1 | \mathsf{V}_{2\gamma} | \mathbf{k}_2\mathbf{k}_3\mathbf{k}_4 \rangle = Z^2 \left(\frac{\pi\alpha\hbar^2}{mV} \right) \frac{\epsilon_3 \cdot \epsilon_4}{(k_3k_4)^{1/2}} \delta_{\mathbf{k}_1, \mathbf{k}_2+\mathbf{k}_3+\mathbf{k}_4} \delta_{n_1n_2}, \tag{11-11}$$

$$\langle \mathbf{k}_1\mathbf{k}_2 | \mathsf{V}_{2\gamma} | \mathbf{k}_3\mathbf{k}_4 \rangle = Z^2 \left(\frac{\pi\alpha\hbar^2}{mV} \right) \frac{\epsilon_2 \cdot \epsilon_4}{(k_2k_4)^{1/2}} \delta_{\mathbf{k}_1+\mathbf{k}_2, \mathbf{k}_3+\mathbf{k}_4} \delta_{n_1n_3}. \tag{11-12}$$

Exercise. Verify Eqs. (11-10)–(11-12).

It is important to observe in the derivation of Eq. (11-10) that the second term has been written in normal form, commensurate with the discussion following Eq. (9-53). Thus, one must actually *put* the second term in this contribution into the form $\mathsf{V}_{2\gamma}^{\dagger}$ as if the commutators were zero. Note also that the second term in Eq. (11-10) has the same *form* as the matrix element of V_2, the two-body interaction. We shall, however, always include this photon-charged particle interaction in V_γ, rather than in V_2, consistent with Eq. (11-2).

Finally, evaluation of the matrix element in the third line of Eq. (11-4) involves substitution of the Coulomb interaction,

$$\mathsf{V}_{\mathrm{C}}(\mathbf{x}_1, \mathbf{x}_2) = \frac{Z_1Z_2e^2}{|\mathbf{x}_1 - \mathbf{x}_2|}, \tag{11-13}$$

into Eq. (10-52). Thus,

$$\begin{aligned} \langle \mathbf{k}_1\mathbf{k}_2 | \mathsf{V}_{\mathrm{C}} | \mathbf{k}_3\mathbf{k}_4 \rangle &= \frac{Z_1Z_2e^2}{V^2} \iint \exp(-i\mathbf{k}_1 \cdot \mathbf{x}_1) \exp(-i\mathbf{k}_2 \cdot \mathbf{x}_2) \frac{d^3x_1\, d^3x_2}{|\mathbf{x}_1 - \mathbf{x}_2|} \\ &\quad \times \exp(i\mathbf{k}_3 \cdot \mathbf{x}_1) \exp(i\mathbf{k}_4 \cdot \mathbf{x}_2). \end{aligned} \tag{11-14}$$

Upon change of integration variables to $\mathbf{r} = \mathbf{x}_1 - \mathbf{x}_2$, the double integral becomes

$$\int \frac{\exp(-i\mathbf{q} \cdot \mathbf{r})}{r} d^3r \int \exp(-i\mathbf{K} \cdot \mathbf{x}_2)\, d^3x_2,$$

where $\mathbf{q} = \mathbf{k}_1 - \mathbf{k}_3$, $\mathbf{K} = \mathbf{k}_1 + \mathbf{k}_2 - \mathbf{k}_3 - \mathbf{k}_4$. The $\mathbf{x}_2$ integral is clearly a δ function, while the Fourier transform of r^{-1} can be found by inserting a factor $\exp(-\Delta r)$ and allowing $\Delta \to 0$ after the integration. Hence,

$$\begin{aligned} \langle \mathbf{k}_1\mathbf{k}_2 | \mathsf{V}_{\mathrm{C}} | \mathbf{k}_3\mathbf{k}_4 \rangle &= \frac{(2\pi)^3}{V^2} \frac{4\pi Z_1Z_2e^2}{q^2} \delta_{n_1n_3} \delta_{n_2n_4} \delta(\mathbf{k}_1 + \mathbf{k}_2 - \mathbf{k}_3 - \mathbf{k}_4), \end{aligned} \tag{11-15}$$

where spin-conserving factors have again been included, and one should note that the third δ is a δ function rather than a Kronecker δ.* Physically, the quantity $\mathbf{q}$ is the momentum transferred in the interaction.

Thus, in the momentum representation the Fock-space Hamiltonian for nonrelativistic QED, Eqs. (11-2) or (11-4), can be written

$$\mathsf{H} = \sum_{\mathbf{k}} a_{\mathbf{k}}^{\dagger} a_{\mathbf{k}} \omega_{\mathbf{k}} + \mathsf{V}_{1\gamma} + \mathsf{V}_{1\gamma}^{\dagger} + \mathsf{V}_{2\gamma} + \mathsf{V}_{2\gamma}^{\dagger} + \mathsf{V}_{C} \,. \tag{11-16}$$

The first term contains both the free-particle kinetic energy and the energy of the radiation field, *if* we assume $\mathbf{k}$ to run over momentum values for *all* types of particle in the system. In this case, $\omega_{\mathbf{k}} = \hbar^2 k^2/2m$ for charged particles, and $\omega_{\mathbf{k}} = \hbar ck$ for photons.

Actually, Eq. (11-16) is a more general result then appears on the surface, because *any* single-particle basis functions could have been chosen in place of the momentum eigenfunctions, Eq. (11-5). In particular, one could use the wave functions describing an electron bound in an atom, and in what follows this will at times become necessary. In this case, one need only reevaluate integrals such as that in Eq. (11-7), with the plane waves replaced by the appropriate bound-state functions. Equations (11-9), (11-11), (11-12), and (11-15) will then have different forms, depending on the nature of the states there described by $|\,\mathbf{k}\rangle$.

B. THE CALCULATIONAL PROBLEM

The Hamiltonian of Eqs. (11-16) or (11-4) represents the complete system of field plus sources, and, as is the case classically, no exact nontrivial solutions are known. In quantum mechanics it is *practicably* impossible to diagonalize H. It would, in fact, be quite embarrassing were this diagonalization to be carried out, for some sort of perturbation expansion seems to be necessary in order to compare QED with experiment.

Therefore, we shall consider the radiation field on the one hand, and the charged particles interacting via Coulomb forces on the other, as two separate, well-understood systems which are weakly coupled together. In QED the strength of this coupling is described by the *coupling constant*

$$\alpha = e^2/\hbar c \sim 1/137 \,, \tag{11-17}$$

the well-known fine-structure constant. The smallness of this quantity immediately suggests that the coupled system be studied in terms of a power series expansion in α, so that the more complicated processes can be considered as "higher order" in the sense that they contain larger powers of α. For QED this procedure works well, although the resulting expansion of the theory is not analytic in powers of α; terms in $\ln \alpha$ appear in pro-

* According to Eq. (9-26), the appropriate relation is $\delta_{\mathbf{kk}'} = [(2\pi)^3/V]\delta^{(3)}(\mathbf{k} - \mathbf{k}')$.

cesses such as the Lamb shift (see Chapter XV). Of course, the convergence of such a scheme is difficult to determine, and usually the best that can be hoped for is an asymptotic approach to the correct physical situation.

While the smallness of the coupling constant has led to tremendous successes in QED, as well as with low-energy pion–nucleon and nucleon–nucleon phenomena, the general theory of interacting fields has encountered great difficulties because of larger coupling constants. In these latter instances other, less-well-understood techniques have been required in order to make any progress. Thus, it is quite possible that, as our understanding of elementary particles improves, the expansion procedures to be used in the following work will be superseded. On the other hand, the weak coupling between systems seems to be essential to any theory of measurement, so that one should expect the *general* features of the following results to be very nearly correct.

TABLE I

Classification of Some Processes Occurring in Nonrelativistic Quantum Electrodynamics

Order	Process
e^2	Emission, absorption, internal conversion, photoelectric effect
e^4	Dispersion, Raman effect, resonance fluorescence, Thomson scattering
e^6	*Bremsstrahlung*, multiple processes
$\vdots$	

Let us now consider how we may construct a logical classification of the phenomena. A cursory examination of Eqs. (11-9), (11-11), and (11-12) shows that associated with each photon in a process is a factor of e, the electronic charge. But in calculating cross sections, transition rates, etc., squares of matrix elements are always involved, so that a one-photon process will lead to a factor of e^2, two-photon processes to factors of e^4, and so on. Moreover, one sees from Eq. (11-15) that, insofar as its content in powers of e is concerned, a Coulomb interaction is a two-photon process of order e^4. Thus, we can tentatively classify the processes in the radiation problem as first order (e^2), second order (e^4), and so on, and summarize them as in Table I. We shall see that processes of order higher than e^6 are essentially unobservable.

Within the above general classification scheme it is very convenient to further group the problems to be studied according to whether they involve the interactions of free particles, or if transitions between bound states are important. Such a grouping has some value, because the methods of calculation are somewhat different in the two cases. In the former group the problem is essentially that of scattering, and the cross section can be calculated in terms of the S matrix, in which case the calculation is facilitated by using the plane-wave, or momentum, representation. This is not always possible with problems of the second group, because the matrix elements involve bound-state wave functions. In such cases one must follow the procedure discussed briefly following Eq. (11-16).

C. THE INTERACTION PICTURE

In this section we shall construct a general expansion procedure for studying all of the processes arising in nonrelativistic quantum electrodynamics. To develop the method, it is first necessary to introduce a quantum mechanical representation essentially due to Dirac,[1] and which allows us to take maximum advantage of the quantum mechanical operator formalism. Although we have agreed to work primarily in the Schroedinger picture, it is useful to reintroduce the time here as a convenient parameter—it will disappear in our final results.

The development begins by recalling that the equation of motion for the N-body wave function in the Schroedinger picture is

$$i\hbar \frac{d\psi'(t)}{dt} = (\mathsf{H}_0 + \mathsf{V}')\psi'(t)\,, \tag{11-18}$$

where H_0 is the N-body, free-particle Hamiltonian and V' is a (possibly time-dependent) interaction operator. Let us now make a time-dependent, unitary transformation by defining a new state vector

$$\psi(t) = [\exp(i\mathsf{H}_0 t/\hbar)]\psi'(t)\,, \tag{11-19}$$

which coincides with $\psi'(0)$ at $t = 0$, and evolves from the latter. Differentiating $\psi(t)$ with respect to t and using Eq. (11-18), we obtain the equation of motion for ψ:

$$i\hbar \frac{d\psi(t)}{dt} = \mathsf{V}(t)\psi(t)\,, \tag{11-20}$$

where

$$\mathsf{V}(t) = [\exp(i\mathsf{H}_0 t/\hbar)]\, V' \exp(-i\mathsf{H}_0 t/\hbar) \tag{11-21}$$

is the *effective Hamiltonian* for $\psi(t)$.

Thus, the quantum state described by $\psi'(t)$ in the Schroedinger picture is described by $\psi(t)$ in the new picture, called the *interaction picture*.* An arbitrary operator A' in the former description is transformed to the inter-

* This is sometimes referred to as the "Dirac picture," or "intermediate picture."

action picture by the unitary transformation induced by Eq. (11-19):

$$\mathsf{A}(t) = [\exp(i\mathsf{H}_0 t/\hbar)]\mathsf{A}' \exp(-i\mathsf{H}_0 t/\hbar) . \tag{11-22}$$

One sees that the motion of the state vector is governed by $\mathsf{V}(t)$, while the evolution of the operators is dictated by H_0, so that the present description is intermediate between the Schroedinger and Heisenberg pictures.

Exercise. Show that the transformed Hamiltonian for free particles is unchanged:

$$\mathsf{H}_0(t) = \mathsf{H}_0 . \tag{11-23}$$

A particularly important case of the transformation (11-22) is that for the creation and annihilation operators. These reduce to the simple forms

$$a_i(t) = [\exp(i\mathsf{H}_0 t/\hbar)]a_i \exp(-i\mathsf{H}_0 t/\hbar) = a_i \exp(-it\omega_i/\hbar) , \tag{11-24a}$$

$$a_i^\dagger(t) = [\exp(i\mathsf{H}_0 t/\hbar)]a_i^\dagger \exp(-i\mathsf{H}_0 t/\hbar) = a_i^\dagger \exp(it\omega_i/\hbar) , \tag{11-24b}$$

where the ω_i are the free-particle energy eigenvalues of H_0 (including photons). To prove Eq. (11-24a), for example, one first substitutes the first term of Eq. (11-16) for H_0 into the exponentials. If the exponentials are now expanded and the commutation relations (10-30) applied, it is seen that every term in the momentum sums vanishes but one, so that

$$a_i(t) = \{\exp[(it/\hbar)a_i^\dagger a_i\omega_i]\}a_i \exp[-(it/\hbar)a_i^\dagger a_i\omega_i] . \tag{11-25}$$

One now applies the well-known operator identity[2]

$$f(\lambda) = [\exp(\lambda a_i^\dagger a_i)]a_i \exp(-\lambda a_i^\dagger a_i) = e^{-\lambda}a_i \tag{11-26}$$

for arbitrary λ, which immediately proves Eq. (11-24).

Thus, in the interaction picture the entire Hamiltonian of Eq. (11-16) can be written

$$\mathsf{H} = \sum_{\mathbf{k}} a_{\mathbf{k}}^\dagger a_{\mathbf{k}}\omega_{\mathbf{k}} + \mathsf{V}(t) , \tag{11-27}$$

where

$$\mathsf{V}(t) = \mathsf{V}_{1\gamma}(t) + \mathsf{V}_{1\gamma}^\dagger(t) + \mathsf{V}_{2\gamma}(t) + \mathsf{V}_{2\gamma}^\dagger(t) + \mathsf{V}_{\mathrm{C}}(t) , \tag{11-28}$$

and, for example,

$$\begin{aligned} \mathsf{V}_{\mathrm{C}}(t) &= \tfrac{1}{2} \sum_{\mathbf{k}_1\ldots\mathbf{k}_4} a_{\mathbf{k}_1}^\dagger a_{\mathbf{k}_2}^\dagger \langle \mathbf{k}_1\mathbf{k}_2 \,|\, \mathsf{V}_{\mathrm{C}}(t) \,|\, \mathbf{k}_3\mathbf{k}_4\rangle\, a_{\mathbf{k}_4}a_{\mathbf{k}_3} \\ &= \tfrac{1}{2} \sum_{\mathbf{k}_1\ldots\mathbf{k}_4} a_{\mathbf{k}_1}^\dagger a_{\mathbf{k}_2}^\dagger \langle \mathbf{k}_1\mathbf{k}_2 \,|\, \mathsf{V}_{\mathrm{C}} \,|\, \mathbf{k}_3\mathbf{k}_4\rangle\, a_{\mathbf{k}_4}a_{\mathbf{k}_3} \\ &\quad\times \exp\{it(\omega_1 + \omega_2 - \omega_3 - \omega_4)/\hbar\} . \end{aligned} \tag{11-29}$$

Perhaps the most important assumption of the quantum theory is the validity of the linear superposition principle. This implies that the differential equation (11-20) for the state of the system can be solved formally by a linear relation

$$\psi(t) = \mathsf{U}(t, t_0)\psi(t_0)\,, \tag{11-30}$$

so that determination of the *time-evolution operator* $\mathsf{U}(t, t_0)$ gives the state of the system at all times if that state is known at $t = t_0$. Substitution of Eq. (11-30) into (11-20) then yields the differential equation

$$i\hbar\big(d\mathsf{U}(t, t_0)/dt\big) = \mathsf{V}(t)\mathsf{U}(t, t_0)\,, \tag{11-31}$$

which determines $\mathsf{U}(t, t_0)$ subject to the initial condition

$$\mathsf{U}(t_0, t_0) = 1\,. \tag{11-32}$$

The importance of the operator $\mathsf{U}(t, t_0)$ lies with the observation that it depends only on the physical characteristics of the system, and not on the initial state. Moreover, these operators form a group, since

$$\psi(t_2) = \mathsf{U}(t_2, t_1)\psi(t_1) = \mathsf{U}(t_2, t_1)\mathsf{U}(t_1, t_0)\psi(t_0)$$

implies that

$$\mathsf{U}(t_2, t_0) = \mathsf{U}(t_2, t_1)\mathsf{U}(t_1, t_0)\,. \tag{11-33}$$

The inverse operator is

$$\mathsf{U}^{-1}(t, t_0) = \mathsf{U}(t_0, t)\,. \tag{11-34}$$

Exercise. Show that $\mathsf{U}(t, t_0)$ is unitary if V' is Hermitian.

For QED the primary feature of $\mathsf{U}(t, t_0)$ is its relation to the measurable quantities describing the radiation processes. As suggested by Eq. (11-30), the operator U takes the system from an initial state **a** to a final state **b**, and if the particle is free of the interaction a long time before and a long time after the encounter, these states can be measured when $t_0 \to -\infty$ and $t \to +\infty$. Thus, one defines a scattering operator, or S *matrix*, by the matrix elements*

$$\begin{aligned}\langle \mathbf{b} \,|\, \mathsf{S} \,|\, \mathbf{a}\rangle &= \big(\psi_{\mathbf{b}}, \mathsf{U}(\infty, -\infty)\psi_{\mathbf{a}}\big)\\ &= \langle \mathbf{b} \,|\, \mathsf{U}(\infty, -\infty) \,|\, \mathbf{a}\rangle\,.\end{aligned} \tag{11-35}$$

The importance of the S matrix is that it is related to the *scattering*

* The S matrix is often defined only for plane-wave or momentum eigenstates, a restriction we do *not* make here.

amplitude f in the momentum representation by

$$\langle \mathbf{k}' \,|\, \mathsf{S} \,|\, \mathbf{k} \rangle = \delta_{\mathbf{k}\mathbf{k}'} + \frac{4\pi^2 i}{kV}\, \delta(k - k') f_{\mathbf{k}}(\hat{\mathbf{k}}) \,, \tag{11-36}$$

which in turn is related to the differential scattering cross section

$$d\sigma = |f_{\mathbf{k}}(\hat{\mathbf{k}}')|^2 \, d\Omega \,. \tag{11-37}$$

Thus, by determining the operator $\mathsf{U}(t, t_0)$ and taking the limits $t \to +\infty$, $t_0 \to -\infty$, we can readily obtain the cross sections for scattering processes.

Equations (11-35)–(11-37) are generally discussed in both elementary and advanced texts on quantum mechanics, their derivation properly belonging to the formal theory of scattering. In order to have a self-contained development of the calculational procedures, however, a brief discussion of scattering theory and a derivation of the appropriate equations is given in Appendix C. There we also define another quantity, $\mathsf{R}_{\mathbf{ba}}$, known as the **ba** element of the *reaction matrix*, which is related to the S matrix through the relation

$$\langle \mathbf{b} \,|\, \mathsf{S} \,|\, \mathbf{a} \rangle = \delta_{\mathbf{ba}} - 2\pi i\, \delta(E_{\mathbf{b}} - E_{\mathbf{a}})\, \mathsf{R}_{\mathbf{ba}} \,. \tag{11-38}$$

The significance of the R matrix is that it is related to an important physical quantity, the *transition probability per unit time*, the appropriate expression being

$$w_{\mathbf{ba}} = (2\pi/\hbar)\, \delta(E_{\mathbf{b}} - E_{\mathbf{a}})\, |\, \mathsf{R}_{\mathbf{ba}} \,|^2 \,. \tag{11-39}$$

However, one is usually interested in the transition to a group of final states with energy between $E_{\mathbf{b}}$ and $E_{\mathbf{b}} + dE_{\mathbf{b}}$, described in phase space by a density of final states $\rho_f(E_{\mathbf{b}})$. Integration over these final states then yields for the transition rate per unit time

$$w = (2\pi/\hbar)\, |\, \mathsf{R}_{\mathbf{ba}} \,|^2 \, [\rho_f(E_{\mathbf{b}})]_{E_{\mathbf{b}} = E_{\mathbf{a}}} \,, \tag{11-40}$$

which is just *Fermi's golden rule*.[3] One then defines the cross section for the process as w divided by the incident flux. Equations (11-37) and (11-40) are the primary quantities to be calculated in what follows, and the reader is referred to Appendix C for further physical and mathematical details underlying their derivation.

From the preceding paragraphs it is amply clear that the calculational problem for QED can be reduced to the determination of the time-evolution operator $\mathsf{U}(t, t_0)$. An iterative technique for calculating the matrix elements of this operator can be developed by first converting the differential equation (11-31) into an integral equation. Integrating over t and using the initial condition (11-32), we obtain

$$\mathsf{U}(t, t_0) = 1 - (i/\hbar) \int_{t_0}^{t} \mathsf{V}(t_1) \mathsf{U}(t_1, t_0)\, dt_1 \,. \tag{11-41}$$

This is an inhomogeneous integral equation of Volterra type, which in many cases can be solved by iteration. From the discussion at the beginning of Section XI-B, we might expect an iteration in powers of V to be quite useful for our purposes, since the coupling constant is small.* Moreover, under fairly general conditions the iterated series obtained from Eq. (11-41) is known to converge, although we shall not attempt to demonstrate this property for the present situation. One easily verifies that Eq. (11-41) is equivalent to

$$\begin{aligned} \mathsf{U}(t, t_0) = 1 &+ (-i/\hbar) \int_{t_0}^{t} \mathsf{V}(t_1)\, dt_1 + (-i/\hbar)^2 \int_{t_0}^{t} dt_1 \int_{t_0}^{t_1} dt_2\, \mathsf{V}(t_1)\mathsf{V}(t_2) \\ &+ (-i/\hbar)^3 \int_{t_0}^{t} dt_1 \int_{t_0}^{t_1} dt_2 \int_{t_0}^{t_2} dt_3\, \mathsf{V}(t_1)\mathsf{V}(t_2)\mathsf{V}(t_3) + \cdots, \end{aligned} \tag{11-42}$$

which is essentially a power series in the coupling constant. Note carefully that it is not the usual perturbation expansion, but merely an iteration of the integral equation, so that no questions regarding the validity of perturbation theory are involved here. Also, one must note that $\mathsf{V}(t_1)$ and $\mathsf{V}(t_2)$ do not necessarily commute, and therefore the ordering of the V's is important. The time ordering of the products in the integrands can be formally removed by introducing Dyson's time-ordered product,[4] but we shall not find this a useful concept in the present context.

From the previous discussion, then, the procedure to be followed is to now take matrix elements of both sides of Eq. (11-42) in order to obtain an approximation in powers of the interaction for the quantities needed in Eqs. (11-36) and (11-38). It will be seen shortly that the terms on the right-hand side of Eq. (11-42) can be interpreted in terms of all possible (nonrelativistic) electrodynamic processes, and thus, by choosing from this sum those processes we wish to study, the above physical quantities can be calculated. If $|\mathbf{a}\rangle$ is a many-body initial state and $|\mathbf{b}\rangle$ a final state after an interaction, we obtain from Eq. (11-42)

$$\begin{aligned} \langle \mathbf{b} | \mathsf{U}(t, t_0) | \mathbf{a} \rangle &= \delta_{\mathbf{ba}} + (-i/\hbar) \int_{t_0}^{t} \langle \mathbf{b} | \mathsf{V}(t_1) | \mathbf{a} \rangle\, dt_1 \\ &\quad + (-i/\hbar)^2 \int_{t_0}^{t} dt_1 \int_{t_0}^{t_1} dt_2\, \langle \mathbf{b} | \mathsf{V}(t_1)\mathsf{V}(t_2) | \mathbf{a} \rangle + \cdots, \end{aligned} \tag{11-43}$$

assuming the states to be normalized. It is the analysis of the right-hand side of this equation which will occupy the rest of the book.

It is of some interest to examine Eq. (11-43) in general when $t \to \infty$, $t_0 \to -\infty$. These limits must be taken rather carefully, however, because when $|\mathbf{b}\rangle$ and $|\mathbf{a}\rangle$ are plane-wave states, the integrals are divergent at

* This is just the standard Liouville–Neumann series solution of Eq. (11-41).

their lower limits, and, furthermore, the number of ways of taking the limit is equal to the order of the iteration! This difficulty can be overcome by the customary device of introducing a convergence factor through the replacement

$$\mathsf{V}(t) \to \mathsf{V}(t)\, e^{-\Delta|t|}\,, \tag{11-44}$$

and taking the limit $\Delta \to 0$ *after* the integration is performed. Following this procedure, one finds from Eq. (11-43), for $\mathbf{b} \neq \mathbf{a}$,

$$\langle \mathbf{b} \,|\, \mathsf{S} \,|\, \mathbf{a}\rangle = -2\pi i\, \delta(E_{\mathbf{b}} - E_{\mathbf{a}}) \times\left[\langle \mathbf{b} \,|\, \mathsf{V}' \,|\, \mathbf{a}\rangle + \sum_n \frac{\langle \mathbf{b} \,|\, \mathsf{V}' \,|\, n\rangle \langle n \,|\, \mathsf{V}' \,|\, \mathbf{a}\rangle}{E_{\mathbf{a}} - E_n} + \cdots\right], \tag{11-45}$$

where V' is the interaction in the Schroedinger picture, $E_{\mathbf{a}}$ is the relevant energy eigenvalue of H_0, and sums over intermediate states have been introduced by means of the identity

$$\langle \mathbf{b} \,|\, \mathsf{AB} \,|\, \mathbf{a}\rangle = \sum_n \langle \mathbf{b} \,|\, \mathsf{A} \,|\, n\rangle \langle n \,|\, \mathsf{B} \,|\, \mathbf{a}\rangle\,. \tag{11-46}$$

The reader will recognize Eq. (11-45) as the usual result of Rayleigh–Schroedinger perturbation theory.[5] Thus, one can *formally* identify the expression enclosed in square brackets in Eq. (11-45) as the energy-level shift ΔE due to the interaction. This identification is precarious, however, because, as will be shown later, the level shift of Eq. (11-45) is often infinite! In fact, we shall not have too much need for this equation, but the procedure used to perform the time integrations will prove to be of value in the sequel.

D. DIAGRAMMATIC ANALYSIS OF ELECTRODYNAMIC PROCESSES

In order to facilitate identification of the radiation processes described by the right-hand side of Eq. (11-43), and to introduce the reader to some of the highly successful techniques of modern field theory and the many-body problem, we shall undertake a formulation of the theory in terms of certain diagrams to be introduced below. One can really appreciate completely the power of these methods only through hindsight, and by making applications beyond those considered here. However, the techniques to be developed do lend a degree of elegance to nonrelativistic QED, and certainly serve to systematize the theory in a manner consistent with the covariant formulations of field theory, which have been so successful.

We begin by examining the second term on the right-hand side of Eq. (11-43), momentarily ignoring the temporal variables. Substituting from

Eq. (11-28), we have

$$\langle \mathbf{b} \,|\, \mathsf{V} \,|\, \mathbf{a} \rangle = \langle \mathbf{b} \,|\, \mathsf{V}_{1\gamma} + \mathsf{V}_{1\gamma}^\dagger + \mathsf{V}_{2\gamma} + \mathsf{V}_{2\gamma}^\dagger + \mathsf{V}_{\mathrm{C}} \,|\, \mathbf{a} \rangle \,, \tag{11-47}$$

where $|\,\mathbf{b}\rangle$ and $|\,\mathbf{a}\rangle$ are symmetrized Fock-space vectors as given by Eq. (10-36). For demonstration purposes it is convenient to use the momentum representation in what follows, although it is to be remembered that this notation must be changed if $|\,\mathbf{b}\rangle$ or $|\,\mathbf{a}\rangle$, or both, are bound-state vectors. Then consider only the first term in Eq. (11-47), $\mathsf{V}_{1\gamma}$, and substitute from Eq. (11-8):

$$\langle \mathbf{b} \,|\, \mathsf{V}_{1\gamma} \,|\, \mathbf{a} \rangle = \sum_{l_1 l_2 l_3} \langle l_1 \,|\, \mathsf{V}_{1\gamma} \,|\, l_2 l_3 \rangle \, \langle \mathbf{b} \,|\, a_{l_1}^\dagger a_{l_2} a_{l_3}^{(\gamma)} \,|\, \mathbf{a} \rangle \,. \tag{11-48}$$

Now, the state $|\,\mathbf{a}\rangle$ could be an arbitrary N-particle Fock state (in the momentum representation),

$$|\,\mathbf{k}_1 \cdots \mathbf{k}_N\rangle = \prod_{\mathbf{k}} \frac{(a_{\mathbf{k}}{}^\dagger)^{n_{\mathbf{k}}}}{(n_{\mathbf{k}}!)^{1/2}} \,|\,0\rangle \,. \tag{11-49}$$

Consider for a moment only particles of one kind. Then, if N is less than the number of creation or annihilation operators in the matrix element, it follows that the latter quantity vanishes. On the other hand, if N is greater than the number of creation or annihilation operators in the last factor of Eq. (11-48), then a number of Kronecker δ functions will appear as products in the evaluation of this matrix element. For instance, since the Fock-state vectors are symmetrized,

$$\begin{aligned} \langle \mathbf{k}_1 \mathbf{k}_2 \,|\, a_{l_1}^\dagger a_{l_2} \,|\, \mathbf{k}_3 \mathbf{k}_4 \rangle = {} & \delta_{l_2 \mathbf{k}_3} [\delta_{\mathbf{k}_1 l_1} \, \delta_{\mathbf{k}_2 \mathbf{k}_4} + \varepsilon \delta_{\mathbf{k}_1 \mathbf{k}_4} \, \delta_{\mathbf{k}_2 l_1}] \\ & + \delta_{l_2 \mathbf{k}_4} [\delta_{l_2 \mathbf{k}_3} \, \delta_{\mathbf{k}_2 l_1} + \varepsilon \delta_{\mathbf{k}_1 l_1} \, \delta_{\mathbf{k}_2 \mathbf{k}_3}] \,, \end{aligned} \tag{11-50}$$

and, since they have no effect on the sums over $\boldsymbol{l}$ states, the Kronecker δ functions involving only $\mathbf{k}$ variables can be considered extraneous factors. The appearance of these factors is due to the specification of initial or final states involving more particles than the interaction can accommodate, so that a factor such as $\delta_{\mathbf{k}\mathbf{k}'}$ indicates that that particular particle does not participate in the interaction indicated by the matrix element. For the moment we can ignore these extraneous factors and specify only states $|\,\mathbf{b}\rangle$ and $|\,\mathbf{a}\rangle$ which are sufficient to give a nonzero value to the matrix element. We shall see below, however, that these seemingly extraneous factors have an important physical interpretation.

We can now turn to the evalution of the last matrix element in Eq. (11-48) and, for clarity, let us agree to label particle momenta by $\mathbf{p}$ and photon momenta by $\mathbf{k}$. Since a_i and $a_i{}^\dagger$ commute for different kinds of particles, it is clear that the matrix element factors into a particle part and

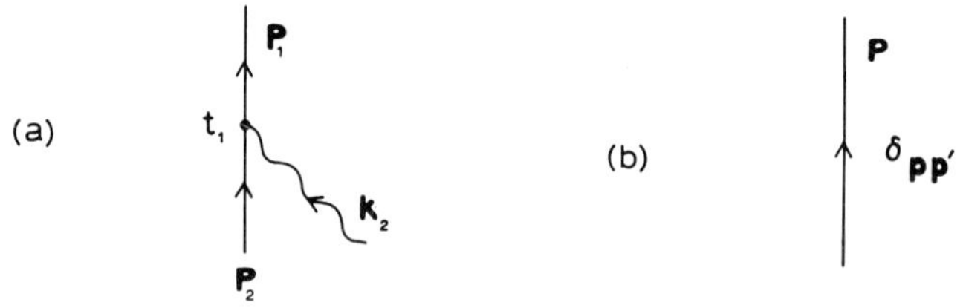

Fig. 25. Diagrammatic representation of (a) absorption of a photon by a charged particle; (b) a noninteracting particle (with mass) with momentum **p**.

a photon part. We examine the case $\langle \mathbf{b}| = \langle \mathbf{p}_1\mathbf{k}_1|$, $|\mathbf{a}\rangle = |\mathbf{p}_2\mathbf{k}_2\rangle$, so that

$$\langle \mathbf{b}|\, a^\dagger_{l_1} a_{l_2} a^{(\gamma)}_{l_3}\, |\mathbf{a}\rangle = \langle \mathbf{p}_1|\, a^\dagger_{l_1} a_{l_2}\, |\mathbf{p}_2\rangle \langle \mathbf{k}_1|\, a^{(\gamma)}_{l_3}\, |\mathbf{k}_2\rangle\,. \tag{11-51}$$

For a nonzero result it is clear that $l_3 = \mathbf{k}_2$ and that $\mathbf{k}_1$ is the vacuum. Likewise, we must have $l_2 = \mathbf{p}_2$ and $l_1 = \mathbf{p}_1$, and Eq. (11-48) becomes

$$\begin{aligned}\langle \mathbf{b}|\, \mathsf{V}_{1\gamma}\, |\mathbf{a}\rangle &= \sum_{l_1 l_2 l_3} \langle l_1|\, \mathsf{V}_{1\gamma}\, |l_2 l_3\rangle\, \delta_{l_1\mathbf{p}_1}\, \delta_{l_2\mathbf{p}_2}\, \delta_{l_3\mathbf{k}_2} \\ &= \langle \mathbf{p}_1|\, \mathsf{V}_{1\gamma}\, |\mathbf{p}_2\mathbf{k}_2\rangle\,. \end{aligned} \tag{11-52}$$

Reinserting the time dependence, we see that the first term of Eq. (11-47), for the specified initial and final states, gives to Eq. (11-43) the contribution

$$-\frac{i}{\hbar}\int_{t_0}^{t} \langle \mathbf{p}_1|\, \mathsf{V}_{1\gamma}(t_1)\, |\mathbf{p}_2\mathbf{k}_2\rangle\, dt_1\,. \tag{11-53}$$

This contribution can now be represented by a *diagram*, in which particle momenta are designated by solid lines and photon momenta by wiggly lines. Each line possesses an arrow pointing in the direction of increasing time, and this direction is considered to be upward from bottom to top of the diagram. The lines join at a vertex, which represents the interaction, and which is labeled by a time variable t. The quantity of Eq. (11-53) is represented in this fashion by the diagram of Fig. 25a, whereas Fig. 25b signifies a noninteracting particle with momentum **p**.

A graphical picture such as that of Fig. 25a suggests an absorption process, which is precisely what the $\mathsf{V}_{1\gamma}$, or $\mathbf{p}\cdot\mathbf{A}$, interaction describes. However, one must be careful with such an interpretation in the momentum representation, which we have been using as an example, for a free electron cannot absorb a photon (see Problem 9-5). Thus, the states $\mathbf{p}_1$ and $\mathbf{p}_2$ must be replaced by bound-state vectors, and the matrix elements describing this process, Eqs. (11-7) and (11-9), must be reevaluated using the correct wave functions. One can then interpret the process of Fig. 25a as the absorption of a photon by an atom, with subsequent excitation of an atomic electron to a higher-energy state. If the incoming photon energy exceeds the ioni-

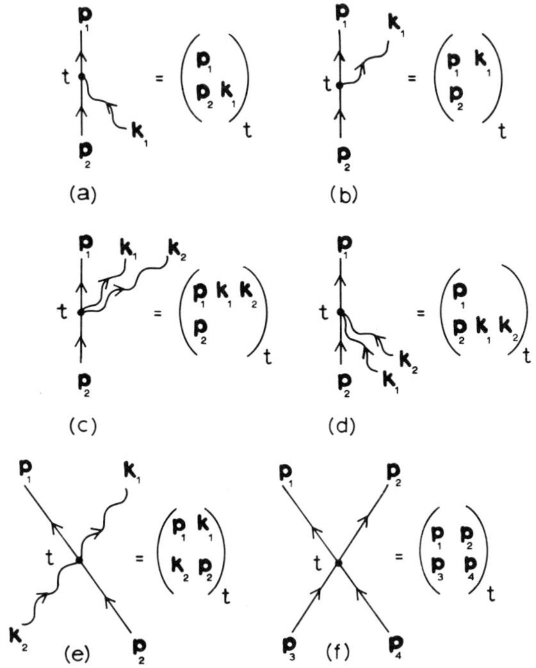

Fig. 26. Diagrammatic representation (a)–(f) of the six *fundamental* processes contained in the matrix element of Eq. (11-47).

zation energy of a particular electron, then this diagram can also describe the atomic photoelectric effect, and the final state $\mathbf{p}_1$ can again be a plane-wave state.

The diagrammatic analysis which we have begun here was first conceived by Feynman[6] for studing QED, and these graphs are generally referred to as *Feynman diagrams*. Not only did Feynman introduce a novel way of analyzing perturbation theory, but he suggested an entirely new method for bookkeeping in studying the complicated behavior of large numbers of identical particles. Many of the diagrammatic techniques which have proliferated in the last twenty years differ in several respects from Feynman's original treatment, as does ours, but all have the same underlying aim of simplifying the calculational problem.

As we shall see below, the diagrammatic method allows us to reinterpret the perturbation-type expansion (11-43) quite physically in terms of all possible (nonrelativistic) electrodynamic processes which can take place for given initial and final states. In fact, one might well interpret the subsequent diagrammatic expansion as a more fundamental view of the theory than that of the preceding sections. However, the use of these techniques here amounts to little more than an elegant way to employ ordinary perturbation theory to any desired order. It is in the study of a true $N(\gg 1)$-body theory that the techniques become overwhelmingly powerful, because in the many-body problem the diagrammatic method allows one to select certain types of processes and (effectively) perform infinite-order perturbation theory. An excellent introductory treatment of the application of diagrammatic techniques to field theory and the many-body problem has been given by Mattuck,[7] although the point of view is strictly that of field-theoretic methods.

The preceding discussion suggests that the right-hand side of Eq. (11-43) can be written entirely as a sum of diagrams. Prior to establishing this, however, it is necessary to study several other terms in the expansion in order to deduce the rules by which such diagrams can be drawn.

The second term in Eq. (11-47), $\mathsf{V}_{1\gamma}^{\dagger}$, yields a diagram similar to that of Fig. 25a, but describing an emission process. In Fig. 26 we have diagrammatically represented the six fundamental processes arising from Eq. (11-47) in the case that the initial and final states are the simplest possible. It is important, however, to make an observation regarding the last term in Eq. (11-47), which corresponds to a Coulomb interaction between two charged particles. From Eq. (10-49)

$$\langle \mathbf{b} | \mathsf{V}_C | \mathbf{a} \rangle = \tfrac{1}{2} \sum_{l_1 l_2 l_3 l_4} \langle l_1 l_2 | \mathsf{V}_C | l_3 l_4 \rangle \langle \mathbf{b} | a_{l_1}^{\dagger} a_{l_2}^{\dagger} a_{l_4} a_{l_3} | \mathbf{a} \rangle , \qquad (11\text{-}54)$$

again ignoring time variables. Suppose now that the particles are identical, so that the simplest two-particle initial and final states can be taken as $\langle \mathbf{b} | = \langle \mathbf{p}_1 \mathbf{p}_2 |$, $| \mathbf{a} \rangle = | \mathbf{p}_3 \mathbf{p}_4 \rangle$. Remembering that these state vectors are symmetrized, we easily find that

$$\langle \mathbf{b} | \mathsf{V}_C | \mathbf{a} \rangle = \langle \mathbf{p}_1 \mathbf{p}_2 | \mathsf{V}_C | \mathbf{p}_3 \mathbf{p}_4 \rangle + \varepsilon \langle \mathbf{p}_1 \mathbf{p}_2 | \mathsf{V}_C | \mathbf{p}_4 \mathbf{p}_3 \rangle , \qquad (11\text{-}55)$$

a direct plus exchange term, as is to be expected for identical particles. If the two particles are not identical, the exchange term does not arise, because in this case the matrix element factors. After substitution into Eq. (11-43), the diagram corresponding to Eq. (11-55) is that of Fig. 26f, and it is clear that when two lines representing identical particles attach to the same vertex at their head ends, an exchange term must be included.

Consider next the two-vertex terms arising from the third term on the right-hand side of Eq. (11-43). Since this is essentially the square of the expression in Eq. (11-47), we expect to have to study 25 terms in second order, and the value of the diagrammatic approach begins to be appreciated! As an example, consider the product $\mathsf{V}_{1\gamma}^{\dagger} \mathsf{V}_{1\gamma}$, and initial states $\langle \mathbf{b} | = \langle \mathbf{p}_1 \mathbf{k}_1 |$,

$|\mathbf{a}\rangle = |\mathbf{p}_2\mathbf{k}_2\rangle$. Again suppressing time variables, we have

$$\langle \mathbf{b}| \mathsf{V}_{1\gamma}^{\dagger}\mathsf{V}_{1\gamma} |\mathbf{a}\rangle = \sum_{l_1l_2l_3l_4l_5l_6} \langle l_1l_2| \mathsf{V}_{1\gamma}^{\dagger} |l_3\rangle \langle l_4| \mathsf{V}_{1\gamma} |l_5l_6\rangle \times \langle \mathbf{p}_1| a_{l_1}^{\dagger}a_{l_3}a_{l_4}^{\dagger}a_{l_5} |\mathbf{p}_2\rangle \langle \mathbf{k}_1| a_{l_2}^{(\gamma)\dagger}a_{l_6}^{(\gamma)} |\mathbf{k}_2\rangle\,. \tag{11-56}$$

The evaluation of the photon matrix element in this last equation is trivial, and one finds the value $\delta_{\mathbf{k}_1l_2}\,\delta_{l_6\mathbf{k}_2}$. The particle matrix element, on the other hand, is an example of how an arbitrary matrix element can appear with the creation and annihilation operators "mixed up," and this strongly suggests that a systematic procedure for evaluating such quantities be adopted. That which proves most convenient is to first put all products of creation and annihilation operators into normal form [recall the discussion following Eq. (9-53)], using the commutation relations (10-30).* This procedure will break up a given matrix element into a sum of terms, but each term can now be evaulated in a simple and systematic manner. For example, the product in Eq. (11-56) can be written (for identical particles)

$$a_{l_1}^{\dagger}a_{l_3}a_{l_4}^{\dagger}a_{l_5} = \delta_{l_3l_4}a_{l_1}^{\dagger}a_{l_5} + \varepsilon a_{l_1}^{\dagger}a_{l_4}^{\dagger}a_{l_3}a_{l_5}\,. \tag{11-57}$$

Before completing the evaluation of the expression in Eq. (11-56), let us digress a bit further as regards this procedure of putting the creation and annihilation operators into normal form prior to evaluation. Since it is unlikely that we shall ever have to calculate diagrams of more than three or four vertices in QED, as discussed in Section B, the straightforward procedure just outlined is certainly adequate. When studying the macroscopic many-body problem with these techniques, though, it is necessary to analyze the general term in the expansion (11-43), and this involves matrix elements containing an arbitrary number of creation and annihilation operators. In this case one needs a general theorem regarding the process of putting matrix elements into normal form, and this was first stated by Wick in 1950.[8] In the present diagrammatic method the procedure would be somewhat simpler than Wick's theorem, because we are not concerned with time-ordered products, and this version of the theorem has been developed by Mohling and Grandy.[9]

We now return to the evaluation of Eq. (11-56), and notice that the second term on the right-hand side of Eq. (11-57) will lead to a zero result, because N is less than the number of annihilation operators, as discussed

* Note carefully that we *do not* here treat the commutators as zero, as when agreeing to always write the canonical operators in normal form. The procedure being discussed here is completely different from that following Eq. (9-53).

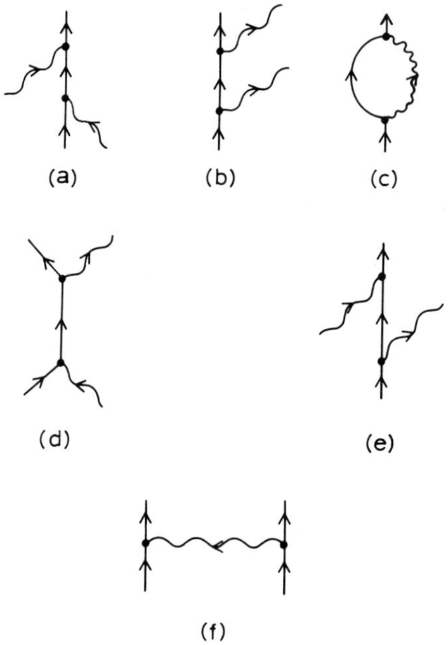

Fig. 27. All possible two-vertex diagrams containing only $\mathbf{p} \cdot \mathbf{A}$ vertices (a)–(f).

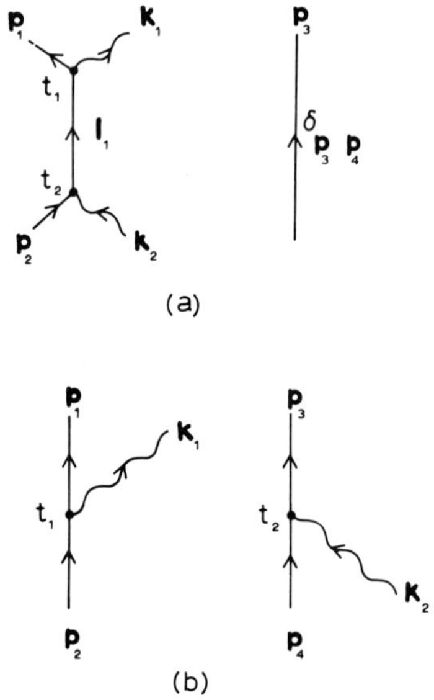

Fig. 28. Two examples (a) and (b) of unconnected diagrams associated with the product $V_{1\gamma}^{\dagger} V_{1\gamma}$.

above. Hence, the equation becomes

$$\langle \mathbf{b} | \mathsf{V}_{17}^{\dagger} \mathsf{V}_{17} | \mathbf{a} \rangle = \sum_{l_3} \langle \mathbf{p}_1 \mathbf{k}_1 | \mathsf{V}_{17}^{\dagger} | l_3 \rangle \langle l_3 | \mathsf{V}_{17} | \mathbf{p}_2 \mathbf{k}_2 \rangle . \qquad (11\text{-}58)$$

The diagram representing this expression* is that of Fig. 27d, and in Fig. 27e one finds the diagram obtained by considering the reverse product, $\mathsf{V}_{17}\mathsf{V}_{17}^{\dagger}$. These diagrams will be found important in studying photon scattering from electrons, but in this connection one again meets the situation of a free electron absorbing or emitting a photon. In this case, however, the process *is* allowed because of the intermediate state of the electron between the two vertices. The interpretation is that the electron interacts with the photon by absorbing it and becoming excited to a *virtual state*. Almost immediately, the electron decays into the original photon and electron, but with different energies and momenta. Although energy and momentum cannot both be conserved in the intermediate state, the lifetime of this virtual state is so short that the uncertainty principle allows for nonconservation of energy or momentum, or both. Such virtual states are unobservable, and, therefore, not subject to the conservation laws, even though these laws are obeyed for the overall process. Thus, the processes represented by Figs. 27d and 27e are indeed physical, even for electrons.

One of the rules for diagrams which we learn from the foregoing analysis is that an *internal line* corresponds to a sum over intermediate states, as in Eq. (11-58). Moreover, there are other important things to be learned from the matrix element $\langle \mathbf{b} | \mathsf{V}_{17}^{\dagger} \mathsf{V}_{17} | \mathbf{a} \rangle$, problems which arise when the initial and final states contain *two* charged particles and a photon. In this case, one obtains terms corresponding to the diagrams of Fig. 28, plus other similar diagrams with the labels interchanged as a result of the symmetrized nature of the state vectors when the two charged particles are identical. The diagrams of Fig. 28 are called *unconnected diagrams*, because they appear as products. This means that their analytical expression contains no sum over intermediate states, so that no internal line connects the two diagrams. On the other hand, the diagrams of Fig. 27 are *connected diagrams*.

Now, the interpretation of Fig. 28 is based on the nature of the initial and final states which led to these diagrams through the interaction $\mathsf{V}_{17}^{\dagger}\mathsf{V}_{17}$. It is only possible for *one* of the initial particles to take part in this *combined* interaction, so that the other must behave in one of the two possible ways indicated. Thus, in Fig. 28a one charged particle interacts with the photon and the other is unaffected, while in Fig. 28b one particle absorbs the photon and the other emits a photon. However, this interpretation is only reasonable if the two processes are truly independent. This is clear for the diagram of Fig. 28a, but not so clear for Fig. 28b when it is remembered that the product diagram is still interconnected by the time integrals in the third term on the right-hand side of Eq. (11-43).

* When no ambiguity can arise, it is convenient to omit time and momentum labels.

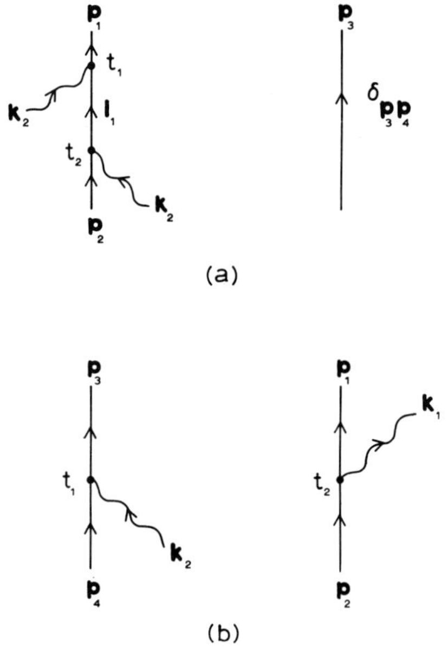

Fig. 29. Examples (a) and (b) of unconnected diagrams associated with the product $\mathsf{V}_{1\gamma}\mathsf{V}_{1\gamma}^{\dagger}$.

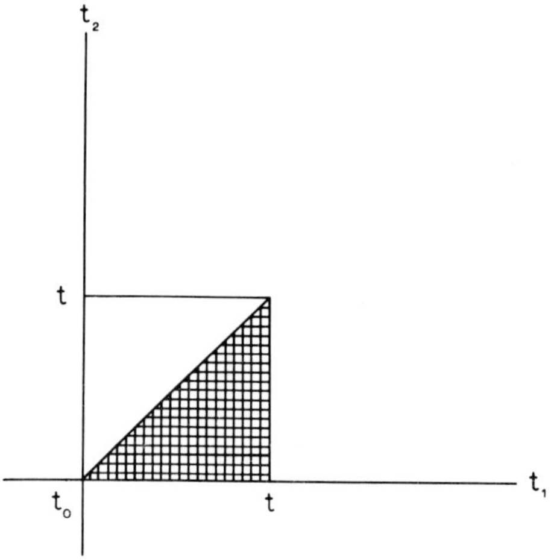

Fig. 30. The checkered region corresponds to the domain of integration in Eq. (11-59).

The resolution of this difficulty, and the proof that unconnected diagrams such as those of Fig. 28b can correctly be evaluated as products of *independent* diagrams, constitutes one of the major advances in mathematical physics of recent years. The problem is resolved by remembering that these diagrams came from a matrix element involving $V_{1\gamma}^{\dagger}V_{1\gamma}$, and that in the same order of the expansion (11-43) there will also occur a matrix element of the reverse interaction, $V_{1\gamma}V_{1\gamma}^{\dagger}$. For initial and final states involving two charged particles and a photon, the latter matrix element corresponds to the unconnected diagrams of Fig. 29. Again, Fig. 29a contains truly independent products, whereas Fig. 29b does not. However, the unconnected diagram of Fig. 29b is almost identical to that of Fig. 28b, except for the ordering of the time variables—and this is the key to the solution of the problem.

Exercise. Show that, except for the time variables, the analytical expressions corresponding to Figs. 28b and 29b are identical.

With the result of this exercise, we can now completely unconnect these diagrams by denoting the matrix element by $f(t_1, t_2)$ and adding together the two time orderings of the diagrams:

$$\int_{t_0}^{t} dt_1 \int_{t_2}^{t_1} dt_2 \,[f(t_1, t_2) + f(t_2, t_1)]$$
$$= \int_{t_0}^{t} dt_1 \int_{t_0}^{t_1} dt_2\, f(t_1, t_2) + \int_{t_0}^{t} dt_1 \int_{t_0}^{t_1} dt_2\, f(t_2, t_1)\,. \tag{11-59}$$

In the second term on the right-hand side of this equation we interchange the order of integration, as indicated in Fig. 30, and then relabel the dummy variables of integration, to obtain finally

$$\int_{t_0}^{t} dt_1 \int_{t_0}^{t_1} dt_2 \,[f(t_1, t_2) + f(t_2, t_1)]$$
$$= \int_{t_0}^{t} dt_1 \int_{t_0}^{t_1} dt_2\, f(t_1, t_2) + \int_{t_0}^{t} dt_1 \int_{t_1}^{t} dt_2\, f(t_1, t_2)$$
$$= \int_{t_0}^{t} dt_1 \int_{t_0}^{t} dt_2\, f(t_1, t_2)\,. \tag{11-60}$$

Thus, the temporal interrelations of the product diagrams have been entirely removed by summing over all possible ways of ordering the time variables. As a consequence, the product diagrams of Fig. 28b can be considered truly unconnected, and evaluated independently. The diagrams of Fig. 29b, therefore, are redundant and no longer occur, because they have been absorbed by the process of completely unconnecting the product. Although we have given the proof only for a particular case, it should be clear that the procedure goes through to all orders, and that, in writing down the processes relative to given initial and final states, one must also include simultaneous processes in which the particles can participate, by including all different unconnected diagrams. In this sense, the product of Fig. 28b *or* Fig. 29b must only be included once.

In passing, it is worth noting that this procedure of summing over all possible time orderings to obtain truly unconnected diagrams is also a necessary ingredient of many-body theory.* But, in this context it turns out that unconnected diagrams can then be eliminated from the theory and they do not appear in the expression for the thermodynamic quantities of interest. In the approach to the many-body problem from a field-theoretic point of view, the dependence of the ground-state energy of the system on connected diagrams only is known as the linked-cluster theorem,[10] whereas in the approach of quantum statistical mechanics the grand potential of the system depends only on connected diagrams as a direct consequence of the Ursell cluster equations.[9]

The preceding, rather lengthy discussion of some particular cases has now prepared us to deduce the general rules for constructing all such diagrams. In order to facilitate this discussion, *vertex functions* have been associated with each of the one-vertex diagrams in Fig. 26, since these are the basic structures from which all other diagrams are constructed. Explicit expressions for these vertex functions are

$$\begin{pmatrix} \mathbf{p}_1 & \\ \mathbf{p}_2 & \mathbf{k}_1 \end{pmatrix}_t = \langle \mathbf{p}_1 \,|\, V_{1\gamma}(t) \,|\, \mathbf{p}_2\mathbf{k}_1 \rangle \,, \tag{11-61a}$$

$$\begin{pmatrix} \mathbf{p}_1 & \mathbf{k}_1 \\ \mathbf{p}_2 & \end{pmatrix}_t = \langle \mathbf{p}_1\mathbf{k}_1 \,|\, V^\dagger_{1\gamma}(t) \,|\, \mathbf{p}_2 \rangle \,, \tag{11-61b}$$

$$\begin{pmatrix} \mathbf{p}_1 & \mathbf{k}_1 & \mathbf{k}_2 \\ \mathbf{p}_2 & & \end{pmatrix}_t = \langle \mathbf{p}_1\mathbf{k}_1\mathbf{k}_2 \,|\, V^\dagger_{2\gamma}(t) \,|\, \mathbf{p}_2 \rangle + \langle \mathbf{p}_1\mathbf{k}_2\mathbf{k}_1 \,|\, V^\dagger_{2\gamma}(t) \,|\, \mathbf{p}_2 \rangle \,, \tag{11-61c}$$

$$\begin{pmatrix} \mathbf{p}_1 & & \\ \mathbf{p}_2 & \mathbf{k}_1 & \mathbf{k}_2 \end{pmatrix}_t = \langle \mathbf{p}_1 \,|\, V_{2\gamma}(t) \,|\, \mathbf{p}_2\mathbf{k}_1\mathbf{k}_2 \rangle + \langle \mathbf{p}_1 \,|\, V_{2\gamma}(t) \,|\mathbf{p}_2\mathbf{k}_2\mathbf{k}_1 \rangle \,, \tag{11-61d}$$

$$\begin{pmatrix} \mathbf{p}_1 & \mathbf{k}_1 \\ \mathbf{p}_2 & \mathbf{k}_2 \end{pmatrix}_t = 2\langle \mathbf{p}_1\mathbf{k}_1 \,|\, V_{2\gamma}(t) \,|\, \mathbf{p}_2\mathbf{k}_2 \rangle \,, \tag{11-61e}$$

$$\begin{pmatrix} \mathbf{p}_1 & \mathbf{p}_2 \\ \mathbf{p}_3 & \mathbf{p}_4 \end{pmatrix}_t = \langle \mathbf{p}_1\mathbf{p}_2 \,|\, V_C(t) \,|\, \mathbf{p}_3\mathbf{p}_4 \rangle + \varepsilon \langle \mathbf{p}_1\mathbf{p}_2 \,|\, V_C(t) \,|\, \mathbf{p}_4\mathbf{p}_3 \rangle \quad \text{for identical particles,} \tag{11-61f}$$

$$= \langle \mathbf{p}_1\mathbf{p}_2 \,|\, V_C(t)\, \mathbf{p}_3\mathbf{p}_4 \rangle \quad \text{for nonidentical particles.} \tag{11-61g}$$

A Qth order (or Q-vertex) *interaction graph* (or I graph) is defined to be a set of Q vertices which are entirely interconnected by directed lines

* In this case, of course, the time variable is replaced by an inverse temperature variable.

equipped with arrows, such that these arrows always point from bottom to top of the graph, the direction in which time increases. Lines corresponding to particles with mass are solid lines and are labeled with momentum (and spin) variables $\mathbf{p}$, while those corresponding to photons are wiggly lines and are labeled with a variable $\mathbf{k}$. The rules for connecting the Q vertices of an I graph, together with the prescription for writing down the corresponding expression, are as follows:

1. Every line is attached to a vertex at either its head end or tail end, or both. If it is attached at both ends, then the temporal variable t_i at the tail end must be *less than* the temporal variable t_j at the head end.
2. Associate with each line an integer i and a corresponding momentum variable $\mathbf{p}_i$ or $\mathbf{k}_i$, according to the scheme described above.
3. Two I graphs are different if they cannot be topologically (including the relative positions of the labels) deformed into one another.
4. A factor of $\frac{1}{2}$ is included for each *identical-particle* double bond, where a *double bond* is a structure in which two lines connect the same two vertices (there may even be a third, different line in the case of photon double bonds).
5. Associate with each vertex a factor of $(-i/\hbar)$, and a step-function factor $\theta(t_j' - t_j)$, where t_j' is the temporal variable at the vertex where the outgoing line from t_j attaches at its head end.* If there is more than one outgoing line, then assign a step function for each line. If the head end of a line is free, then $t_j' = t$.
6. When the I graph is written in terms of its associated vertex functions using the symbols of Eqs. (11-61), a factor $\prod_i(\varepsilon^{P_B})_i$ is included, where $(P_B)_i$ is the total permutation among identical particles of the bottom-row momenta with respect to the top-row momenta.
7. Time integrations are performed over the Q time variables associated with the product of Q vertex functions, and all integration limits are from t_0 to t.
8. For all lines connected at *both* ends the sum over momentum (and spin) coordinates is performed, and for some particles is may also be necessary to sum over internal states.

It would seem appropriate to make some brief comments regarding these rules, the first of which is that exchange terms, when applicable, have been included in the vertex functions of Eq. (11-61), so that according to rule 3 no distinct diagrams appear for these exchange terms. The factor of 2 in Eq. (11-61e) arises because the term appears twice in Eq. (11-16). One should also observe that rule 1 prohibits diagrams with loops, such as those in Fig. 35. This is an important observation, because such diagrams would

* The step function $\theta(x)$ is unity if $x > 0$, and vanishes if $x \leq 0$.

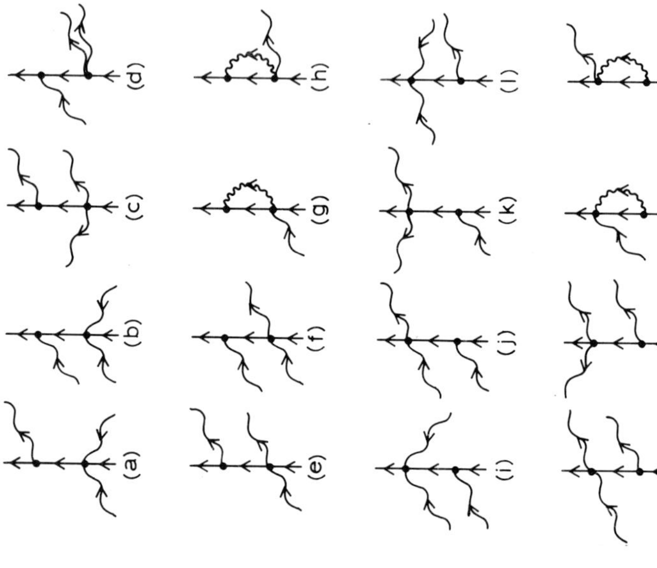

Fig. 31. All of the two-vertex diagrams containing one $\mathbf{p} \cdot \mathbf{A}$ vertex and one $\mathbf{A}^2$ vertex (a)–(p).

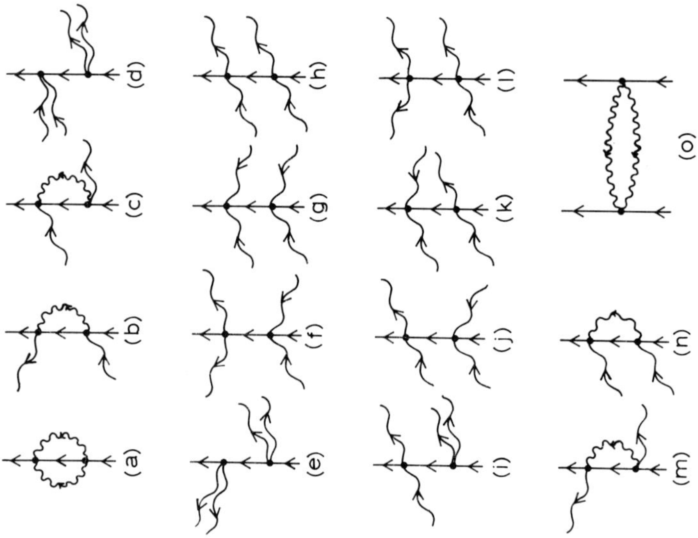

Fig. 32. All of the two-vertex diagrams containing two $\mathbf{A}^2$ vertices (a)–(o).

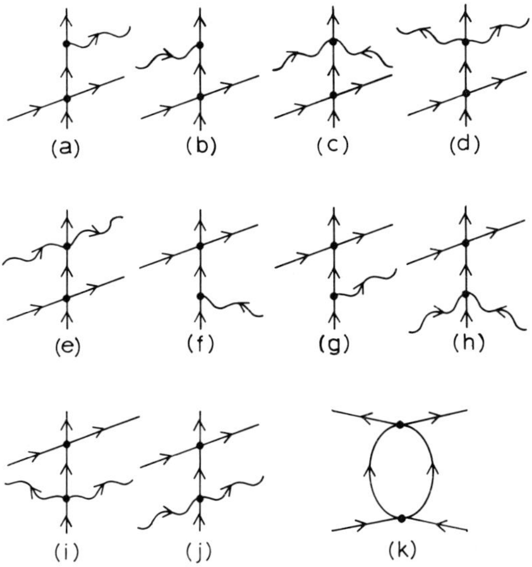

Fig. 33. All of the two-vertex diagrams containing at least one Coulomb vertex (a)–(k).

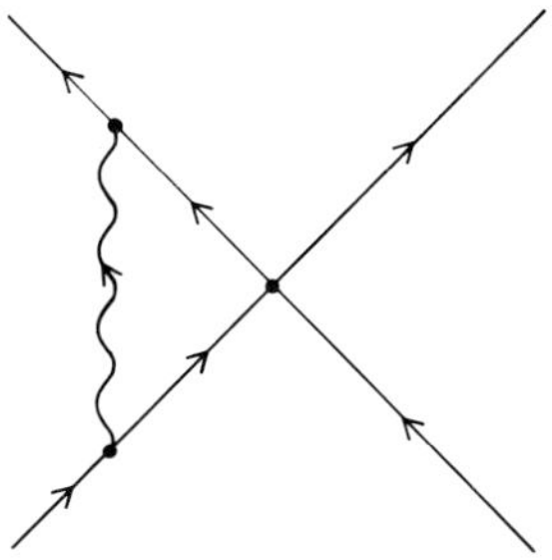

Fig. 34. Example of a three-vertex diagram.

correspond to an identically zero momentum transfer, which is no interaction at all. This sometimes leads to difficulties in other diagrammatic theories. Examples of identical-particle double bonds are provided by the diagrams of Figs. 32a and 33k. The reason for rule 6 stems from the factors of ε arising in the evaluation procedure discussed in conjunction with Eq. (11-57). Finally, one must remember that momentum labels were used in Eqs. (11-61) and in the above rules only for definiteness of notation, and that when a process calls for the use of other quantum numbers suitable changes must be made in the labeling.

The diagrams of Figs. 25–27 and 31–33 represent all possible one- and two-vertex I graphs, and Fig. 34 provides an example of a three-vertex

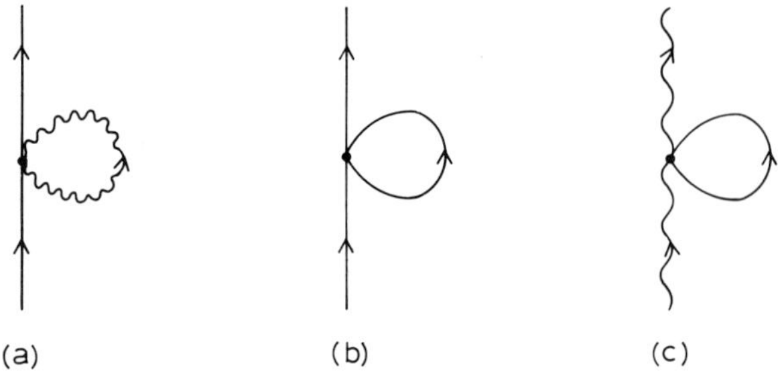

Fig. 35. Closed-loop diagrams (a)–(c) prohibited by rule 1 for I graphs.

diagram. However, the preceding analysis has been very general, in that one probably need not consider in the nonrelativistic theory any diagrams beyond those already exhibited. In fact, we shall see that almost all higher-order diagrams diverge, so that one is forced at any rate to reorganize the theory later. Nevertheless, it is now possible to write in place of Eq. (11-43)

$$\langle \mathbf{b} | \mathsf{U}(t, t_0) | \mathbf{a} \rangle = \delta_{\mathbf{ba}} + \sum \left\{ \begin{array}{l} \text{all different } Q\text{th-order connected} \\ \textit{and} \text{ unconnected interaction} \\ \text{graphs for initial states } |\mathbf{a}\rangle \text{ and} \\ \text{final states } |\mathbf{b}\rangle \end{array} \right\} . \tag{11-62}$$

The proof of this diagrammatic expansion is constructed by generalizing the foregoing discussion of this section and then proving rules 1–8 to all orders. After studying several of the lower-order terms, it is usually possible to satisfy oneself that the general terms behave in the way indicated. The most complicated rule to prove is rule 6, although one can again verify this in a satisfactory manner by examining a number of examples; or, one can construct a systematic evaluation procedure along the lines of Wick's theorem. We shall now adopt Eq. (11-62) as valid, and devote the remaining chapters to a detailed analysis of its right-hand side.

PROBLEMS

11-1. In many of the calculational problems of QED it is necessary to perform sums over polarization indices for photons. Prove the following very useful formulas:

(a) $$\sum_{i=1}^{2} (\mathbf{p} \cdot \boldsymbol{\epsilon}^{(i)})^2 = p^2 - (\mathbf{p} \cdot \mathbf{k})^2/k^2 . \tag{11-63}$$

(b) $$\sum_{i=1}^{2} (\mathbf{p}_1 \cdot \boldsymbol{\epsilon}^{(i)})(\mathbf{p}_2 \cdot \boldsymbol{\epsilon}^{(i)}) = (\mathbf{p}_1 \cdot \mathbf{p}_2) - [(\mathbf{p}_1 \cdot \mathbf{k})(\mathbf{p}_2 \cdot \mathbf{k})/k^2] . \tag{11-64}$$

(c) For two photons with wave vectors $\mathbf{k}_1$ and $\mathbf{k}_2$

$$\sum_{\substack{i=1\\j=1}}^{2} (\epsilon_1^{(i)} \cdot \epsilon_2^{(j)})^2 = 1 + (\mathbf{k}_1 \cdot \mathbf{k}_2)^2/k_1^2 k_2^2 = 1 + \cos^2\theta\,, \tag{11-65}$$

where $\cos\theta = \hat{\mathbf{k}}_1 \cdot \hat{\mathbf{k}}_2$.

REFERENCES

1. P. A. M. Dirac, *Proc. Roy. Soc.* (*London*) **A112,** 661 (1926); **A114,** 243 (1927).
2. See, e.g., E. Merzbacher, "Quantum Mechanics," p. 162. Wiley, New York, 1961.
3. E. Fermi, "Nuclear Physics," 2nd ed., p. 142. Univ. of Chicago Press, Chicago, Illinois, 1950.
4. F. J. Dyson, *Phys. Rev.* **75,** 486, 1736 (1949).
5. See, e.g., L. I. Schiff, "Quantum Mechanics," p. 154. McGraw-Hill, New York, 1955.
6. R. P. Feynman, *Phys. Rev.* **76,** 749, 769 (1949).
7. R. D. Mattuck, "A Guide to Feynman Diagrams in the Many-Body Problem." McGraw-Hill, New York, 1967.
8. G. C. Wick, *Phys. Rev.* **80,** 268 (1950).
9. F. Mohling and W. T. Grandy, Jr., *J. Math. Phys.* **6,** 348 (1965).
10. J. Goldstone, *Proc. Roy. Soc.* (*London*) **A239,** 267 (1957).

XII APPLICATION TO CLASSICAL PROCESSES

The preceding chapters were devoted to a careful development of a (nonrelativistic) quantum theory of radiation, and in this chapter we begin at long last to apply the theory to a calculation of the measurable quantities describing radiation processes. Formally, we wish to extract the maximum amount of physics possible from the right-hand side of Eq. (11-62). In order to begin on familiar ground, we shall restrict the present discussion to primarily classical processes.

A. THOMSON SCATTERING

As a first application, let us examine the scattering of a photon from a free electron. This is identical to the problem studied in Chapters VI and VII, if the photon is interpreted as an electromagnetic plane wave. Since we are interested only in the nonrelativistic effect, we invoke the criterion that the photon not have sufficient energy to create electron–positron pairs*:

$$\hbar ck \ll 2mc^2 \,. \tag{12-1}$$

This is equivalent to stating that the photon energy is much greater than the electron kinetic energy, for the same wave number:

$$\hbar^2k^2/2m \ll \hbar ck \,. \tag{12-2}$$

We now examine the diagrams in the figures of the last chapter to find the leading-order contributions to scattering of a photon from a charged particle. These are clearly the $O(\alpha)$ diagrams of Figs. 26e, 27d, and 27e. According to the rules for I graphs, the contribution from the

* See Chapter XVI.

first diagram can be written

$$\text{Fig. 26e:}\quad -(i/\hbar)\int_{t_0}^{t}\begin{pmatrix}\mathbf{p}_1 & \mathbf{k}_1\\ \mathbf{p}_2 & \mathbf{k}_2\end{pmatrix}_{t_1} dt_1$$
$$= -(2i/\hbar)\langle \mathbf{p}_1\mathbf{k}_1 \,|\, \mathsf{V}_{2r} \,|\, \mathbf{p}_2\mathbf{k}_2\rangle$$
$$\times \int_{t_0}^{t} \exp\{it_1[\omega_1 + \hbar ck_1 - \omega_2 - \hbar ck_2]/\hbar\}\, dt_1 . \qquad (12\text{-}3)$$

The objective here is to calculate the scattering cross section, and therefore the scattering matrix, and so we must take the limits $t \to +\infty$, $t_0 \to -\infty$. There is no difficulty with the time integral in this particular case, since it is clearly a δ function, and with the aid of Eq. (11-12) we have

$$\text{Fig. 26e:}\quad -4\pi i\left[\frac{\hbar^2\pi\alpha}{mV}\,\frac{\epsilon_1\cdot\epsilon_2}{(k_1k_2)^{1/2}}\right]\delta_{n_1n_2}\,\delta_{\mathbf{p}_1+\mathbf{k}_1,\mathbf{p}_2+\mathbf{k}_2}$$
$$\times\, \delta(\omega_1 + \hbar ck_1 - \omega_2 - \hbar ck_2)\,, \qquad (12\text{-}4)$$

with $Z = -1$ for electrons.

The δ functions ensure conservation of spin, momentum, and energy, and the first is duly noted and discarded. The product of the second two δ functions implies the relation

$$\hbar ck_1 = \hbar ck_2 - \frac{(\hbar\mathbf{k}_1 - \hbar\mathbf{k}_2)^2}{2m} + \frac{(\hbar\mathbf{k}_1 - \hbar\mathbf{k}_2)\cdot\hbar\mathbf{p}_2}{m}\,, \qquad (12\text{-}5)$$

so that in the nonrelativistic limit of Eq. (12-1),

$$\hbar ck_1 \simeq \hbar ck_2\,. \qquad (12\text{-}6)$$

This result demonstrates that the photon momentum and energy remain unchanged in the nonrelativistic approximation, and that electron recoil is completely negligible. This bit of physics can be expressed formally by writing for the δ functions

$$\delta_{\mathbf{p}_1+\mathbf{k}_1,\mathbf{p}_2+\mathbf{k}_2}\,\delta\left(\frac{\hbar^2p_1^{\,2}}{2m} + \hbar ck_1 - \frac{\hbar^2p_2^{\,2}}{2m} - \hbar ck_2\right)$$
$$\simeq \frac{1}{\hbar c}\,\delta(k_1 - k_2)\,. \qquad (12\text{-}7)$$

The contribution to the matrix element of $\mathsf{U}(\infty, -\infty)$ is then

$$\text{Fig. 26e:}\quad -4\pi i\left[\frac{\hbar\pi\alpha}{mcV}\,\frac{\epsilon_1\cdot\epsilon_2}{k_1}\right]\delta(k_1 - k_2)\,. \qquad (12\text{-}8)$$

The diagram of Fig. 27d yields a contribution

$$\text{Fig. 27d:}\quad \varepsilon(-i/\hbar)^2 \int_{t_0}^{t} dt_1 \int_{t_0}^{t_1} dt_2 \sum_{l} \begin{pmatrix} l & \\ \mathbf{p}_1 & \mathbf{k}_1 \end{pmatrix}_{t_2} \begin{pmatrix} \mathbf{p}_2 & \mathbf{k}_2 \\ l & \end{pmatrix}_{t_1}$$
$$= \hbar^2 \sum_{l} \langle \mathbf{p}_2\mathbf{k}_2 \,|\, \mathsf{V}_{1\gamma}^{\dagger} \,|\, l\rangle \langle l \,|\, \mathsf{V}_{1\gamma} \,|\, \mathbf{p}_1\mathbf{k}_1\rangle$$
$$\times \int_{-\infty}^{\infty} dt_1 \int_{-\infty}^{t_1} dt_2 \exp\{it_1[\omega_2 + \hbar ck_2 - \omega_l]/\hbar\}$$
$$\times \exp\{it_2[\omega_l - \omega_1 - \hbar ck_1]/\hbar\} \tag{12-9}$$

in the limits $t \to +\infty$, $t_0 \to -\infty$.

The t_2 integral is evaluated as follows:

$$\lim_{\Delta\to 0} \int_{-\infty}^{t_1} \exp(-\Delta\,|\,t_2\,|) \exp\{it_2[\omega_l - \omega_1 - \hbar ck_1]/\hbar\}\, dt_2$$
$$= \frac{\hbar \exp\{it_1[\omega_l - \omega_1 - \hbar ck_1]/\hbar\}}{i[\omega_l - \omega_1 - \hbar ck_1]}. \tag{12-10}$$

The t_1 integral then yields an energy-conserving δ function which, with the momentum-conserving Kronecker δ, gives

$$2\pi\hbar\,\delta\left(\frac{\hbar^2 p_2^2}{2m} + \hbar ck_2 - \frac{\hbar^2 p_1^2}{2m} - \hbar ck_1\right) \simeq \frac{2\pi}{c}\,\delta(k_2 - k_1) \tag{12-11}$$

in the nonrelativistic limit. The energy denominator can also be approximated in this limit after the l sum is performed by means of the momentum δ functions in the matrix elements, and one finds

$$\frac{\hbar^2}{2m}(\mathbf{p}_2 + \mathbf{k}_2)^2 - \frac{\hbar^2 p_1^2}{2m} - \hbar ck_1 \simeq -\hbar ck_1 + \frac{\hbar^2}{m}\mathbf{k}_2 \cdot \mathbf{p}_2. \tag{12-12}$$

The contribution from this diagram is then

$$\text{Fig. 27d:}\quad -i\frac{4\pi^2\hbar^3\alpha}{Vm^2ck_1}(\mathbf{p}_2 \cdot \epsilon_2)^2 \frac{\delta(\mathbf{k}_2 - \mathbf{k}_1)}{[-\hbar ck_1 + (\hbar^2/m)\mathbf{k}_2 \cdot \mathbf{p}_2]}. \tag{12-13}$$

If the contribution from the diagram of Fig. 27e is now calculated in the nonrelativistic limit, one finds a term exactly equal to that of Eq. (12-13), but of opposite sign.

Exercise. Verify this last assertion.

Thus, in this limit the contributions from the two-vertex, $(\mathbf{p} \cdot \mathbf{A})$, processes cancel one another, and the scattering is due entirely to the $\mathbf{A}^2$

interaction. From Eqs. (11-43), (11-35), and (11-36), one identifies the scattering amplitude in Eq. (12-8) as

$$f_{\mathbf{k}_2}(\hat{\mathbf{k}}_1) \simeq (-e^2/mc^2)(\epsilon_1 \cdot \epsilon_2)\,, \tag{12-14}$$

and from Eq. (11-37) we have for the differential scattering cross section

$$d\sigma/d\Omega \simeq (e^2/mc^2)^2(\epsilon_1 \cdot \epsilon_2)^2\,. \tag{12-15}$$

Equation (12-15) is just the classical Thomson cross section of Eq. (6-17) (see Fig. 19), and integration over all angles yields the total cross section of Eq. (6-18). By summing over all polarizations of the scattered photon and averaging over all polarizations of the initial photon, one obtains the Thomson formula of Eq. (6-21).

Exercise. Use Eq. (11-65) of Problem 11-1 to obtain Eq. (6-21) from Eq. (12-15).

The preceding calculation shows that the leading-order quantum mechanical theory agrees exactly with the classical result, while also demonstrating an inability to deal with the high-energy problem leading to the Compton effect. The discussion in Chapter VI describes this difficulty quite well, and the necessary approximations based on Eq. (12-1) transparently indicate why we cannot obtain the Klein–Nishina formula here. On the other hand, we *can* obtain higher-order quantum corrections to Eq. (12-15) by considering diagrams such as that of Fig. 32b, and we shall do this in Chapter XV.

B. NONRELATIVISTIC *BREMSSTRAHLUNG*

In Chapter VI the classical theory of *Bremsstrahlung* was studied nonrelativistically, and a generalization to the quantum mechanical case was made by means of the uncertainty principle. It is the purpose of this section to place the quantum aspects of this process on a more rigorous basis and obtain a cross section which agrees quite well with experiment.

The physical discussion of Chapter VI is still pertinent, and we consider a nonrelativistic electron deflected by the Coulomb field of a nucleus, with the concomitant emission of radiation. There are two possible diagrams contributing to this process in lowest order, and these are given by Figs. 33a and 33g. Photon emission by a free electron is permissible in the Coulomb field of a nucleus when the electron is in a virtual state following its interaction with the scattering center. From the rules for I graphs a now-familiar calculation gives the S matrix element between an initial state

$|\mathbf{a}\rangle$ and a final state $\langle\mathbf{b}|$ as

$$\langle\mathbf{b}|\,\mathsf{S}\,|\mathbf{a}\rangle = \delta_{\mathbf{ba}} + 2\pi i\delta(\omega_1 + \omega_2 - \omega_3 - \omega_4 + \hbar c k_1)\frac{\hbar^2}{mV}\left(\frac{2\pi\alpha}{Vk_1}\right)^{1/2}$$
$$\times\,\delta_{n_1 n_3}\frac{4\pi Ze^2}{|\mathbf{p}_1 + \mathbf{k}_1 - \mathbf{p}_3|^2}\left[\frac{\mathbf{p}_3\cdot\boldsymbol{\epsilon}_1}{\omega(\mathbf{p}_3 - \mathbf{k}_1) + \hbar c k_1 - \omega(\mathbf{p}_3)}\right.$$
$$\left.+\,\frac{\mathbf{p}_1\cdot\boldsymbol{\epsilon}_1}{\omega(\mathbf{p}_1 + \mathbf{k}_1) + \omega(\mathbf{p}_2) - \omega(\mathbf{p}_3) - \omega(\mathbf{p}_4)}\right]$$
$$\times\,\frac{(2\pi)^3}{V}\,\delta(\mathbf{k}_1 + \mathbf{p}_1 + \mathbf{p}_2 - \mathbf{p}_3 - \mathbf{p}_4)\,, \tag{12-16}$$

where we have included the contributions from both diagrams, and have taken the view that it is the electron that emits the photon.

Now, one generally makes the approximation that the nucleus remains stationary throughout the interaction, which is quite reasonable in this nonrelativistic approach. Then, it is not possible to conserve momentum in the process, because the stationary nucleus can absorb any amount of momentum. Therefore, we *assume* $\mathbf{p}_2 = \mathbf{p}_4$ and drop the last two factors on the right-hand side of Eq. (12-16), an approximation resulting in an error of about one part in 1000. The energy-conserving δ function is then approximately

$$\delta(\omega_1 - \omega_3 + \hbar c k_1)\,. \tag{12-17}$$

Note, however, that this procedure implies that the energy denominators are very small.

After making these approximations, a comparison of Eq. (12-16) with Eq. (11-38) leads to the following expression for the reaction matrix element:

$$\mathsf{R}_{\mathbf{ba}} \simeq -\frac{\hbar^2}{mV}\left(\frac{2\pi\alpha}{Vk_1}\right)^{1/2}\frac{4\pi Ze^2\delta_{n_1 n_3}}{|\mathbf{p}_1 + \mathbf{k}_1 - \mathbf{p}_3|^2}$$
$$\times\left[\frac{\mathbf{p}_1\cdot\boldsymbol{\epsilon}_1}{\omega(\mathbf{p}_1 + \mathbf{k}_1) - \omega(\mathbf{p}_3)} + \frac{\mathbf{p}_3\cdot\boldsymbol{\epsilon}_1}{\omega(\mathbf{p}_3 - \mathbf{k}_1) - \omega(\mathbf{p}_1)}\right], \tag{12-18}$$

where we have used Eq. (12-2).

As in the classical case, we are primarily interested in studying the continuous soft X-ray region, so that we can take

$$k_1 \ll p_1\,. \tag{12-19}$$

This approximation, along with Eqs. (12-17) and (12-2), shows that the energy denominators in Eq. (12-18) have magnitudes proportional to $\hbar c k_1$,

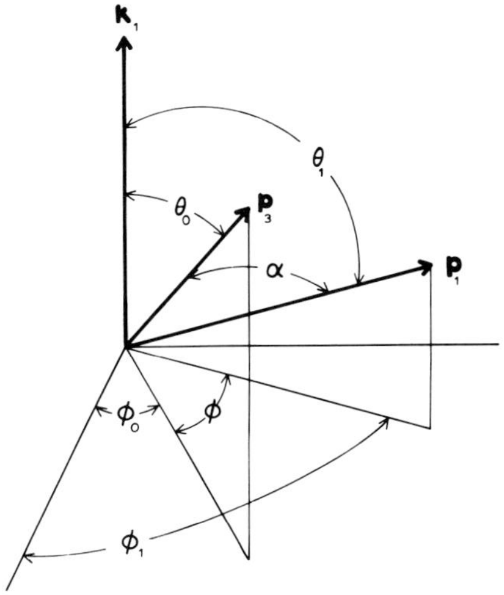

Fig. 36. Coordinate system in which the evaluation (12-22) is performed.

and are indeed small. Referring to Eq. (11-40), we find for the transition rate per unit time

$$w = \frac{64\pi^4\hbar^3 Z^2}{m^2 V^3 k_1^{\,3}} \frac{\alpha^3 \delta_{n_1 n_3}}{|\mathbf{p}_1 - \mathbf{p}_3|^4} (\mathbf{p}_1 \cdot \epsilon_1 - \mathbf{p}_3 \cdot \epsilon_1)^2 \rho_f \,. \tag{12-20}$$

It is now necessary to calculate the density of final states, and, since the final electron and photon are independent,* ρ_f is merely a product of the individual densities. The photon factor was essentially calculated in Eq. (8-28), and the electron density is obtained in the same way, remembering that ρ_f is the number of states in momentum space divided by the energy interval dE. Hence,

$$\rho_f = \frac{mV^2}{\hbar^2} \frac{p_1\, d\Omega_{p_1} k_1^{\,2}\, dk_1\, d\Omega_{k_1}}{(2\pi)^6} \,. \tag{12-21}$$

To obtain the cross section, we need the incident electron flux, which is clearly the incident velocity divided by the volume: $\hbar p_3/mV$.

The polarization factor in Eq. (12-20) is evaluated by noting that we shall not measure the polarization of the radiation, so that we sum over all polarization states. We shall denote the various angles as in Fig. 36.

* This is due to the approximation of a stationary nucleus.

Exercise. With the aid of Problem 11-1 show that

$$\sum_{i=1}^{2} (\mathbf{p}_1 \cdot \epsilon_1^{(i)} - \mathbf{p}_3 \cdot \epsilon_1^{(i)})^2 = \mathbf{p}_1^2 \sin^2\theta + \mathbf{p}_3^2 \sin^2\theta_0 - 2p_1p_3 \sin\theta \sin\theta_0 \cos\phi . \tag{12-22}$$

Show also that the products $d\phi_0\, d\phi_1$ from the solid angle elements can effectively be replaced by $2\pi\, d\phi$.

Combining all these results, one can now write for the differential cross section

$$d\sigma \simeq \frac{Z^2\alpha^3}{\pi^2} \frac{dk_1}{k_1} \left(\frac{p_1}{p_3}\right) \frac{\sin\theta\, d\theta \sin\theta_0\, d\theta_0\, d\phi}{|\mathbf{p}_1 - \mathbf{p}_3|^4} \times [p_1^2 \sin^2\theta + p_3^2 \sin^2\theta_0 - 2p_1p_3 \sin\theta \sin\theta_0 \cos\phi] . \tag{12-23}$$

In this equation we have summed over the final spin states of the electron and averaged over initial states, since we are not interested in the spin, and this results in a factor of unity. Also, the differential cross section of Eq. (12-23) refers to a fixed value of k_1, since that is what would be measured.

One learns from Eq. (12-23) that, for a given angle of deflection of the electron, the maximum photon intensity is perpendicular to the plane of motion of the electron. The total cross section for emission of the photon k_1 is obtained by integrating Eq. (12-23) over all angles. This is an elementary calculation and is left for the problems at the end of the chapter. Moreover, the total cross section is usually written relative to the variation of photon energy with respect to the incident electron energy, $d(\hbar c k_1)/\omega(p_3)$. Hence, we finally obtain the *Bethe–Heitler formula*[1] for nonrelativistic *Bremsstrahlung*,

$$\sigma_{\text{BH}} = \frac{d\sigma}{dk_1} = \frac{Z^2}{\pi} \alpha\sigma_{\text{T}} \left(\frac{mc^2}{\hbar c k_1}\right) \ln\left(\frac{p_3 + p_1}{p_3 - p_1}\right), \tag{12-24}$$

in terms of the Thomson cross section, σ_{T}.

Exercise. Show that σ_{BH} can also be written as

$$\sigma_{\text{BH}} = \frac{Z^2}{\pi} \alpha\sigma_{\text{T}} \left(\frac{mc^2}{\hbar c k_1}\right) \ln \frac{[\sqrt{\omega_3} + (\omega_3 - \hbar c k_1)^{1/2}]^2}{\hbar c k_1} . \tag{12-25}$$

The Bethe–Heitler formula is compared with the data in Fig. 37, along with the cross sections calculated in Chapter VI. One sees that σ_{BH} gives quite good agreement, indicating that the process cannot, in fact, be considered completely classical. Actually, the data are sparse in the region of validity of Eq. (12-24). For an excellent review of both the theoretical

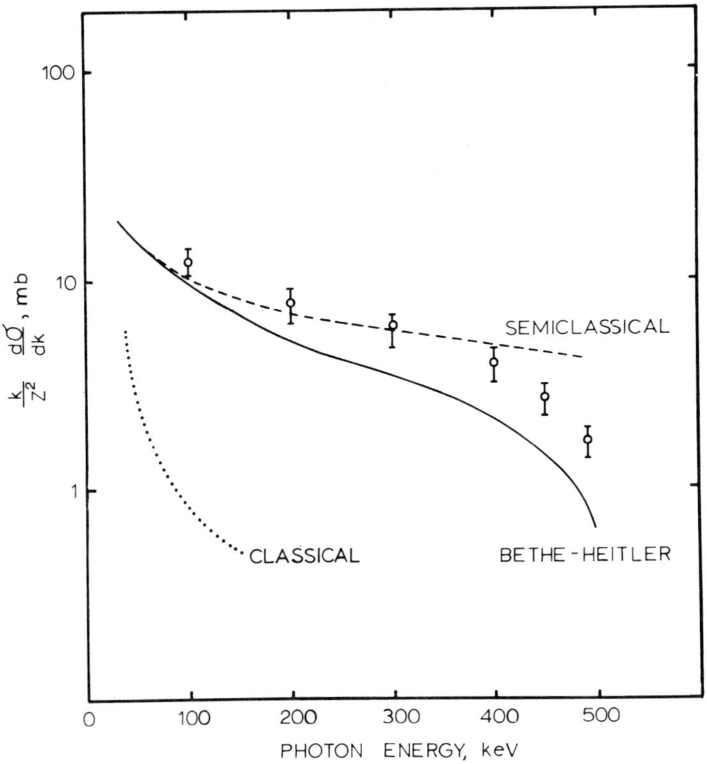

Fig. 37. Comparison of the Bethe–Heitler formula, Eq. (12-24), with the classical and semiclassical cross sections of Chapter VI. The experimental results are due to Motz,[2] and indicate the X-ray yield from 500-keV electrons incident on an aluminum target. The deviations are due to relativistic and screening effects, and, for very low incident energies, the Sommerfeld correction discussed in Chapter VI.

and experimental situations for *Bremsstrahlung*, the reader is referred to the article by Koch and Motz.[3]

As a final point, the careful reader will note that the cross section σ_{BH} of Eq. (12-24) is proportional to $(\hbar ck_1)^{-1}$, so that the probability for emission of a very soft photon is quite large. The cross section would actually be cut off at extremely low photon momenta, because the limitations on the energy resolution of any experimental setup negates the possibility of observing arbitrarily small energies. Thus, one must actually admit to the *Bremsstrahlung* process diagrams in which an arbitrarily large number of "infrared" photons are emitted with energies lower than some maximum value $\hbar ck_m$, all of which are entirely unobservable. Nevertheless, this possibility presents a thorny theoretical problem because there in fact exist many processes in which an arbitrarily large number of infrared photons can be emitted, and for which the cross sections diverge as $k \to 0$. We shall examine this *infrared problem* in detail later, and, fortunately, find that it is only a formal difficulty which in no way affects the physical predictions of the theory. In the present situation, the conclusion to be drawn from these observations is that there is no such thing as a *pure* Coulomb interaction, for the process can always be accompanied by emission of an arbitrary number of (unobservable) infrared photons.

C. COULOMB SCATTERING

It is of some interest to examine the scattering of two charged particles in order to determine what changes, if any, are introduced into the classical results by QED. The leading-order diagram, of course, is that of Fig. 26f, and a now straightforward evaluation results in the expression

Fig. 26f:

$$-\frac{i}{\hbar}\int_{t_0}^{t} dt_1 \begin{pmatrix} \mathbf{p}_1 & \mathbf{p}_2 \\ \mathbf{p}_3 & \mathbf{p}_4 \end{pmatrix}_{t_1}$$

$$= -2\pi i(4\pi Z_1 Z_2 e^2)\,\delta(\omega_1 + \omega_2 - \omega_3 - \omega_4)$$

$$\times \left[\frac{\delta_{n_1 n_3}\,\delta_{n_2 n_4}}{|\mathbf{p}_1 - \mathbf{p}_3|^2} + \varepsilon\frac{\delta_{n_1 n_4}\,\delta_{n_2 n_3}}{|\mathbf{p}_1 - \mathbf{p}_4|^2}\frac{(2\pi)^3}{V^2}\right]\delta(\mathbf{p}_1 + \mathbf{p}_2 - \mathbf{p}_3 - \mathbf{p}_4)\,. \qquad (12\text{-}26)$$

The second term in this equation is the exchange term, and, of course, is only included if the particles are identical. We will wish to compare the cross section for this process with the classical results, and so it is convenient to work in the center-of-mass (CM) coordinate system, a choice which also simplifies the calculations. Thus, we write

$$\mathbf{p}_3 = -\mathbf{p}_4\,, \qquad \mathbf{p}_1 = -\mathbf{p}_2\,, \qquad |\mathbf{p}_1| = |\mathbf{p}_3|\,, \qquad (12\text{-}27)$$

and one easily verifies the relations

$$|\mathbf{p}_1 - \mathbf{p}_3|^2 = 4p_1^2 \sin^2(\theta/2)\,, \qquad (12\text{-}28a)$$

$$|\mathbf{p}_1 - \mathbf{p}_4|^2 = 4p_1^2 \cos^2(\theta/2)\,, \qquad (12\text{-}28b)$$

where θ is the angle between $\mathbf{p}_1$ and $\mathbf{p}_3$ in the CM system.

The reaction matrix element can now be identified by means of Eqs. (11-35) and (11-38), so that

$$\mathsf{R}_{ba} = \frac{\pi}{V}\frac{Z_1 Z_2 e^2}{p_1^2}\left[\frac{\delta_{n_1 n_3}\,\delta_{n_2 n_4}}{\sin^2(\theta/2)} + \varepsilon\,\delta_{12}\frac{\delta_{n_1 n_4}\,\delta_{n_2 n_3}}{\cos^2(\theta/2)}\right], \qquad (12\text{-}29)$$

where the factor of δ_{12} sets the second term to zero unless the particles are identical. The transition rate and, therefore, the differential cross section are now found from Eq. (11-40). An additional bonus is obtained here due to our choice of the CM coordinate system, because a little thought shows that the density of final states is just twice that for one of the charged particles. Thus,

$$\rho_f = 2\,\frac{p_1^2\,dp_1\,d\Omega}{(2\pi)^3}\,\frac{V}{dE} = \frac{mVp_1\,d\Omega}{4\pi^3\hbar^2}, \tag{12-30}$$

and for identical particles there can be no confusion over the value of the mass. The total flux is just twice the velocity of one of the incident particles per unit volume. Finally, we shall not be interested in the spins of the particles, so we average over initial and sum over final spin states, *after* squaring the matrix element $\mathbf{R}_{\mathbf{ba}}$. A small bit of algebra now yields

$$\frac{d\sigma}{d\Omega} = \frac{Z_1^2 Z_2^2 e^4}{4(2T_0)^2}\left[\frac{1}{\sin^4(\theta/2)} + \frac{\delta_{12}}{\cos^4(\theta/2)} + \frac{\varepsilon\delta_{12}}{(2S+1)}\,\frac{2}{\sin^2(\theta/2)\cos^2(\theta/2)}\right], \tag{12-31}$$

where T_0 is the initial particle kinetic energy in the CM system, and S is the spin of the identical particles.

The first term in the cross section (12-31) is just the Rutherford formula,[4] and the equality of the classical and quantum results of this process for nonidentical particles is well known.* The second two (exchange) terms were first derived by Mott,[5] and these represent the leading-order corrections to the classical scattering. The quantities δ_{12} are to remind one that these terms vanish if the particles are not identical.

As is the situation classically, if one attempts to calculate the total cross section from Eq. (12-31), a divergent integral results, due to the long range of the Coulomb interaction. Thus, in practice it is usually necessary to introduce a cutoff based on the experimental conditions. A similar phenomenon occurs in the many-body problem when one considers large systems of charged particles, but in this case the charges undergo many successive Coulomb interactions, as suggested by the diagram of Fig. 33k. The cross section for this process is very small compared with that of Eq. (12-31), but the investigation of this and the n-fold Coulomb interaction is not academic in the many-body problem. By summing together all diagrams corresponding to successive Coulomb interactions (the so-called ring and ladder diagrams), the basic interaction is replaced by a screened Coulomb interaction, which no longer exhibits a long-range divergence.[6]

* It is generally considered a complete coincidence that the quantum mechanical first Born approximation gives the Rutherford cross section exactly. This is another of those results emphasizing the peculiarity of the Coulomb potential.

Further corrections to the Rutherford cross section, such as those due to retardation effects, can be obtained by studying diagrams like that of Fig. 34. We shall take up this problem in Chapter XV.

PROBLEMS

12-1. Complete the verification of the Bethe–Heitler formula, Eq. (12-24), by showing that the angular integral of Eq. (12-23) yields a factor

$$\frac{8\pi^2}{3}\frac{1}{p_1p_3}\ln\left|\frac{p_1+p_3}{p_1-p_3}\right|.$$

12-2. Calculate the differential scattering cross section for double Coulomb scattering of nonidentical particles as suggested by the diagram of Fig. 33k, and compare the result with Eq. (12-31).

REFERENCES

1. H. Bethe and W. Heitler, *Proc. Roy. Soc.* (*London*) **A146,** 83 (1934).
2. J. W. Motz, *Phys. Rev.* **100,** 1560 (1955).
3. H. W. Koch and J. W. Motz, *Rev. Mod. Phys.* **31,** 920 (1959).
4. E. Rutherford, *Phil. Mag.* **21,** 669 (1911).
5. N. F. Mott, *Proc. Roy. Soc.* (*London*) **A126,** 259 (1930).
6. See, e.g., F. Mohling and W. T. Grandy, Jr., *J. Math. Phys.* **6,** 348 (1965). This problem was also investigated in the context of quantum electrodynamics by R. H. Dalitz, *Proc. Roy. Soc.* (*London*) **A206,** 509 (1951).

XIII QUANTUM MECHANICAL PROCESSES

A. PHOTON EMISSION AND ABSORPTION

In this chapter we take up a study of radiation processes which are inherently quantum mechanical, and also investigate the necessary changes in the theory when composite particles are considered. Perhaps the simplest, and yet most fundamental, of these processes is that in which an atom in an excited state, $E_{\mathbf{a}}$, say, emits radiation and makes a transition to a lower state $E_{\mathbf{b}}$. The frequency of the emitted radiation is then

$$\omega_{\mathbf{ab}} = (E_{\mathbf{a}} - E_{\mathbf{b}})/\hbar\,. \tag{13-1}$$

As discussed in Chapter XI, the diagram describing this emission process is that of Fig. 26b, and the matrix element must now be evaluated using the appropriate states of the atom. In the usual manner, the transition rate is then given by Fermi's golden rule:

$$w = (2\pi/\hbar)\,|\langle\mathbf{b}\,|\,\mathsf{V}_{1\gamma}\,|\,\mathbf{a}\rangle\,|^2\,\rho_f\,. \tag{13-2}$$

For reasons of generality, we shall consider the initial state to contain n_λ photons of momentum $\hbar\mathbf{k}_\lambda$ and polarization ϵ_λ, such as a beam of photons incident upon the atom. Since our interest is with the one-photon emission process, the pertinent part of the Hamiltonian is the $\hbar\mathbf{p}\cdot\mathbf{A}$ interaction, and there will be $(n_\lambda + 1)$ photons in the final state. The needed matrix element is then

$$\begin{aligned}\langle\mathbf{b}\,|\,\mathsf{V}_{1\gamma}\,|\,\mathbf{a}\rangle &= \frac{e}{mc}\,\langle\mathbf{b}, (n_\lambda + 1)\mathbf{k}_\lambda\,|\,\hbar\mathbf{p}\cdot\mathbf{A}\,|\,\mathbf{a}, n_\lambda\mathbf{k}_\lambda\rangle \\ &= \frac{e}{mc}\left(\frac{2\pi\hbar c}{Vk}\right)^{1/2}(n_\lambda + 1)^{1/2}\langle\mathbf{b}\,|\,\hbar\mathbf{p}\cdot\epsilon_\lambda(\mathbf{k})\exp(-i\mathbf{k}\cdot\mathbf{r})\,|\,\mathbf{a}\rangle\,,\end{aligned} \tag{13-3}$$

where we have used the expansion (9-21) for **A**, Eqs. (10-26) and (10-28) to evaluate that part of the matrix element involving photon variables, and retained only that term in the expansion of **A** corresponding to creation of a photon. Physically, Eq. (13-3) illustrates that preparation of a beam of photons in the state $(\mathbf{k}_\lambda, \epsilon_\lambda)$ filters from **A** only that state for emission.

It is almost always the case that the wavelength of emitted radiation is much greater than the atomic dimensions, say,

$$\lambda \gg |\mathbf{r}| . \tag{13-4}$$

In this case we can make the *dipole approximation* of replacing the exponential in Eq. (13-3) by unity. Also, we can use the operator equation of motion,

$$\frac{d\mathbf{r}}{dt} = \frac{i}{\hbar}[\mathsf{H}, \mathbf{r}] + \frac{\partial \mathbf{r}}{\partial t} , \tag{13-5}$$

along with Ehrenfest's theorem, to eliminate the electron momentum $\hbar\mathbf{p}$. Since **r** does not depend on time explicitly, we have, to within an error of order e,*

$$\langle \mathbf{b} | \hbar\mathbf{p} | \mathbf{a} \rangle = -im\omega_{\mathbf{ab}} \langle \mathbf{b} | \mathbf{r} | \mathbf{a} \rangle . \tag{13-6}$$

Exercise. Verify Eq. (13-6).

The dipole-moment operator is defined by the relation $\boldsymbol{\mu} = e\mathbf{r}$. If, however, the atom contains a number of electrons, then Eq. (13-3) would contain a sum of terms, one for each electron. The end result would be to introduce a total magnetic moment operator

$$\boldsymbol{\mu} = \sum_i e_i \mathbf{r}_i , \tag{13-7}$$

where the sum goes over all atomic electrons.

The matrix element of Eq. (13-3) can now be written in the dipole-moment approximation as

$$\langle \mathbf{b} | \mathsf{V}_{1\gamma} | \mathbf{a} \rangle = -\frac{i\omega_{\mathbf{ab}}}{c} \left(\frac{2\pi\hbar c}{Vk} \right)^{1/2} (n_\lambda + 1)^{1/2} \langle \mathbf{b} | \boldsymbol{\mu} \cdot \epsilon_\lambda | \mathbf{a} \rangle . \tag{13-8}$$

As usual, the density of final states is the number of states in phase space divided by the energy interval,

$$\rho_f = k^2 V \, d\Omega / (2\pi)^3 \hbar c . \tag{13-9}$$

* It is to be remembered that the momentum of the electron is the electromagnetic momentum $\mathbf{p} - (e/c)\mathbf{A}$. Thus, Eq. (13-6) is correct only to leading order in e.

Thus, the transition rate for emission is

$$w_e = \frac{\omega_{ab}^3}{2\pi\hbar c^3} (n_\lambda + 1) \,|\, \langle \mathbf{b} \,|\, \boldsymbol{\mu} \cdot \epsilon_\lambda \,|\, \mathbf{a} \rangle \,|^2 \, d\Omega \,, \tag{13-10}$$

where the relation $\omega = ck$ has been used.

We are interested in the rate for all polarizations, and the sum over photon polarization states can be performed with the aid of Problem 11-1. This sum results in a factor of $\sin^2\theta$, where θ is the angle between $\mathbf{k}$ and $\boldsymbol{\mu}$. The rate for emission into all space is obtained by integrating over all angles, and we finally find

$$w_e = \frac{4}{3} \frac{\omega_{ab}^3}{\hbar c^3} (n_\lambda + 1) \,|\, \boldsymbol{\mu}_{ba} \,|^2 \,. \tag{13-11}$$

This last expression is a very general result, and we can deduce from it two historically and physically important quantities. In the first place, suppose there were no photons present in the initial state, so that $n_\lambda = 0$. Equation (13-11) still gives a finite rate of photon emission from the atom; that is, *spontaneous emission* is predicted. The right-hand side of Eq. (13-11) with $n_\lambda = 0$ is the famous Einstein A coefficient for spontaneous dipole radiation.[1]

Suppose now that $n_\lambda \neq 0$, and there is a beam containing a large number of photons incident upon the atom. In this case $(n_\lambda + 1) \simeq n_\lambda$, and this number can also be written in terms of the intensity of the beam $I(\nu)$ as a function of the frequency $\nu = \omega/2\pi$. The energy of the beam in a frequency interval $(\nu, \nu + d\nu)$, including all angles and polarizations, is then

$$I(\nu)\, d\nu = \frac{n_\lambda h\nu}{V} \frac{8\pi\nu^2\, d\nu}{c^3} V \,, \tag{13-12}$$

and, therefore,

$$n_\lambda = \frac{c^3}{8\pi h \nu^3} I(\nu) \,. \tag{13-13}$$

One can now rewrite Eq. (13-11) in terms of ν and replace the factor $(n_\lambda + 1)$ by means of Eq. (13-13):

$$w_e = (2\pi/3\hbar^2) \,|\, \boldsymbol{\mu}_{ba} \,|^2 \, I(\nu) \,. \tag{13-14}$$

This expression predicts a *stimulated emission* of radiation, and the coefficient of $I(\nu)$ is just the Einstein B coefficient.[1]

In a similar manner, the transition rate for absorbing a photon from the incident beam can be calculated, and the pertinent diagram is that of Fig. 26a. In this case, the transition is from a photon state n_λ to a state $(n_\lambda - 1)$. The matrix element obtained is the same as above, except that the factor $(n_\lambda + 1)^{1/2}$ is replaced by $\sqrt{n_\lambda}$. Thus, the ratio of the two probabilities is

$$w_e/w_a = (n_\lambda + 1)/n_\lambda\,, \tag{13-15}$$

and this is just the condition for thermal equilibrium in a gas emitting and absorbing radiation.

Exercise. Show that for dipole absorption

$$w_a = (\pi/3\hbar^2)\,|\boldsymbol{\mu}_{ba}|^2 I(\nu)\,, \tag{13-16}$$

if, instead of summing over all polarizations, one averages over all orientations of the atom with respect to the incident beam.

B. THE ATOMIC PHOTOELECTRIC EFFECT

If in the above absorption problem there is an incident photon with energy greater than the ionization energy I of an electron, then the photon can be absorbed and the electron can make a transition to the continuous spectrum. The kinetic energy of the ejected electron is given by Einstein's photoelectric equation[2]

$$T = \hbar\omega - I\,, \tag{13-17}$$

which is merely a statement of energy conservation.

According to Problem 9-5 we expect the probability for ejection of a K-shell electron to be dominant. Thus, the final state can be taken as a plane wave, but the initial state is a K-shell wave function. The incident photon energy is assumed large compared to the ionization energy, but such that the nonrelativistic limit is still applicable:

$$I \ll \hbar ck \ll mc^2\,. \tag{13-18}$$

A K-shell electron is the closest to the nucleus, and therefore is minimally affected by the other atomic electrons. Thus, an excellent approximation to the wave function is the normalized state vector for hydrogen,

$$\phi_b(\mathbf{x}) = (\pi a^3)^{-1/2}\exp(-r/a)\,, \tag{13-19}$$

where $a = a_B/Z$, $a_B = \hbar^2/me^2$ is the first Bohr radius, and Z is the nuclear charge number. The integral of Eq. (10-51) must now be reevaluated by replacing the plane-wave state $\phi_{\lambda\mathbf{k}}$ by ϕ_b, thereby obtaining the correct form of the matrix element in Eq. (11-8). The integral of importance is then

$$(V\pi a^3)^{-1/2}\int [\exp(-i\mathbf{k}_1\cdot\mathbf{r})](\hbar\mathbf{p}\cdot\mathbf{A})\exp(-r/a)\,d^3r\,. \tag{13-20}$$

One again substitutes the expansion (9-21) for **A**, and it is convenient to define a momentum transfer vector

$$\mathbf{q} = \mathbf{k} - \mathbf{p}\,, \tag{13-21}$$

where **k** is the photon momentum obtained from the vector potential. The resulting integral can now be done in a straightforward manner:

$$\int \exp(i\mathbf{q}\cdot\mathbf{r})\exp(-r/a)\,d^3r = 8\pi a/(a^{-2}+q^2)^2\,. \tag{13-22}$$

Since there is one photon in the initial state and none in the final state, we can immediately write down the desired S-matrix element:

$$\langle\mathbf{b}\,|\,\mathsf{S}\,|\,\mathbf{a}\rangle = \delta_{\mathbf{ba}} - 2\pi i\,\delta(\omega_{\mathbf{p}} - \omega_{\mathrm{b}} - \hbar ck) \times\left[-\frac{8\pi e}{mca}\left(\frac{2\pi\hbar c}{V^2\pi a^3}\right)^{1/2}\frac{\hbar\mathbf{p}\cdot\boldsymbol{\epsilon}}{k^{1/2}(a^{-2}+q^2)^2}\right], \tag{13-23}$$

where ω_{b} is the energy of the bound electron. In the sense of the formalism developed in Chapter X, the bound electron must be considered a different kind of particle than a free electron.

The transition rate can now be identified from Eqs. (11-38) and (11-39),

$$w = \frac{2^8\pi^3}{V^2a^5}\frac{\hbar\alpha}{m^2}\frac{(\hbar\mathbf{p}\cdot\boldsymbol{\epsilon})^2}{k(a^{-2}+q^2)^4}\rho_f\,, \tag{13-24}$$

and the density of final states is

$$\rho_f = \frac{p^2\,dp\,d\Omega}{(2\pi)^3}\frac{V}{dE} = \frac{mpV\,d\Omega}{8\pi^3\hbar^2}\,. \tag{13-25}$$

The incident photon flux is clearly c/V, so that the differential cross section obtained from Eq. (13-24) is

$$\frac{d\sigma}{d\Omega} = \frac{32\hbar\alpha}{mca^5}\left(\frac{p}{k}\right)\frac{(\mathbf{p}\cdot\boldsymbol{\epsilon})^2}{(a^{-2}+q^2)^4}\,. \tag{13-26}$$

The angular distribution exhibited by this last expression can be made more transparent by simplifying the denominator. Use of the definition of a, along with the expression for the ground-state energy of hydrogen, allows us to write

$$I = e^2Z/2a = \hbar^2a^{-2}/2m\,. \tag{13-27}$$

From Eq. (13-17),

$$\hbar^2p^2/2m = \hbar ck - I\,,$$

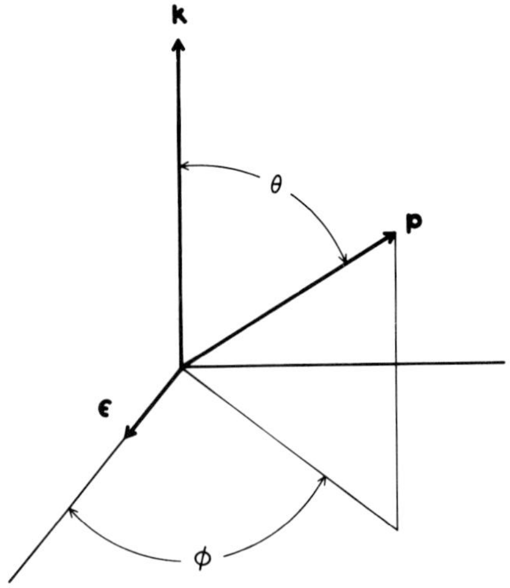

Fig. 38. Orientation of momentum and polarization vectors for the *K*-shell photoelectric effect.

or

$$\hbar ck = (\hbar^2/2m)(a^{-2} + p^2) \ll mc^2 , \tag{13-28}$$

in the nonrelativistic limit (13-18). Then, with the definition (13-21),

$$\begin{aligned} a^{-2} + q^2 &= a^{-2} + p^2 + k^2 - 2pk \cos\theta \\ &= (k/\hbar c)(2mc^2 + \hbar ck - 2\hbar cp \cos\theta) \\ &\simeq (2mc/\hbar)k[1 - (\hbar cp/mc^2)\cos\theta] , \end{aligned} \tag{13-29}$$

where θ is the angle between $\mathbf{p}$ and $\mathbf{k}$, as in Fig. 38.

Rather than sum over polarization states, it is customary to exhibit the angular distribution with respect to the polarization of the incident photon. The factor $(\mathbf{p} \cdot \epsilon)^2$ in Eq. (13-26) is easily evaluated with the aid of Fig. 38, and with the result of Eq. (13-29) the differential cross section can now be written

$$\frac{d\sigma}{d\Omega} \simeq 4\sqrt{2}\, r_0^2 \alpha^4 Z^5 \left(\frac{mc}{\hbar k}\right)^{7/2} \frac{\sin^2\theta \cos^2\phi}{[1 - (\hbar cp/mc^2)\cos\theta]^4} . \tag{13-30}$$

Exercise. Verify this form by use of the inequality (13-18).

Thus, we see that the photoelectrons are emitted primarily in the direction of polarization of the incident photon. As the photon energy is increased, the momentum of the ejected electron increases, and the de-

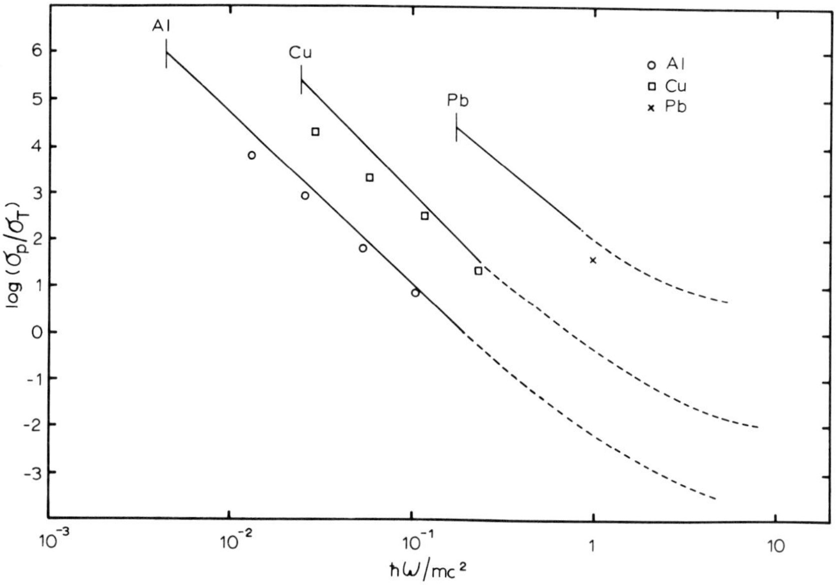

Fig. 39. Logarithmic plot of the photoelectric cross section, (13-32), in units of the Thomson cross section, σ_T, and comparison with some of the data.[3] The dashed portions indicate the projection into the relativistic region due to Sauter,[4] and the vertical lines indicate the edge of the absorption band. [See also Eq. (16-11).]

nominator factor tends to tilt the distribution in the direction of **k**, much as was the case in the classical, relativistic problems of Chapter V.

Having made this observation, however, we can now omit the factor of $\cos\theta$ for the nonrelativistic process, and this renders the angular integration trivial. To obtain the correct total cross section, one must multiply by a factor of two in order to account for *both* *K*-shell electrons. It is also convenient to again introduce the Thomson cross section as a measure of the importance of the process. Then, one readily finds for the total (nonrelativistic) photoelectric cross section:

$$\sigma_p \simeq 4\sqrt{2}\,\alpha^4 Z^5 \sigma_T (mc/\hbar k)^{7/2}\,. \tag{13-31}$$

Exercise. Show that σ_p can be written equivalently as

$$\sigma_p \simeq (64/Z^2\alpha^3)\sigma_T(I/\hbar c k)^{7/2} \tag{13-32}$$

in the nonrelativistic limit.

The nonrelativistic equations, (13-31) and (13-32), are excellent results for most elements, and are compared with some of the data[3] in Fig. 39.

For the very heavy atoms, however, it is possible that the incident photon energy $\hbar ck$ is small enough to make the kinetic energy of the ejected electron of order I. This indicates a close proximity to the edge of the absorption band, and use of the plane wave is no longer a valid approximation for the ejected-electron wave function. In this case the correct continuum wave functions must be used, and the process has been treated by Stobbe.[5]

In the relativistic region the cross section is altered by a multiplicative function of $\gamma = (1 - v^2/c^2)^{-1/2}$ and the exponent of 7/2 in Eqs. (13-31) and (13-32) is changed to 5. The resulting cross section differs surprisingly little in numerical value from the above expressions.[4]

C. SCATTERING OF LIGHT FROM ATOMS

As a final process for consideration in this chapter, we shall discuss photon scattering from an atom. On the surface, the system is very similar to that of Section XII-A for Thomson scattering, and, in fact, the same diagrams contribute to leading order. The physics of the situation, however, is quite different, because the electron is now in a bound state of the atom both before and after the interaction, and the process can be interpreted as an absorption of the incident photon by the atom followed by an immediate emission. Moreover, we shall see that, unlike Thomson scattering, the $(\mathbf{p} \cdot \mathbf{A})$ diagrams do not cancel, but contribute strongly to the process.

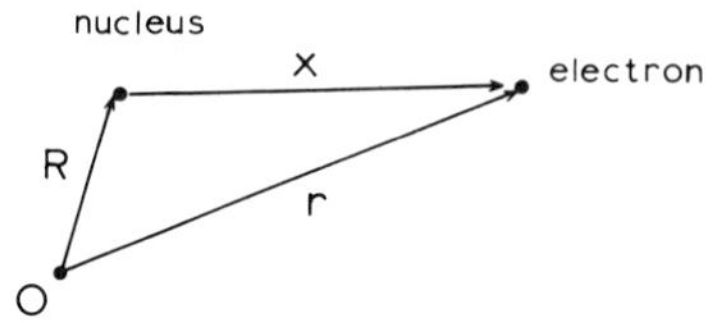

Fig. 40. Appropriate coordinate relations for the evaluation of Eq. (13-34).

The physical assumption which will be made is that the incident photon energy $\hbar ck_1$ is of the same order of magnitude as the binding energy of the electron; for larger energies the problem reduces once again to Thomson scattering, as will be seen below. This assumption covers a spectrum from the visible up to soft X-ray frequencies, and also allows us to consider the wavelengths of both incident and scattered photons as large compared to the atomic dimensions.

We denote the states of the electron in the atom by n, with energy E_n, and for the moment assume that $\hbar ck_1$ is not near a resonant frequency of the atom, $E_n - E_m$. The diagrams of Figs. 26e, 27d, and 27e, give the leading-order contribution, but it is necessary to reevaluate the matrix elements of the interaction using bound-state wave functions before proceeding with the calculations. Thus, the single-particle basis functions in Fock space, Eq. (10-5), are now those of the complete set describing the electron in the atom, and we designate them by $\psi_{n_j}(\mathbf{x})$. If we designate the initial electronic state by n_i and the final state by n_f, then the V_{2r} interaction is readily seen to contribute a matrix element

$$\langle n_f\mathbf{k}_2 | V_{2r} | n_i\mathbf{k}_1\rangle = \frac{e^2}{2mc^2}\int \psi_{n_f}^*(\mathbf{x})\mathbf{A}^2\psi_{n_i}(\mathbf{x})\, d^3x$$
$$= \frac{e^2}{2mc^2}\left(\frac{2\pi\hbar c}{Vk_1^{1/2}k_2^{1/2}}\right)(\epsilon_1 \cdot \epsilon_2)$$
$$\times \int \psi_{n_f}^*(\mathbf{x})\{\exp[i(\mathbf{k}_2 - \mathbf{k}_1)\cdot \mathbf{r}]\}\psi_{n_i}(\mathbf{x})\, d^3x\,. \qquad (13\text{-}33)$$

The coordinate system describing the nucleus and electron is shown in Fig. 40. Because the photon wavelengths are large compared to atomic dimensions, it is again in order to make the dipole approximation and replace all exponentials involving $\mathbf{x}$ by unity. Therefore, the integral in Eq. (13-33) can be written

$$\int \psi_{n_f}^*(\mathbf{x})\{\exp[i(\mathbf{k}_2 - \mathbf{k}_1)\cdot \mathbf{r}]\}\psi_{n_i}(\mathbf{x})\, d^3x$$
$$= \{\exp[i(\mathbf{k}_2 - \mathbf{k}_1)\cdot \mathbf{R}]\}\int \psi_{n_f}^*(\mathbf{x})\{\exp[i(\mathbf{k}_2 - \mathbf{k}_1)\cdot \mathbf{x}]\}\psi_{n_i}(\mathbf{x})\, d^3x$$
$$\simeq \{\exp[i(\mathbf{k}_2 - \mathbf{k}_1)\cdot \mathbf{R}]\}\,\delta_{n_f n_i}\,. \qquad (13\text{-}34)$$

Hence,

$$\langle n_f\mathbf{k}_2 | V_{2r} | n_i\mathbf{k}_1\rangle$$
$$\simeq \frac{e^2}{2mc^2}\left(\frac{2\pi\hbar c}{Vk_1^{1/2}k_2^{1/2}}\right)(\epsilon_1 \cdot \epsilon_2)\,\delta_{n_f n_i}\exp[i(\mathbf{k}_2 - \mathbf{k}_1)\cdot \mathbf{R}]\,, \qquad (13\text{-}35)$$

in the dipole approximation.

The matrix elements for the diagrams of Figs. 27d and 27e are evaluated in the manner of Eqs. (13-3)–(13-8). These calculations are easily performed in a straightforward manner, and when the contributions from all three diagrams are added together, we can identify the reaction matrix element from Eq. (11-38):

$$\mathsf{R}_{ba} \simeq \frac{\hbar}{mc}\frac{2\pi}{Vk_1^{1/2}k_2^{1/2}}\{\exp[i(\mathbf{k}_2 - \mathbf{k}_1)\cdot \mathbf{R}]\}$$
$$\times \left[m\sum_n \omega_{n_f n}\omega_{n n_i}\left(\frac{\langle n_f|\boldsymbol{\mu}\cdot\epsilon_2|n\rangle\langle n|\boldsymbol{\mu}\cdot\epsilon_1|n_i\rangle}{E_{n_i} - E_n + \hbar c k_1}\right.\right.$$
$$\left.\left. + \frac{\langle n_f|\boldsymbol{\mu}\cdot\epsilon_1|n\rangle\langle n|\boldsymbol{\mu}\cdot\epsilon_2|n_i\rangle}{E_{n_i} - E_n - \hbar c k_2}\right) + e^2\,\delta_{n_f n_i}(\epsilon_1\cdot\epsilon_2)\right], \qquad (13\text{-}36)$$

where, for instance, $\omega_{n_f n} = (E_{n_f} - E_n)/\hbar$, and

$$\langle n_f | \boldsymbol{\mu} \cdot \epsilon_1 | n \rangle = \int \psi_{n_f}^*(\mathbf{x})(\boldsymbol{\mu} \cdot \epsilon_1)\psi_n(\mathbf{x})\, d^3x\,. \tag{13-37}$$

One notes that the vector **R** in the phase factor merely locates the position of the atom with respect to the arbitrary origin, and that the sum over states in Eq. (13-36) must include both the bound states and the continuum. In the visible portion of the spectrum the latter part of the sum contributes negligibly.

Exercise. Perform the necessary calculations as indicated, and verify Eq. (13-36).

It is customary to exhibit the cross section in terms of the angles describing the polarizations of the photons, so that we shall *not* sum over polarization states, but take the angle between the polarization vectors to be θ. The density of final states is again given by Eq. (13-9), while the incident flux is c/V. Thus, we obtain for the differential scattering cross section for a given polarization:

$$\frac{d\sigma}{d\Omega} \simeq r_0^2\left(\frac{k_2}{k_1}\right)\left[\frac{m}{e^2}\sum_n \omega_{n_f n}\omega_{n n_i}\left(\frac{\langle n_f | \boldsymbol{\mu}\cdot\epsilon_2 | n\rangle\langle n | \boldsymbol{\mu}\cdot\epsilon_1 | n_i\rangle}{E_{n_i} - E_n + \hbar c k_1}\right.\right.$$
$$\left.\left. + \frac{\langle n_f | \boldsymbol{\mu}\cdot\epsilon_1 | n\rangle\langle n | \boldsymbol{\mu}\cdot\epsilon_2 | n_i\rangle}{E_{n_i} - E_n - \hbar c k_2}\right) + \delta_{n_f n_i}\cos\theta\right]^2, \tag{13-38}$$

which is valid in the dipole approximation and not near a resonance, such as $\hbar c k_1 \sim E_n - E_{n_i}$. The differential cross section of Eq. (13-38) is very general, in that it forms the basis for analyzing several physical situations. If $n_f = n_i$ and the atomic electron is left in the same state, the scattering is *coherent*, since there is no shift of frequency in the scattered photon ($k_2 = k_1$). In this case the differential cross section (13-38) is known as the *Kramers–Heisenberg dispersion formula.*[6] However, the term in $\cos\theta$ was first found by Waller[7] and is dominant if $\hbar c k_1 \gg E_n - E_{n_i}$. The cross section then reduces to the Thomson value of Eq. (12-15) for a free electron, if one analyzes the meaning of the angles correctly.

When $n_f \neq n_i$, Eq. (13-38) predicts a frequency shift in the scattered radiation proportional to the energy difference between the initial and final electronic states. This is the well-known *Raman effect* predicted by Smekal,[8] and which was discovered experimentally by Landsberg and Mandelstamm in solids,[9] and by Raman and Krishnan in liquids.[10]

Finally, an examination of Eq. (13-38) indicates that the first term becomes very large if $\hbar c k_1 \simeq E_n - E_{n_i}$, implying a resonance phenomenon. In this case the process must be reanalyzed, and this was first done by

Weisskopf.[11] The resulting expression is quite similar to the classical result, Eq. (7-65), exhibiting the phenomenon of *resonance fluorescence*. As was the case classically, the resonance is prevented from becoming infinite by taking into account the radiative damping force in a correct way. The exposition of this procedure given by Heitler is excellent.[12]

PROBLEMS

13-1. It sometimes happens that in a particular atom, and for certain initial and final states, the matrix element of the dipole moment vanishes; that is, the transition is forbidden. In these cases the dipole approximation is inadequate and higher multipoles must be examined. Calculate the transition rate for spontaneous quadrupole radiation.

13-2. An important process for studying the nature of multiple effects is represented by the diagrams of Figs. 26c and 27b. Calculate the transition rate for double emission,[13] assuming the energy difference between atomic states to be shared more or less equitably between the two photons, and compare the result with that obtained for single spontaneous emission. Comment on the reduction of this result to that for single emission when one of the photon momenta is allowed to become vanishingly small.

13-3. Calculate the differential cross section for the L-shell photoelectric effect.[14]

REFERENCES

1. A. Einstein, *Mitt. Phys. Ges. Zurich* **16,** 47 (1916); *Verhandl. Deut. Physik. Ges.* **18,** 318 (1916); *Z. Physik* **18,** 121 (1917).
2. A. Einstein, *Ann. Physik* **17,** 132 (1905).
3. S. J. M. Allen, *Phys. Rev.* **27,** 266; **28,** 907 (1926); L. H. Gray, *Proc. Cambridge Phil. Soc.* **27,** 103 (1931).
4. F. Sauter, *Ann. Physik* **9,** 217; **11,** 454 (1931).
5. M. Stobbe, *Ann. Physik* **7,** 661 (1930).
6. H. A. Kramers and W. Heisenberg, *Z. Physik* **31,** 681 (1925).
7. I. Waller, *Z. Physik* **51,** 213 (1928).
8. A. Smekal, *Naturwissenschaften* **11,** 873 (1923).
9. G. Landsberg and L. Mandelstamm, *Naturwissenschaften* **16,** 557, 772 (1928).
10. C. V. Raman and K. S. Krishnan, *Nature* **121,** 501 (1928).
11. V. Weisskopf, *Ann. Physik* **9,** 23 (1931); *Z. Physik* **85,** 451 (1933). See also, W. Heitler and S. T. Ma, *Proc. Roy. Irish Acad.* **52,** 109 (1949).
12. W. Heitler, "The Quantum Theory of Radiation," 3rd ed., Section V-20. Oxford Univ. Press, London and New York, 1954.
13. M. Göppert-Mayer, *Naturwissenschaften* **17,** 932 (1929); *Ann. Physik* **9,** 273 (1931).
14. H. Hall, *Rev. Mod. Phys.* **8,** 358 (1936).

XIV SELF-ENERGIES AND RENORMALIZATION

A. HIGHER-ORDER CALCULATIONS

Investigation of the various processes in the preceding work involved only the leading-order contributions in α, and only the simplest diagrams describing a particular process were included. Suppose, now, that more accuracy were required in, say, the determination of the Thomson scattering cross section. This suggests that we study the diagram of Fig. 41, which can be calculated quite easily following the rules for I graphs. Upon pursuing this calculation, one readily finds that it is necessary to evaluate the sum

$$\sum_{\mathbf{k}_3} \frac{(\epsilon_2 \cdot \epsilon_3)(\epsilon_1 \cdot \epsilon_3)}{k_3[\omega(\mathbf{p}_1 + \mathbf{k}_2 - \mathbf{k}_3) + \hbar c k_3 - \hbar c k_1 - \omega(\mathbf{p}_1)]}, \tag{14-1}$$

where $\mathbf{k}_1$ describes the incident photon and $\mathbf{k}_2$ the scattered photon. If the sum is now converted to an integral, in the usual manner, one observes that the integral diverges at the upper limit. This "ultraviolet" divergence must surely be unphysical, since the theory is nonrelativistic, and we can have some confidence that the singularity has nothing to do with the observable physical properties of the process. On the other hand, it is here, and must be handled in some acceptable manner.

Unfortunately, the problem just uncovered is not unique; in fact, it is quite ubiquitous in the theory. If similar higher-order terms for other electrodynamic processes are calculated in the same manner, it is found that they, too, lead to divergence difficulties of the same kind. Thus, it appears that while the leading-order calculations produce very nice results in agreement with experiment, the theory breaks down in higher order.

This state of affairs is reminiscent of the problems encountered with the classical theory in Chapter VII, where the presence of structure-dependent

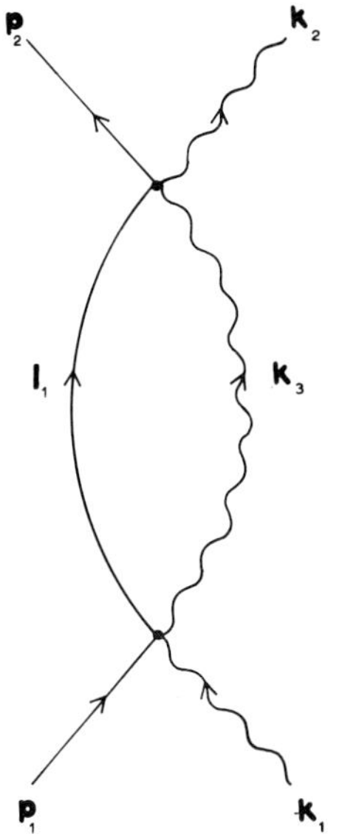

Fig. 41. Diagram representing a higher-order contribution to Thomson scattering.

terms in the equations of motion led to divergence difficulties. In the classical theory of the electron these unpleasant problems were associated with the electromagnetic self-energy of the charged particle. Since it is not obvious that such phenomena should not be present in the quantum theory, the implication is that we should examine this possibility as the source of the above divergences.

Reference to Chapter VII shows that the classical divergence difficulties are associated with the notion of a point electron, in that if the radius of the charge distribution is allowed to approach zero, the electromagnetic self-energy and the electromagnetic mass become infinite. Formally, these structure-dependent terms are small in terms of the physical parameters, and constitute essentially negligible radiative corrections to the equations of motion. Nevertheless, they lead to theoretical difficulties. It will be recalled that the equation of motion gives physical results that are confirmed experimentally if we let the structure constant $a \to 0$. On the other hand, divergences arise unless this parameter is actually cut off at r_0, the classical electron radius. Thus, the divergences arise on a length scale associated with $r_0 = e^2/mc^2$; for instance, as $r_0 \to 0$, the self-energy diverges as r_0^{-1}.

In quantum mechanics there seems to be no change in the above behavior at first glance, for, if one recalculates the self-energy using $e^2|\psi(\mathbf{r})|^2$ in place of the classical charge distribution, there is still a linear divergence in the structure constant a. However, this is a purely nonrelativistic effect, and is modified considerably when one remembers to include in the analysis other processes which can occur in the same order. To correctly treat these other processes, such as the fluctuations of the vacuum, it is necessary to go to a relativistic theory, and therefore the self-energy should also be calculated relativistically for consistency. Thus, one performs the calculation with the Dirac electron–positron theory and finds that the Pauli principle induces a spreading of the electron charge distribution, which, in turn, reduces the linear divergence in a to a logarithmic divergence:

$$\ln(\hbar/mca)\,. \tag{14-2}$$

This behavior of the electron self-energy was first demonstrated by Weisskopf,[1] who later showed that the other processes contributing in this order also diverge logarithmically.[2] We shall have more to say about these contributions in Chapter XVI.

The point of the preceding discussion was to show by means of Eq. (14-2) that the quantum mechanical self-energy problem actually differs qualitatively, as well as quantitatively, from the classical problem. We see that the cutoff here is of the order of an electron Compton wavelength, $\lambda_C = h/mc$. Since the classical electron radius is of order α *less* than λ_C, it is clear that quantum divergence difficulties become important long before the corresponding classical effects. Thus, one expects from the beginning that an independent approach to the self-energy problem must be developed in the quantum theory. Let us hasten to point out, however, that these observations have no bearing on the lengths down to which QED may be valid, a question we will take up in more detail in Chapter XVI.

B. SELF-ENERGY OF THE ELECTRON

Examination of the diagrams exhibited in Chapter XI yields two that immediately give the appearance of self-energy processes: Figs. 27c and 32a. We shall confine our discussion to the former at this time, since it is of lower order in α. This diagram represents the emission of a photon by an electron and its subsequent, almost immediate, reabsorption. Such a photon is referred to as *virtual*, and the process strongly suggests a self-interaction term. If this observation is true, then the interaction should constitute a change in the energy of the free electron. This energy change, ΔE, is most readily evaluated by means of Eq. (11-45), in the form

$$\langle \mathbf{b} | \mathsf{S} | \mathbf{a} \rangle = -2\pi i\, \delta(E_b - E_a)\, \Delta E\,. \tag{14-3}$$

Applying the rules for I graphs, we find from the diagram of Fig. 27c

$$\begin{aligned}\Delta E &= -\sum_{\mathbf{k}\boldsymbol{l}} \frac{\langle \mathbf{p}_2 | \mathsf{V}_{1\gamma} | \boldsymbol{l}\mathbf{k}\rangle \langle \mathbf{k}\boldsymbol{l} | \mathsf{V}_{1\gamma}^{\dagger} | \mathbf{p}_1\rangle}{\omega(\boldsymbol{l}) + \hbar ck - \omega(p_1)} \\ &= -\frac{\alpha\hbar^4}{4\pi^2 m^2}\, p_1^{\,2} \int \frac{\sin^2\theta}{\omega(\mathbf{p}_1 - \mathbf{k}) + \hbar ck - \omega(\mathbf{p}_1)}\, \frac{d^3k}{k}\,. \end{aligned} \tag{14-4}$$

This last integral diverges linearly at the upper limit, whether or not we make any approximations. However, to be consistent with the nonrelativistic nature of the theory, it is proper to apply the inequality (12-2) and approximate the denominator by $\hbar ck$. Then,

$$\Delta E \simeq -\frac{4\alpha}{3\pi}\left(\frac{\hbar^2 p_1{}^2}{2m}\right)\int_0^\infty \frac{\hbar}{mc}\,dk\,. \tag{14-5}$$

One cannot, of course, take the indicated ultraviolet divergence too seriously in the nonrelativistic case, and it is reasonable to assume the existence of a cutoff. In order to obtain a formal physical estimate of ΔE, we shall impose the nonrelativistic quantum criterion that $k_{\max} = \lambda_C^{-1}$, the inverse Compton wavelength (divided by 2π) of the electron. Thus,

$$\Delta E \simeq -\frac{4\alpha}{3\pi}\left(\frac{\hbar^2 p_1{}^2}{2m}\right), \tag{14-6}$$

which has the form of a correction to the unperturbed free-electron energy. A brief reference to Section IX-B suggests that this perturbation may be due to the interaction of the electron with the fluctuating vacuum.

As an aside, it should be pointed out that the above process is not completely self-consistent in the nonrelativistic domain, because the electron could emit a photon of arbitrarily high energy and thereby acquire a relativistic recoil momentum. However, this recoil occurs in the virtual state and is probably unobservable. Also, the interaction could be made nonlocal and cut off at the radius of the charge distribution so that only wave numbers $k < \lambda_C^{-1}$ contribute. We shall see below that this observation is actually of no consequence, so that it is safe to accept Eqs. (14-5) and (14-6) at face value.

Equation (14-6) contains the key to the resolution of the self-energy problem, and this solution was first uncovered by Kramers[3] (although Stückelberg[4] had anticipated a good deal of the method). The key is that the energy shift ΔE is proportional to the kinetic energy of the electron. Therefore, let us introduce a change of notation and label the mass by m_0, and then write the total energy of the electron in the formally divergent form

$$E = \frac{\hbar^2 p_1{}^2}{2m_0}\left(1 - \frac{4\alpha}{3\pi}\frac{\hbar}{m_0 c}\int_0^\infty dk\right), \tag{14-7}$$

so that we have introduced *no* cutoff of any kind.

We now introduce the *true* mass of the particle, m, by writing

$$E = \frac{\hbar^2 p_1{}^2}{2m} = \frac{\hbar^2 p_1{}^2}{2(m_0 + \delta m)}, \tag{14-8}$$

where δm is defined to be the electromagnetic mass the particle has acquired over and above its mechanical mass. This additional mass arises due to the

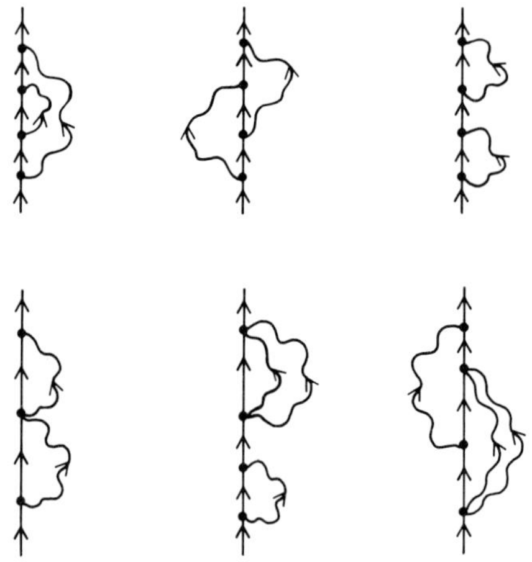

Fig. 42. Higher-order contributions to the self-energy of a charged particle.

interaction of the particle with the radiation field. If we now *assume* that δm is small, the denominator in Eq. (14-8) can be expanded to leading order:

$$E \simeq \frac{\hbar^2 p_1^{\,2}}{2m_0}\left(1 - \frac{\delta m}{m_0}\right). \tag{14-9}$$

Comparing this expression with Eq. (14-7), we can make the *identification*

$$\delta m = \frac{4\alpha}{3\pi}\frac{\hbar}{c}\int_0^\infty dk\,, \tag{14-10}$$

and, since this is (formally) of order α, the expansion in Eq. (14-9) may be reasonable.

The interpretation to be made now is that $m = m_0 + \delta m$ is the true, observable mass of the electron, and this is motivated by the fact that no experiment can possibly differentiate between m and m_0. The *bare mass* is completely unobservable, and only the *dressed*, or *experimental mass* has any physical significance. Thus, the self-energy difficulty has been resolved by *renormalizing* the energy eigenvalue of Eq. (14-8), much as was implicitly the case in the classical theory. It should be pointed out, though, that Eq. (14-10) represents only the leading-order contribution to δm, and that this quantity is really the sum of all self-energy parts of diagrams such as those of Figs. 32a and 42.

The problems which have just been uncovered impeded progress in QED until about twenty years ago. Although the theory was thought to be correct, it was essentially impossible to perform any calculation beyond those of the preceding chapters. While it is not too difficult to remedy the nonrelativistic theory, as demonstrated above, the problems became much more complex in the high-energy domain. The monumental contributions of Feynman,[5] Tomonaga,[6] and Schwinger[7] involved showing that quantum electrodynamics could, and must, be formulated in a manifestly covariant way, that renormalization can be done in a relativistically consistent manner, and that systematic techniques could be found for calculating physically important quantities. Dyson[8] capped these successes by proving that a formal renormalization program can be carried out to *all* orders in the coupling constant. He did this, though, for the covariant theory only, and we are here *assuming* that one can construct a similar formal proof for nonrelativistic QED. The basis for such a proof has been given in another context.[9]

Although we shall discuss it in more detail in Chapter XVI, the renormalization program can be illustrated quite simply in the nonrelativistic theory by first noting that the calculations of Eqs. (14-8)–(14-10) are not sufficient in themselves. This is because the Hamiltonian with which we began still contains the bare mass:

$$\mathsf{H} = (\hbar^2 p^2/2m_0) + \mathsf{H}' \,, \tag{14-11}$$

where H' is the interaction term. If we now substitute $m_0 = m - \delta m$ and expand to leading order in α, we find

$$\mathsf{H} \simeq \frac{\hbar^2 p^2}{2m} + \left[\frac{\hbar^2 p^2}{2m}\frac{\delta m}{m} + \mathsf{H}'\right]. \tag{14-12}$$

The first term in the brackets is referred to as the *mass renormalization counterterm*. Clearly, to order α this quantity will cancel any self-energy terms arising from H', so that if the Hamiltonian of Eq. (14-12) is substituted throughout the formalism of Chapters X and XI, the entire nonrelativistic theory is renormalized to leading order. The extension of this statement to all orders constitutes the renormalization program.

It must be emphasized that the renormalization procedure which we have outlined depends only upon the fact that the self-energy terms are unmeasurable. Therefore, these quantities are unphysical in themselves, and it is a consistent procedure to cancel them with a counterterm. The fact that these terms are infinite has no bearing on this procedure, for even a convergent theory must be expressed in terms of the observed masses and other quantities. The worst that can be said of the theory is that it cannot account for the masses of charged particles, but that one has to always substitute the experimentally determined numbers, and this we are ready to admit. When a higher level of understanding is reached with regard to elementary particles, then it is possible that the peculiarities of QED as currently formulated will become transparent.

Thus, the difficulties first encountered in classical electrodynamics have only been partially resolved, for the occurrence of infinite self-energies is not avoided in the quantum theory either. These infinities always (and only) occur when one attempts to actually calculate something, in which case an approximation procedure has to be adopted, such as the method we have chosen in Chapter XI. As a consequence, it is tempting to inquire whether the ultraviolet divergences arise because of the inadequate mathematical techniques forced upon us by our ignorance, or whether these maladies are actually inherent in the structure of the theory. The latter view seems to be the more nearly accepted state of affairs, and, in fact, Källén[10] has demonstrated that the renormalization constants must be infinite, regardless of the calculational technique adopted. As we have seen above, however, the infinite nature of the self-energies is of no consequence, since such quantities can be absorbed into the observable mass. What still remains is to reexamine the theory in order to ascertain the possible introduction of any new physics by the renormalization program, and this question we shall take up in the following chapter.

It is instructive to loqk a bit further into the way in which our calculational procedure has forced upon us the infinite electromagnetic contributions to the mass of a charged particle. In order to perform any calculations, it has been found *necessary* to separate the Hamiltonian into two distinct parts:

$$\mathsf{H} = \mathsf{H}_0 + \mathsf{V}\,.$$

Nevertheless, such a separation is far from trivial in the case of electromagnetic interactions, because the masses of the free particles in H_0 actually depend on V in this situation. Thus, it would appear as if a nonlinear theory of electrodynamics were needed, and such a theory has indeed been found by Born and Infeld.[11] However, the quantization of this theory has proved to be intractable. It seems, therefore, as if we must be content to first remove the radiation field accompanying the particle, leaving a bare electron, and then add the field back again, *photon by photon*.

As a final observation, let us point out that the divergence difficulties in QED are actually more extensive than those encountered above, and further renormalizations and even more delicate techniques are required to render the theory meaningful. These problems arise in connection with processes which we have not studied as yet, and we shall discuss these affairs as we encounter them in the next chapter.

PROBLEMS

14-1. Consider the process indicated in Fig. 43, which represents a self-energy correction to a Thomson scattering diagram. Demonstrate explicitly the

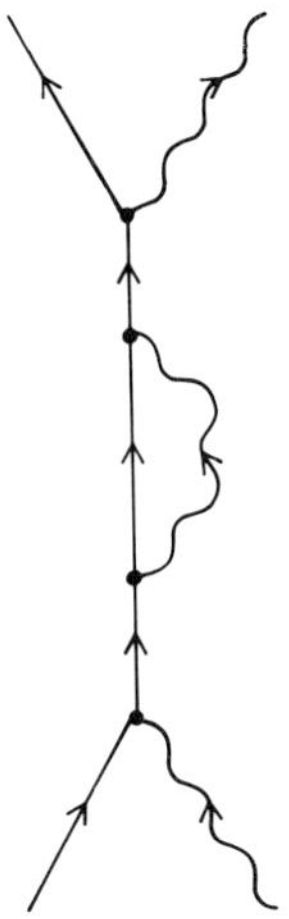

Fig. 43. A possible self-energy correction to Thomson scattering.

mass renormalization of this contribution in the nonrelativistic limit, and thereby show that there is no observable effect from this process over and above that from the original diagram of Fig. 27d.

REFERENCES

1. V. Weisskopf, *Z. Physik* **89,** 27; **90,** 817 (1934).
2. V. Weisskopf, *Phys. Rev.* **56,** 72 (1939).
3. H. A. Kramers, *Rappt. Conseil Solvay 1948*, p. 241. Brussels (1950).
4. E. C. G. Stückelberg, *Helv. Phys. Acta* **14,** 15 (1941); **17,** 3 (1944); **18,** 195 (1945); **19,** 242 (1946).
5. R. P. Feynman, *Phys. Rev.* **74,** 1430 (1948); **76,** 769 (1949).
6. S. Tomonaga, *Progr. Theoret. Phys.* (*Kyoto*) **1,** 27 (1946).
7. J. Schwinger, *Phys. Rev.* **73,** 416; **74,** 1439 (1948); **75,** 651; **76,** 790 (1949).
8. F. J. Dyson, *Phys. Rev.* **75,** 486, 1736 (1949).
9. F. Mohling and W. T. Grandy, Jr., *J. Math. Phys.* **6,** 348 (1965); Section IV.
10. G. Källén, *Helv. Phys. Acta* **25,** 417 (1952); *Kgl. Danske Videnskab. Selskab* **27,** 12 (1953).
11. M. Born, *Proc. Roy. Soc.* (*London*) **A143,** 410 (1934); M. Born and L. Infeld, *ibid.* **A144,** 425; **A147,** 522 (1934); **A150,** 141 (1935).

XV RADIATIVE CORRECTIONS AND THE INFRARED PROBLEM

In the preceding chapter we have seen that the processes contributing to the self-energy of a free electron are unobservable, being absorbed into the experimentally determined mass. If, however, the particle is placed in an external field, this electromagnetic contribution to the mass is no longer completely unobservable. Only a part of this mass correction contributes to the self-energy of the particle, while the remainder can be interpreted as a manifestation of the particle structure. That is, in an external field the structure of the electron plays an important role in the interaction with the radiation field, which leads to an observable change in its total energy. These small energy changes are called *radiative corrections*, and their experimental observance constitutes one of the most brilliant successes of quantum electrodynamics.

A. THE LAMB SHIFT

As an example of how these corrections fit into the nonrelativistic theory, and also to provide an example of how the renormalization program removes the self-energy divergences, we shall first calculate one of the radiative corrections to the energy levels of the hydrogen atom. In 1947 Lamb and Retherford[1] observed an energy difference between the 2S and 2P states in hydrogen. According to the Dirac theory of the electron, no such level displacement should be present, and the only possible explanation of this effect was to be found in the interaction of the electron with the radiation field. Adopting Kramers' renormalization ideas, Bethe[2] calculated this level displacement, now called the *Lamb shift*, and we shall essentially reproduce his nonrelativistic calculation here.

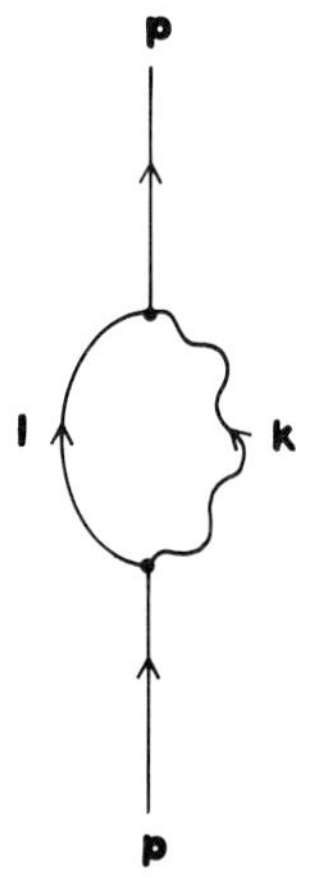

Fig. 44. Leading-order diagrammatic contribution to the Lamb shift. The momentum **p** is now that of an electron in a hydrogenic bound state.

Let us consider an electron located in the nth level of a hydrogen atom, and again calculate the level shift due to the self-interaction, as in Eq. (11-47). The diagram corresponding to the process is reproduced in Fig. 44, and from Eq. (11-45) and the rules for I graphs we easily find in the nonrelativistic limit

$$\Delta E_n = -\frac{2\alpha}{3\pi}\sum_i \frac{|\langle E_n|\hbar\mathbf{p}|E_i\rangle|^2}{m^2c^2}\int_0^\infty \frac{(\hbar c)^2 k\,dk}{\hbar ck + E_i - E_n} + \frac{\delta m}{m}\frac{\langle E_n|\hbar^2\mathbf{p}^2|E_n\rangle}{2m}, \tag{15-1}$$

where we have now used the renormalized Hamiltonian of Eq. (14-12). Aside from the mass renormalization counterterm, the essential difference between Eq. (11-47) for the free electron and Eq. (15-1) is that the latter involves the bound states of the electron in an external field. Thus, the counterterm does not completely cancel the first term in Eq. (15-1).

We now introduce the identity

$$\frac{1}{E_i - E_n + \hbar ck} = \frac{1}{\hbar ck} - \frac{E_i - E_n}{\hbar ck(E_i - E_n + \hbar ck)}. \tag{15-2}$$

Substituting this into Eq. (15-1), one finds

$$\Delta E_n = -\frac{2\alpha}{3\pi c^2}\int_0^\infty (\hbar c)\,dk \sum_i \frac{|\langle E_n|\hbar\mathbf{p}|E_i\rangle|^2}{m^2} + \frac{\delta m}{m}\frac{\langle E_n|\hbar^2\mathbf{p}^2|E_n\rangle}{2m}$$
$$-\frac{2\alpha}{3\pi c^2}\frac{\hbar c}{m^2}\int_0^\infty \sum_i \frac{|\hbar\mathbf{p}_{ni}|^2(E_i - E_n)}{E_i - E_n + \hbar ck}\,dk, \tag{15-3}$$

with the notation $\mathbf{p}_{ni} = \langle E_n | \mathbf{p} | E_i \rangle$. If one now makes the observation that

$$\sum_i | \langle E_n | \mathbf{p} | E_i \rangle |^2 = \sum_i \mathbf{p}_{ni} \cdot \mathbf{p}_{in}$$
$$= | \mathbf{p}^2 |_{nn} = \langle E_n | \mathbf{p}^2 | E_n \rangle \,, \qquad (15\text{-}4)$$

and recalls Eq. (14-10), then it is clear that the first term in Eq. (15-3) is precisely canceled by the counterterm.

The integral in the third term of Eq. (15-3) is easily performed, but contains an apparent divergence at the upper limit. However, the calculation is nonrelativistic, and it must be assumed that a relativistic treatment would yield a proper cutoff for the integral. We choose this cutoff to be K, and find that

$$\Delta E_n = \frac{2\alpha}{3\pi c^2} \sum_i \frac{| \hbar \mathbf{p}_{ni} |^2}{m^2} (E_i - E_n) \ln \left| \frac{K}{E_i - E_n} \right| . \qquad (15\text{-}5)$$

An essential observation made by Bethe was that the choice of cutoff K need not be made with great precision, because it appears in a logarithm, which renders the energy shift quite insensitive to small errors in K. We shall choose $K \simeq mc^2$, which must certainly be of the right order of magnitude.

The remainder of the calculation involves evaluation of the sum in Eq. (15-5), and we follow Bethe by *defining* the average value of $(E_i - E_n)$ as follows:

$$\sum_i | \hbar \mathbf{p}_{ni} |^2 (E_i - E_n) \ln | E_i - E_n |$$
$$\equiv \ln | E - E_n |_{\text{av}} \sum_i | \hbar \mathbf{p}_{ni} |^2 (E_i - E_n) \,. \qquad (15\text{-}6)$$

The sum on the right-hand side of this equation can be evaluated by writing

$$\sum_i (E_i - E_n) | \langle i | \hbar \mathbf{p} | n \rangle |^2 = - \sum_i \langle n | [\mathsf{H}, \hbar \mathbf{p}] | i \rangle \langle i | \hbar \mathbf{p} | n \rangle$$
$$= \tfrac{1}{2} \langle n | \big[\hbar \mathbf{p}, [\mathsf{H}, \hbar \mathbf{p}] \big] | n \rangle \,, \qquad (15\text{-}7)$$

where

$$\mathsf{H} = \frac{\hbar^2 p^2}{2m} + \mathsf{V}(\mathbf{r}) = \frac{\hbar^2 p^2}{2m} - \frac{e^2}{r} \qquad (15\text{-}8)$$

is the Hamiltonian for the electron in the hydrogen atom. The commutator can be evaluated as follows:

$$[\mathsf{H}, \hbar \mathbf{p}] = - \hbar \mathbf{p} \mathsf{V}(\mathbf{r}) \,, \qquad (15\text{-}9a)$$
$$[\hbar \mathbf{p}, -(\hbar \mathbf{p} \mathsf{V})] = - \hbar^2 p^2 \mathsf{V}(\mathbf{r}) \,. \qquad (15\text{-}9b)$$

Exercise. Verify Eqs. (15-7) and (15-9) explicitly.

Thus, Eq. (15-7) becomes

$$\begin{aligned}\sum_i (E_i - E_n) \,|\, \langle i \,|\, \hbar\mathbf{p} \,|\, n\rangle|^2 &= \tfrac{1}{2}\langle n \,|\, -\hbar^2 p^2 \mathrm{V} \,|\, n\rangle \\ &= 2\pi\hbar^2 e^2 \langle n \,|\, \delta(\mathbf{r}) \,|\, n\rangle \\ &= 2\pi\hbar^2 e^2 \,|\, \psi_n(\mathbf{0}) \,|^2 , \end{aligned} \qquad (15\text{-}10)$$

where $\psi_n(\mathbf{0})$ is the hydrogen atom wave function evaluated at the origin, and we have used the relation $\nabla^2(1/r) = -4\pi\delta(\mathbf{r})$. Hence, in the nonrelativistic approximation there is a level shift only for S states. For an S state with quantum number n

$$|\,\psi_n(\mathbf{0})\,|^2 = 1/\pi n^3 a_B{}^3 , \qquad (15\text{-}11)$$

where $a_B = \hbar^2/me^2$ is the first Bohr radius. Substitution of Eqs. (15-6), (15-10), and (15-11) into Eq. (15-5) yields

$$\Delta E_n = \frac{8\alpha^3}{3\pi n^3}\frac{e^2}{2a_B} \ln\left|\frac{K}{E - E_n}\right|_{\text{av}} . \qquad (15\text{-}12)$$

The calculation of $(E - E_n)_{\text{av}}$ is quite tedious, although not too mysterious. In his first publication Bethe[2] obtained the value 17.8 Ry (a surprisingly large number), but this was refined in a later calculation[3] to 16.646 ± 0.007 Ry. Consequently, we shall use the value

$$(E - E_n)_{\text{av}} \simeq 8.3\alpha^2(mc^2) . \qquad (15\text{-}13)$$

One readily finds that

$$\frac{\alpha^3}{3\pi}\frac{e^2}{2a_B} \simeq 136 \quad \text{MHz} , \qquad (15\text{-}14)$$

and the Lamb shift in energy due to the interaction of the electron with the radiation field for hydrogenic S states is

$$(\Delta E)_{\text{LS}} \simeq 1049 \quad \text{MHz} . \qquad (15\text{-}15)$$

This value is in excellent agreement with the experimental result of 1057.77 ± 0.10 MHz obtained by Triebwasser *et al.*[4] In fact, our result is a little too good to be true, and a more precise numerical evaluation would reduce this number somewhat. Nevertheless, Eq. (15-15) represents an outstanding success for the theory, and indicates that the great bulk of the Lamb shift is due to nonrelativistic effects, as guessed by Bethe. When one performs the calculation relativistically, and accounts for finite proton mass and higher-order radiative corrections, the theoretical result is within 0.4 MHz of the observed value (see Chapter XVI). It is difficult to believe that such accuracy is fortuitous, and one must conclude that the expansion in α upon which QED rests must be at least asymptotic, if not convergent. As added support for this belief, there has for some time existed a discrepancy between the theoretical and experimental values of the hyperfine splitting in the ground state of atomic hydrogen.[5,6] However, with the discovery of the Josephson effect[7] it has been possible to make a much more delicate determination of the fine-structure constant. This new value[8] for α corrects the theoretical results such that QED is in *exact* agreement with observation within the experimental errors. Although the renormalization program is esthetically unpleasing and logically unsatisfactory, the present state of the theory must in some sense be "right"!

B. RADIATIVE CORRECTIONS TO THOMSON SCATTERING

In Section XII-A we obtained the differential cross section for Thomson scattering, Eq. (12-15). There are, however, radiative corrections to this process, corresponding to emission and reabsorption of a photon by the electron during the basic scattering interaction. Such processes also constitute quantum mechanical corrections to Thomson scattering, although it should be remembered that there also exist classical corrections to this process due to radiation damping, as exhibited by Eq. (7-64). We shall see in this section that both types of correction can be calculated from our nonrelativistic theory.

Radiative corrections to the basic Thomson scattering diagrams of Figs. 26 and 27 are given by the seven diagrams of Fig. 45. Note that diagrams representing photon emission and reabsorption by a single parcle line have *not* been included, since it is assumed that all simple mass renormalizations have been carried out. When one writes down the analytical expression corresponding to each of these diagrams, following the rules given in Chapter XI, it is found that only one of them need be evaluated. That is, diagrams a–d of Fig. 45 are all of order (v^2/c^2), possess no momentum-integral divergences, and can therefore be discarded as not contributing to the nonrelativistic cross section. Diagrams 45e and 45f, on the other hand, exhibit infrared divergences when the photon-momentum sums are converted to integrals, but the divergences cancel when the two expressions are combined, and the resulting sum is of order (v^2/c^2) and can be neglected.

Exercise. Verify the preceding analysis of the diagrammatic contributions from Figs. 45a–45f.

Finally, the scattering amplitude for the process depicted in Fig. 45g can be calculated straightforwardly in a manner similar to that of Section

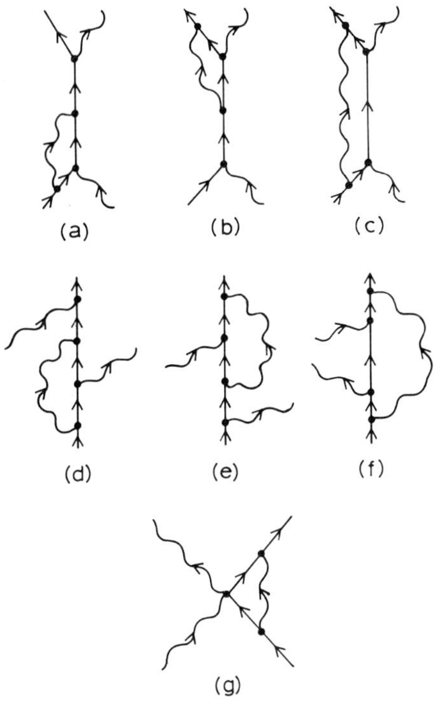

Fig. 45. Diagrams (a)–(g) representing the radiative corrections to Thomson scattering.

XII-A. One finds in the nonrelativistic limit

$$f_{\mathbf{k}_2}(\hat{\mathbf{k}}_1)_{\text{rad}} \simeq \frac{1}{8\pi^2}\,\alpha r_0(\epsilon_1 \cdot \epsilon_2)\,\frac{\hbar^2}{m^2c^2}\int \frac{d^3l}{l}\left[(\mathbf{p}_1 \cdot \mathbf{p}_2) - \frac{(\mathbf{p}_1 \cdot \boldsymbol{l})(\mathbf{p}_2 \cdot \boldsymbol{l})}{l^2}\right], \tag{15-16}$$

where we have averaged over initial and summed over final electron-spin states, introduced the classical electron radius r_0, and applied the identity (11-64).

Now, one must recall that we cannot conserve *both* energy and momentum in the Thomson effect, because we assume the photon frequency to be unshifted in the process. Therefore, we set $\mathbf{p}_1 = \mathbf{p}_2$ and note that energy conservation requires

$$p_2^2 = (\mathbf{k}_1 - \mathbf{k}_2)^2 = 2k_1^2(1 - \cos\theta)\,, \tag{15-17}$$

where θ is the angle between the incident and scattered photons. Also, one observes that the integral in Eq. (15-16) diverges at both limits. The ultraviolet divergence is merely due to the nonrelativistic approximation, so that it is permissible to introduce a cutoff at the inverse Compton wavelength λ_C^{-1}. At the lower limit, however, there is no obvious physical cutoff, and thus we shall insert one temporarily at $l = a$ and study the

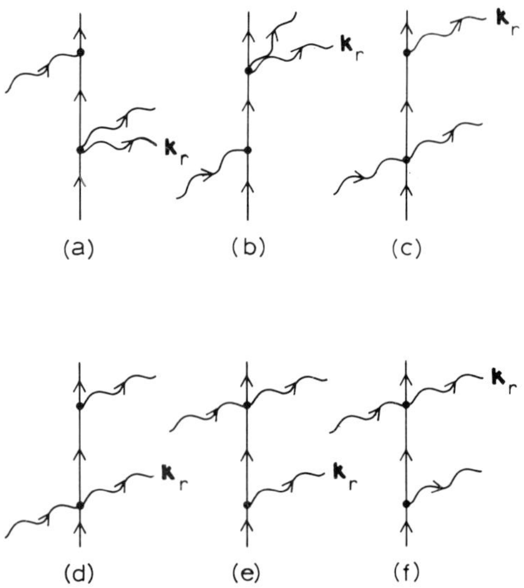

Fig. 46. The six diagrams (a)–(f) with $\mathbf{A}^2$ vertices contributing to double Thomson scattering.

limit $a \to 0$ shortly. Equation (15-16) can now be rewritten as

$$f_{\mathbf{k}_2}(\hat{\mathbf{k}}_1)_{\text{rad}} \simeq \frac{2}{3\pi}\,\alpha r_0(\epsilon_1 \cdot \epsilon_2)(1 - \cos\theta)\left(\frac{\hbar c k_1}{mc^2}\right)^2 \int_a^{\lambda_C^{-1}} \frac{dl}{l}. \tag{15-18}$$

In order to obtain the differential cross section via Eq. (11-37), one must note that the *total* amplitude for Thomson scattering to this order in α is the sum of the amplitude in Eq. (12-14) and that of Eq. (15-18). Hence,

$$\begin{aligned}\left(\frac{d\sigma}{d\Omega}\right)_T &\simeq |f_{\mathbf{k}_2}(\hat{\mathbf{k}}_1) + f_{\mathbf{k}_2}(\hat{\mathbf{k}}_1)_{\text{rad}}|^2 \\ &\simeq r_0^2(\epsilon_1 \cdot \epsilon_2)^2\left[1 - \frac{4}{3\pi}\,\alpha(1 - \cos\theta)\left(\frac{\hbar c k_1}{mc^2}\right)^2 \int_a^{\lambda_C^{-1}} \frac{dl}{l}\right],\end{aligned} \tag{15-19}$$

through leading order in α. The leading factor in this result is just the usual Thomson cross section.

The result of Eq. (15-19) is apparently unsatisfactory in the sense that it possess an infrared divergence as the mathematical cutoff vanishes. This type of divergence difficulty abounds in QED, and it might be recalled that it arose previously and was discussed briefly at the end of Section XII-B. We noticed there that the cross section for *Bremsstrahlung* diverged for emission of a very "soft" photon. In general, this is known as the *infrared problem*, and it is the resolution of this problem to which we must now turn.

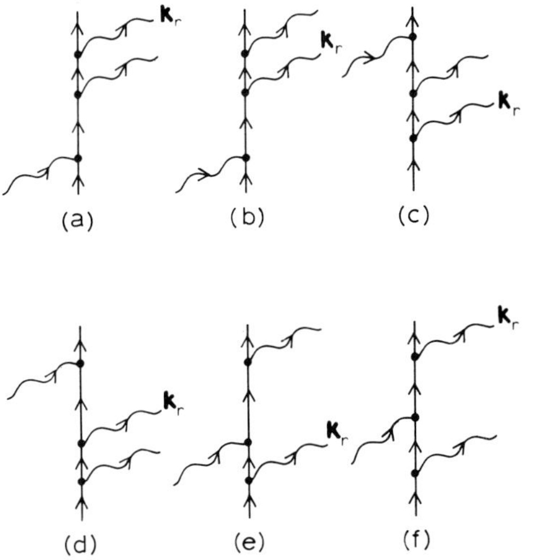

Fig. 47. The six diagrams (a)–(f) with only $\mathbf{p} \cdot \mathbf{A}$ vertices contributing to double Thomson scattering.

As might be suspected, the infrared problem arises from the necessity of having to expand the S matrix in terms of all possible processes in order to calculate anything, and it is possible that the many processes arising in QED which exhibit infrared divergences just might cancel one another. In the next section we shall return to the problem encountered with *Bremsstrahlung*, but at this time it is imperative to continue the study of Eq. (15-19).

There does indeed exist a canceling process for Thomson scattering, and that is *double* Thomson scattering. A photon is absorbed by the electron, and *two* photons are emitted in the final state. Multiple processes in general have very small cross sections[9] and it is experimentally quite difficult to be sure that one is observing an actual multiple event. Some evidence has been seen in cosmic-ray studies,[10] and elsewhere.[11] Nevertheless, our immediate interest in multiple processes is a formal one only, to help resolve the infrared problem. In leading order there are 12 diagrams contributing to double Thomson scattering, and these are shown in Figs. 46 and 47. When the analytical expressions corresponding to these diagrams are written down, it is an easy matter to show that the parametric dependence of the cross sections is

$$\left(\frac{d\sigma}{d\Omega}\right)_{46} \propto \alpha r_0^2(\hbar ck/mc^2)^2\,, \qquad \left(\frac{d\sigma}{d\Omega}\right)_{47} \propto \alpha r_0^2(\hbar ck/mc^2)^4\,, \tag{15-20}$$

in keeping with our claim that the cross sections are very small. In obtaining Eqs. (15-20), one assumes that the energies of the two emitted quanta are of the same order of magnitude.

However, our current interest is with the case in which one of the photons in the final state has very small energy, and therefore one of the outgoing photons in each diagram has been labeled $\mathbf{k}_r$ to denote an infrared photon. We shall want to examine the behavior of these processes as $k_r \to 0$. It is necessary to laboriously write out the analytical expression for each diagram in Figs. 46 and 47, and this explicit task is left as a problem. After carrying out this evaluation, one observes that the diagrams 46e, 47c, and 47e are clearly dominant as $k_r \to 0$. Moreover, the latter two expressions are of order (v^2/c^2) smaller than that for Fig. 46e, so that the leading-order contribution in the nonrelativistic limit comes from only one diagram:

(Fig. 46e):

$$2\pi i\,\delta(\omega_{\mathbf{p}_2} + \hbar c k_2 + \hbar c k_r - \hbar c k_1 - \omega_{\mathbf{p}_1})\,\delta_{n_1 n_2}\,\delta_{\mathbf{p}_1+\mathbf{k}_1,\mathbf{p}_2+\mathbf{k}_2+\mathbf{k}_r}$$
$$\times \frac{(\mathbf{p}_1 \cdot \epsilon_r)(\epsilon_1 \cdot \epsilon_2)}{(k_1 k_2 k_r)^{1/2}} \frac{(2\pi\alpha/V)^{3/2}(\hbar^2/m)^2}{\omega(\mathbf{p}_1 - \mathbf{k}_r) + \hbar c k_r - \omega(\mathbf{p}_1)}. \qquad (15\text{-}21)$$

Now, if one wishes to observe double Thomson scattering, there is plainly going to exist a value of k_r below which the extra photon cannot be detected. This observation merely points out that there is a finite energy resolution in any experiment, and so we can indicate this region of unobservability by $k_r < k_m$, where k_m corresponds to the minimum observable photon momentum. Thus, we should separate Eq. (15-21) into an observable process for soft photons with observable momenta, and into a term containing an unobservable infrared photon. In the latter term we now integrate over all $k_r < k_m$ and again introduce the mathematical cutoff a. One then finds for the cross section corresponding to this unobservable process

$$\left(\frac{d\sigma}{d\Omega}\right)_{\mathrm{DT,ir}} = \frac{4}{3\pi}\alpha r_0^2(\epsilon_1 \cdot \epsilon_2)^2 \left(\frac{\hbar c k_1}{mc^2}\right)^2 (1 - \cos\theta) \int_a^{k_m} \frac{dk_r}{k_r}. \qquad (15\text{-}22)$$

Combining Eqs. (15-19) and (15-22), we now see that the mathematical cutoff a has been eliminated, and one obtains the finite result for radiative corrections to Thomson scattering

$$\left(\frac{d\sigma}{d\Omega}\right)_{\mathrm{T,rad}} \simeq r_0^2(\epsilon_1 \cdot \epsilon_2)^2$$
$$\times \left[1 + \frac{4}{3\pi}\alpha\left(\frac{\hbar c k_1}{mc^2}\right)^2 (1 - \cos\theta)\ln\left|\frac{\hbar c k_m}{mc^2}\right|\right]. \qquad (15\text{-}23)$$

Thus, the infrared divergence arising in radiative corrections to Thomson

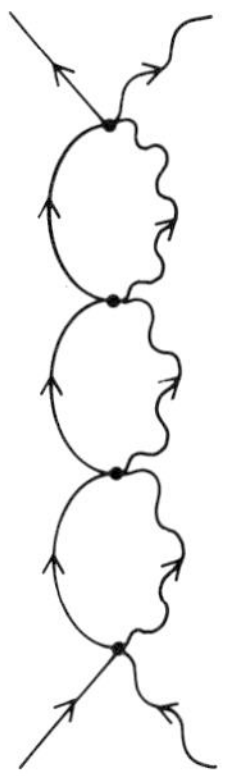

Fig. 48. Iteration of the double-bond structure of Fig. 41.

scattering is canceled completely by the corresponding propensity for double Thomson scattering to emit an arbitrarily soft photon, at least to leading order in the fine-structure constant. This cancellation of infrared divergences was first observed over 30 years ago,[12] although it was not then possible to also eliminate the associated ultraviolet divergences without a completely covariant formulation of QED. Bethe and Oppenheimer[13] first pointed out that this cancellation might occur to all orders in α, and the modern proof of this conjecture should be credited to Jauch and Rohrlich.[14] Although the proof has only been given for covariant QED, we shall assume that the corresponding phenomenon is implied for the nonrelativistic theory. In the following section a further explicit example of canceling infrared divergences will be studied.

There are two important points to be noted regarding the above result, the first of which is that the value of k_m must be determined by the energy resolution of any experiment designed to verify Eq. (15-23); the charged particle does not have infinite space–time available to it, so that the laboratory cuts off long-wavelength photons. Secondly, the entire second term in the brackets is of order (v^2/c^2), and thus is actually a relativistic correction to the classical Thomson cross section. Consequently, it appears that there are really *no* radiative corrections to Thomson scattering in the nonrelativistic limit. This is a special case of a more general result first found by Thirring,[15] who proved the statement exactly.

Up to this point we have discussed only the quantum mechanical radiative corrections to Thomson scattering, as exhibited by Eq. (15-23). But, as mentioned at the beginning of this section, there are also corrections to this process which correspond to the classical radiation damping. It is realized that such corrections are contained in the present theory when one notices that we have overlooked diagrams of the type shown in Fig. 41, which, in its overall aspects, resembles Thomson scattering. This diagram could also be interpreted as a self-energy diagram, as discussed in Section XIV-B, and we shall see that this aspect of such diagrams is also contained in the present analysis.

Unfortunately, the diagram of Fig. 41, which corresponds to the divergent expression (14-1), is only a sample of the difficulty being uncovered. The diagram can be iterated, and therefore the divergence also, as indicated by the structure of Fig. 48. This diagram, too, possesses the overall characteristics of the Thomson scattering process. Consequently, it is necessary to consider all such diagrams, and it is most convenient to consider their sum. In fact, it is this procedure of summing these diagrams together which eliminates the divergences. Such a technique has become known as *peratization*,[16] and we represent it diagrammatically in Fig. 49.

In covariant QED an infinite series of such diagrams is most easily studied by means of an integral equation formulation, resulting in the so-called Heitler integral equation.[17] However, for the present purposes this elegant treatment is unnecessary, because the nonrelativistic terms in the series are so simple that we can actually sum the series in a straightforward manner. It is by now a simple matter to write down the expression for each diagram in the series of Fig. 49, and we find the corresponding analytical expression for the sum to be

$$
\begin{aligned}
S = & -2\pi i\,\delta(\omega_{\mathbf{p}_2} + \hbar c k_2 - \omega_{\mathbf{p}_1} - \hbar c k_1) \\
& \times \delta_{\mathbf{p}_1+\mathbf{k}_1,\,\mathbf{p}_2+\mathbf{k}_2}\,\delta_{n_1 n_2}\,\frac{2\pi\alpha\hbar^2 Z^2/mV)}{(k_1 k_2)^{1/2}}\left\{(\epsilon_1\cdot\epsilon_2) + \varepsilon Z^2\left(\frac{2\pi\alpha\hbar^2}{mV}\right)\right. \\
& \times \sum_{l_3} \frac{(\epsilon_2\cdot\epsilon_3)(\epsilon_3\cdot\epsilon_1)}{l_3[\omega(\mathbf{p}_1+\mathbf{k}_1-\boldsymbol{l}_3) + \hbar c l_3 - \hbar c k_1 - \omega(\mathbf{p}_1)]} + \varepsilon^2 Z^4\left(\frac{2\pi\alpha\hbar^2}{mV}\right)^2 \\
& \times \sum_{l_3 l_4} \frac{(\epsilon_2\cdot\epsilon_4)(\epsilon_4\cdot\epsilon_3)}{l_3 l_4[\omega(\mathbf{p}_2+\mathbf{k}_2-\boldsymbol{l}_3) + \hbar c l_3 - \hbar c k_1 - \omega(\mathbf{p}_1)]} \\
& \times \left.\frac{(\epsilon_3\cdot\epsilon_1)}{[\omega(\mathbf{p}_2+\mathbf{k}_2-\boldsymbol{l}_4) + \hbar c l_4 - \hbar c k_1 - \omega(\mathbf{p}_1)]} + \cdots\right\} \\
\simeq & -2\pi i\,\delta(\omega_{\mathbf{p}_2} + \hbar c k_2 - \omega_{\mathbf{p}_1} - \hbar c k_1)\,\delta_{\mathbf{p}_1+\mathbf{k}_1,\,\mathbf{p}_2+\mathbf{k}_2}\,\delta_{n_1 n_2}\,\frac{(\epsilon_1\cdot\epsilon_2)}{(k_1 k_2)^{1/2}}\,\frac{2\pi\alpha\hbar^2 Z^2}{mV} \\
& \times \left[1 - \left(\varepsilon Z^2\,\frac{2\alpha\hbar^2}{3\pi m}\right)\int_0^{\bar{\lambda}_C^{-1}} \frac{l_3^2\,dl_3}{\hbar c l_3(l_3 - k_1)}\right. \\
& + \left.\left(\varepsilon Z^2\,\frac{2\alpha\hbar^2}{3\pi m}\right)^2 \int_0^{\bar{\lambda}_C^{-1}} \frac{l_3^2\,dl_3}{\hbar c l_3(l_3 - k_1)} \int_0^{\bar{\lambda}_C^{-1}} \frac{l_4^2\,dl_4}{\hbar c l_4(l_4 - k_1)} + \cdots\right],
\end{aligned}
\tag{15-24}
$$

in the nonrelativistic approximation. In obtaining this last form for S we have used the following result.

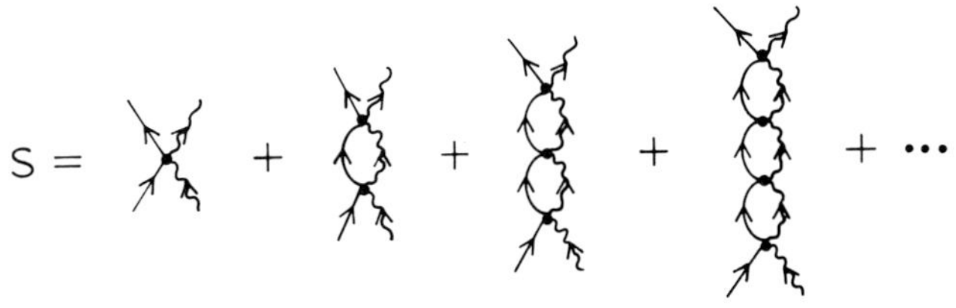

Fig. 49. Diagrammatic representation of the sum in Eq. (15-24).

Exercise. With the aid of Eq. (11-64), show that

$$\sum_{l_3 l_4 \cdots l_n} \frac{(\epsilon_2 \cdot \epsilon_n)(\epsilon_n \cdot \epsilon_{n-1}) \cdots (\epsilon_3 \cdot \epsilon_1)}{\hbar c l_3(l_3 - k_1)\hbar c l_4(l_4 - k_1) \cdots \hbar c l_n(l_n - k_1)}$$
$$= \left(-\frac{V}{3\pi^2}\right)^n \int_0^\infty \frac{l_3^2\, dl_3}{\hbar c l_3(l_3 - k_1)} \cdots \int_0^\infty \frac{l_n^2\, dl_n}{\hbar c l_n(l_n - k_1)}, \tag{15-25}$$

when all the sums are converted to integrals.

Thus, we can now write for the sum

$$S \simeq -2\pi i\, \delta(\omega_{\mathbf{p}_2} + \hbar c k_2 - \hbar c k_1 - \omega_{\mathbf{p}_1})\, \delta_{\mathbf{p}_1+\mathbf{k}_1,\, \mathbf{p}_2+\mathbf{k}_2}\, \delta_{n_1 n_2}$$
$$\times \frac{(\epsilon_1 \cdot \epsilon_2)}{(k_1 k_2)^{1/2}} \frac{2\pi\hbar^2 \alpha Z^2}{mV} X(k_1)\,, \tag{15-26}$$

where

$$X(k_1) \equiv \sum_{n=0}^{\infty} \left(-\varepsilon Z^2 \frac{2\alpha\hbar^2}{3\pi m}\right)^n \left(\int_0^{\lambda_C^{-1}} \frac{x^2\, dx}{\hbar c x(x - k_1)}\right)^n$$
$$\equiv \frac{1}{1 + A(k_1)} \tag{15-27}$$

and

$$A(k_1) \equiv \varepsilon Z^2 \frac{2\alpha\hbar}{3\pi m c} \int_0^{\lambda_C^{-1}} \frac{x\, dx}{x - k_1}\,. \tag{15-28}$$

Now, one must be quite careful in evaluating the integral in Eq. (15-28). Formally,

$$I(15\text{-}28) = \lambda_C^{-1}\left[1 + k_1\lambda_C \ln\left(1 - \frac{1}{k_1\lambda_C}\right)\right]. \tag{15-29}$$

For certain values of $k_1\lambda_C$ the argument of the logarithm may be a negative real number, in which case one must observe that

$$\ln x = \ln|x| - i\pi, \qquad x < 0. \tag{15-30}$$

With this observation we see that

$$I(15\text{-}28) = O(k_1/k_1^2\lambda_C^2) \ll 1\,, \qquad k_1\lambda_C \gg 1\,;$$
$$= \lambda_C^{-1} - i\pi k_1\,, \qquad k_1\lambda_C \ll 1\,.$$

Therefore, in the nonrelativistic limit

$$A(k_1) = \varepsilon Z^2\left(\frac{2}{3}\frac{\alpha}{\pi} - i\,\frac{2\alpha\hbar k_1}{3mc}\right),$$

and for the electron

$$X(k_1) \simeq 1/(1 + i\tau_0 ck_1)\,, \tag{15-31}$$

where $\tau_0 \equiv 2e^2/3mc^3$ was defined in Eq. (7-2), and we have neglected α compared to unity in the real part of the denominator.

When Eq. (15-31) is substituted into Eq. (15-26), the scattering amplitude can be identified in the usual way, and one obtains precisely the differential scattering cross section of Eq. (7-64) for the classical radiation damping effect in Thomson scattering. Thus, while each diagram in Fig. 49 corresponds to a quantum mechanical process, the sum contains both a classical (imaginary) and a quantum mechanical (real) part. This latter part, however, corresponds precisely to mass renormalization when the classical term is neglected.

In Section VII-C it was observed that the effect of radiation damping was essentially negligible, because it was only significant for very-high-energy photons. The same comment applies here, since an observable effect of this kind is outside the region of validity of our theory. However, it turns out that such damping effects are *always* negligible, even from the point of view of covariant QED, and are probably unobservable. This was first demonstrated by Wilson.[18]

C. RADIATIVE CORRECTIONS TO COULOMB SCATTERING

The Coulomb scattering of two charged particles can also be accompanied by the emission and subsequent reabsorption of one or more photons, which leads to radiative corrections to the Rutherford cross section. We shall now attempt to calculate this corrected cross section, and, in order to simplify the problem, it is convenient to focus on electron scattering from a proton. There is then no exchange term, the relevant lowest-order diagram is that of Fig. 50, and again we do not consider those diagrams involved in simple mass renormalization.

Following the same procedure we have been using throughout, we find the amplitude for this process to be

$$(\mathbf{R}_{\mathbf{ba}})_{\mathrm{rad}} = 2\pi e^2\left(\frac{2\pi\alpha}{V^2}\right)\frac{\hbar^2}{m^2c^2}\,|\,\mathbf{p}_1 - \mathbf{p}_3\,|^{-1}\sum_{l_3}\frac{-2(\mathbf{p}_1\cdot\epsilon_3)(\mathbf{p}_3\cdot\epsilon_3)}{l_3^2}\,, \tag{15-32}$$

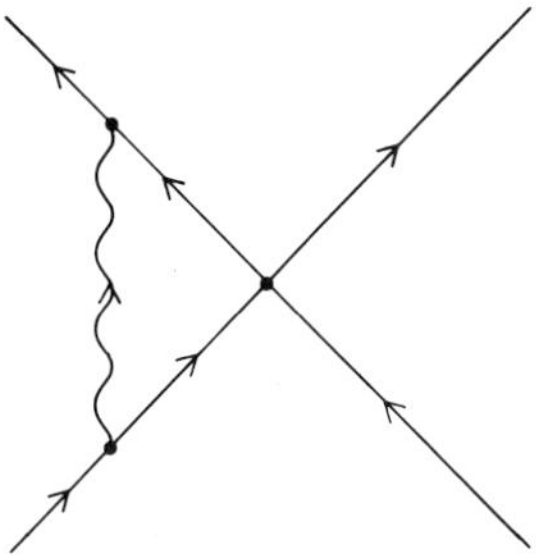

Fig. 50. Leading-order radiative correction to Coulomb scattering.

in the nonrelativistic approximation. We have also summed over final and averaged over initial spin states of the electron and proton. The recoil of the proton can be neglected nonrelativistically, which means that we cannot strictly conserve momentum in the process; i.e., the proton is assumed to remain stationary throughout the interaction. It is also convenient to adopt the center-of-mass coordinate system defined by Eqs. (12-27) and (12-28). Thus, we can write

$$\begin{aligned} -2(\mathbf{p}_1 \cdot \epsilon_3)(\mathbf{p}_3 \cdot \epsilon_3) &= (\mathbf{q} \cdot \epsilon_3)^2 - (\mathbf{p}_2 \cdot \epsilon_3)^2 - (\mathbf{p}_4 \cdot \epsilon_3)^2 \\ &\simeq (\mathbf{q} \cdot \epsilon_3)^2 , \end{aligned} \tag{15-33}$$

where $\mathbf{q} \equiv \mathbf{p}_1 - \mathbf{p}_3$ is the momentum transferred in the process. The total amplitude for the process of Fig. 50 is then the sum of that for the usual Coulomb scattering,

$$(\mathbf{R}_{\mathbf{ba}})_{\mathrm{C}} = -\frac{4\pi e^2}{|\mathbf{p}_1 - \mathbf{p}_3|^2} \frac{1}{V},$$

and that of Eq. (15-32). Using Eq. (15-33), we find for the differential cross section

$$\left(\frac{d\sigma}{d\Omega}\right)_{\mathrm{C,rad}} \simeq \left(\frac{d\sigma}{d\Omega}\right)_{\mathrm{C}} - \frac{1}{\pi^2}\frac{\alpha^3}{q^4}\int_a^{\lambda_{\mathrm{C}}^{-1}} \frac{dl}{l} \int d\Omega \sum_i (\mathbf{q} \cdot \epsilon^i)^2 , \tag{15-34}$$

to leading order in α. We have again introduced the mathematical cutoff a to emphasize the indicated infrared divergence, and $(d\sigma/d\Omega)_{\mathrm{C}}$ is the Rutherford cross section.

Equation (15-34) exhibits the infrared problem again, and it is now necessary to recall the discussion at the end of Section XII-B. There we saw that the *Bremsstrahlung* process could result in an infrared photon. As pointed out in the last section, a very-low-energy photon cannot be observed below some finite experimental resolution, so that we can cut off the infrared part of the *Bremsstrahlung* spectrum for $k_1 < k_m$. One now integrates over the photon momentum from a to k_m and, in this case, also over all angles of the emitted photon. The (unobservable) cross section for infrared *Bremsstrahlung* is then

$$\left(\frac{d\sigma}{d\Omega}\right)_{\mathrm{B,ir}} \simeq \frac{1}{\pi^2}\frac{\alpha^3}{q^4}\int_a^{k_m}\frac{dl}{l}\int d\Omega \sum_i (\mathbf{q}\cdot\epsilon^i)^2 . \tag{15-35}$$

Comparison of Eqs. (15-34) and (15-35) thus shows that the infrared divergence is canceled when the two results are combined. One can now do the indicated integrals trivially, and we obtain for the Coulomb cross section with leading-order radiative correction

$$\frac{d\sigma}{d\Omega} \simeq \left(\frac{d\sigma}{d\Omega}\right)_{\mathrm{C}}(1-\delta), \tag{15-36}$$

with

$$\begin{aligned}\delta &= \frac{2\alpha}{3\pi}\left(\frac{\hbar}{mc}\right)^2 |\mathbf{p}_1 - \mathbf{p}_3|^2 \ln\frac{mc^2}{\hbar c k_m}\\ &= \frac{8\alpha}{3\pi}\left(\frac{v}{c}\right)^2 \sin^2(\theta/2)\ln\frac{mc^2}{\hbar c k_m}.\end{aligned} \tag{15-37}$$

The correction δ is seen to actually be of order (v^2/c^2), and therefore relativistic. To this order, Eq. (15-37) is not completely correct, because there is a constant associated with the logarithm which arises from polarization of the vacuum and the covariant cutoff procedure necessary in a correct relativistic treatment. To order (v^2/c^2), the logarithm should be replaced by[19]

$$\ln\frac{mc^2}{2\hbar c k_m} + \frac{19}{30}. \tag{15-38}$$

It is, perhaps, worth inquiring at this point as to what may be expected from the typical solutions (15-23) and (15-37) to the infrared problem in the hypothetical limit of perfect experimental resolution. It can be shown[20] that in the limit $(\hbar c k_m) \to 0$ the cross sections for all the processes we have considered go over into the usual cross sections for elastic scattering. However, it appears that in order to explicitly demonstrate this behavior, one must examine the complete solution of the infrared problem from the standpoint of covariant QED.[14] The basic reason for the cancellation of the infrared divergences is that the amplitudes for emission and absorption of an infrared photon turn out to be the same, which follows from a theorem due to Lee and Nauenberg.[21]

PROBLEMS

15-1. Examine the analytical expressions associated with the diagrams of Figs. 46 and 47 and demonstrate the validity of Eqs. (15-20).

REFERENCES

1. W. E. Lamb and R. C. Retherford, *Phys. Rev.* **72,** 241 (1947); **75,** 1325 (1949); **79,** 549 (1950); **81,** 222 (1951); **85,** 259 (1952); **86,** 1014 (1952).
2. H. A. Bethe, *Phys. Rev.* **72,** 339 (1947).
3. H. A. Bethe, L. M. Brown, and J. R. Stehn, *Phys. Rev.* **77,** 370 (1950).
4. S. Triebwasser, E. S. Dayhoff, and W. E. Lamb, *Phys. Rev.* **89,** 98, 106 (1953). For the most recent experimental result see the article by R. T. Robiscoe and B. L. Cosens, *Phys. Rev. Letters* **17,** 69 (1966).
5. S. B. Crampton, D. Kleppner, and N. Ramsey, *Phys. Rev. Letters* **11,** 338 (1963).
6. S. D. Drell and J. D. Sullivan, *Phys. Rev.* **154,** 1477 (1967).
7. B. D. Josephson, *Phys. Letters* **1,** 251 (1962); *Advan. Phys.* **14,** 419 (1965).
8. W. H. Parker, B. N. Taylor, and D. N. Langenberg, *Phys. Rev. Letters* **18,** 287 (1967).
9. M. Goeppert-Mayer, *Naturwissenschaften* **17,** 932 (1929); *Ann. Physik* **9,** 273 (1931).
10. J. E. Hooper and D. T. King, *Phil. Mag.* **41,** 1194 (1950).
11. I. F. Boekelheide, Ph. D. Thesis, State Univ. of Iowa, Ames, Iowa, 1952; see also, P. E. Cavanaugh, *Phys. Rev.* **87,** 1131 (1952).
12. F. Bloch and A. Nordsieck, *Phys. Rev.* **52,** 54 (1937); W. Pauli and M. Fierz, *Nuovo Cimento* **15,** 1967 (1938); W. Braunbeck and E. Weinmann, *Z. Physik* **110,** 360 (1938).
13. H. A. Bethe and J. R. Oppenheimer, *Phys. Rev.* **70,** 451 (1946); see also, R. Jost, *ibid.* **72,** 815 (1947).
14. J. M. Jauch and F. Rohrlich, *Helv. Phys. Acta* **27,** 613 (1954); F. Rohrlich, *Phys. Rev.* **98,** 181 (1955). For a more recent treatment, see D. R. Yennie, S. C. Frautschi, and H. Suura, *Ann. Phys.* (*N. Y.*) **13,** 379 (1961); K. T. Mahanthappa, *Phys. Rev.* **126,** 329 (1962).
15. W. Thirring, *Phil. Mag.* **41,** 1193 (1950).
16. See, e.g., G. Feinberg and A. Pais, *Phys. Rev.* **131,** 2724 (1963).
17. W. Heitler, *Proc. Cambridge Phil. Soc.* **37,** 291 (1941); see also, I. Waller, *Z. Physik* **88,** 436 (1934).
18. A. H. Wilson, *Proc. Cambridge Phil. Soc.* **37,** 301 (1941).
19. See, e.g., J. M. Jauch and F. Rohrlich, "The Theory of Photons and Electrons," Section 15.2. Addison-Wesley, Reading, Massachusetts, 1955.
20. J. M. Jauch and F. Rohrlich, "The Theory of Photons and Electrons," Section 16.1. Addison-Wesley, Reading, Massachusetts, 1955.
21. T. D. Lee and M. Nauenberg, *Phys. Rev.* **133B,** 1549 (1964).

XVI COVARIANT QUANTUM ELECTRODYNAMICS

It is in order at this point to summarize briefly the program carried out in the preceding pages. As we have seen, the classical theory of charged particles and electromagnetic fields encountered difficulties which eventually led to the theory of relativity. Owing to the finite value of the speed of light, it became necessary to construct a theory in terms of Lorentz frames of reference. While this theory was successful in treating electromagnetic wave propagation, it, too, led to inconsistencies in the ensuing attempts to formulate theories of charged particles. In particular, great difficulties arose in connection with the structure of these particles. Not until it was realized that a completely relativistic formulation of the theory of charged particles and their interactions with their own and external fields was needed did a satisfactory theory emerge. Moreover, it was necessary to also realize that particle structure was really ouside the domain of classical physics. This work culminated in the Lorentz–Dirac theory of the point electron, which provides a satisfactory classical description.

The second stage in this program was the reformulation of classical electrodynamics in terms of the Hamiltonian description of classical mechanics. A notable point in this development was the nonrelativistic nature of the theory; because the Lagrangian, and therefore the Hamiltonian, is a function of instantaneous particle velocities and coordinates, and because the potentials depend on retarded times, a classical relativistic canonical formulation faces formidable difficulties. These conceptual problems are alleviated by the realization that a quantum mechanical treatment is quite necessary, both to put charged particles and light waves on more or less the same level and to make an attempt at an understanding of the structure of charged particles. The former step is quite necessary to radiation theory, because it leads directly to the notion of photon.

Thus, in the ensuing chapters the quantum theory of electrodynamics was developed from a nonrelativistic point of view, which permitted a certain degree of simplicity and yet led to a good understanding of the essential physics. It was then a fairly straightforward matter to construct the needed many-body theory with which to describe quantum electrodynamics (QED), and many important physically observable quantities could then be calculated. Agreement with experiment on this level appears to be quite good, and the goals established at the beginning have been reached.

Nevertheless, the theory developed up to this point is incapable of describing the considerable high-energy phenomena observable in nature. Therefore, the next step is to generalize this theory to the relativistic domain. While that is not a goal of this book, it is appropriate to indicate the way in which the foregoing work forms a sound basis for this generalization and to mention some of the high-energy processes to which such a treatment gives understanding. We shall outline in this chapter the way in which one must generalize the preceding ideas to obtain a manifestly covariant field theory. It is of interest to study qualitatively several intrinsically relativistic processes, as well as the covariant forms of the equations derived in previous chapters. Finally, we shall give a brief discussion of the present state of QED.

It should be emphasized at this time, however, that the following sections are not meant to be, nor do they succeed in being, a scholarly and exhaustive review of quantum electrodynamics. Rather, the purpose is to provide a kind of transition to the next stage of studying charged particles and their radiation fields, and to stimulate an interest in pursuing further an investigation of relativistic quantum field theory.

A. MANIFESTLY COVARIANT FORMULATION OF QED

The methods of calculation and the *formal* approach to covariant QED are very similar to those of the previous chapters, but the relativistic nature of the procedures leads to important and extensive modifications. One, of course, can no longer employ the Schroedinger equation to describe the motion of the electron. Rather, it is the relativistic Dirac equation[1] which is now correct, at least for fermions, and the consequent electron-positron theory. Not only does the Dirac equation introduce the notion of particle spin in a correct way, but it also leads to the concept of antiparticle. While the positive-energy solutions correspond to the electron, the negative-energy solutions are to be associated with a particle in every way similar to the electron, but with positive charge. This is referred to as the antiparticle of the electron, called a *positron*. Thus, there is already an extra particle to deal with in the relativistic theory.

In Chapter X we developed the Fock space techniques, and these must now be formulated in a covariant manner. The process of second quantization becomes very essential, in that the Dirac equation now governs the Dirac *field*, and QED becomes a theory of interacting fields. This is rather necessary because it is at this point that the calculational procedures change significantly. Although any relativistic theory may be invariant under both Lorentz and gauge transformations, it is conceivable that the inherent divergences of the theory may result in a loss of covariance when one chooses a particular coordinate system or gauge in which to perform a calculation. It became clear that a theory was needed which manifested covariance at each stage of the calculation, and, in different ways, such a theory was constructed by Tomonaga,[2] Schwinger,[3] and Feynman.[4] In this way, one was able to evaluate the equations in a systematic manner devoid of divergences. There is probably no better history of the tortuous paths traversed in arriving at this theory than the Nobel lectures of the above physicists,[5] which should be read by every serious student of QED.

One now proceeds essentially as in Chapter XI, performing an S-matrix expansion in the interaction picture and making a diagrammatic identification of each resulting term, but in a manifestly covariant way. The diagrams, of course, are Feynman diagrams and are to be evaluated by means of a set of rules somewhat different than those we introduced. All of this becomes considerably more complicated, and the details can be found in any one of a number of excellent books devoted primarily to covariant field theory.[6–12]

The major difference in the formalism which one notices immediately is the qualitative difference in the diagrams and the new types of diagram now allowed, corresponding to many new kinds of physical process. Because of the relativistic character of the theory, high-energy interactions are permissible which were prohibited in the preceding theory. For example, the restriction (12-1) is no longer necessary, and the production of electron–position pairs becomes energetically possible. In the presence of an external field, e.g., the field of a nucleus, a photon can create an electron–positron pair, and the pair can then annihilate into two or more photons. At higher energies particle–antiparticle pairs of even more massive particles can be produced if they have a coupling to the electromagnetic field. Moreover, such processes can also occur virtually in intermediate states, and so the numbers and types of diagrams are increased appreciably in the covariant theory. The concepts of creation and annihilation of particles with mass, introduced in Chapter X, now indeed correspond to physically real processes.

An additionally important qualitative and quantitative difference between the covariant and nonrelativistic theories is that now the vacuum acquires a greater significance. First of all, vacuum–vacuum transitions arise, the diagrams for which are characterized by the lack of any external lines. These processes are referred to collectively as *fluctuations of the vacuum.* One sees that the state of no particles is *not* an eigenstate of QED, but the vacuum can be expanded into all possible states of noninteracting particles. These diagonal matrix elements of the vacuum also lead to unconnected and zero-momentum-transfer diagrams such as those encountered in Chapter XI, and they all correspond to infinite mathematical expressions. Nevertheless, these effects are totally unobservable, and, if desired, can be formally eliminated by means of a phase transformation of the S matrix. If one adds together the infinite set of all such diagrams, the sum obtained is merely a phase factor; since one is always interested only in the square of the matrix elements, even the fact that the phase is infinite has no observable consequences.

Second, and more important physically, the vacuum can be polarized by an external field, and it is precisely this *vacuum polarization* which gives the vacuum physical significance. As an example, we consider the scattering of an electron by an external field. During the action of the field on the electron, an electron–positron pair can be created and then annihilate itself in an intermediate state because the field has polarized the vacuum. This is the lowest-order effect, and there exist infinitely many such higher-order processes. The resulting mathematical expression is divergent, but this divergence can be removed (see below) and the remaining finite part is observable. One interprets this polarization as a spreading out of the potential over a region whose radius is of the order of an electron Compton wavelength. For instance, in a hydrogen atom the electron sees not the Coulomb potential of the proton, but the form perturbed due to vacuum polarization:

$$e^2\left[\frac{1}{r} - \frac{\alpha}{15\pi m^2}\,4\pi\,\delta(\mathbf{r})\right]. \tag{16-1}$$

This spreading of the point charge is known as the Uehling effect,[13] after Uehling, who first calculated the leading correction term of Eq. (16-1). This effect reduces the Lamb shift in hydrogen by about 27 MHz and, since the total shift has been measured to much higher accuracy, one feels very certain that vacuum-polarization effects are real.

Several formal difficulties still arise in the covariant theory, some of which lead to important observable effects, and others which do not. Among the latter there is a problem which arises concerning the polarization vector of the photon, now a Lorentz 4-vector. Although one can construct a relativistically invariant formalism in the Coulomb gauge, the covariance is not manifest, and the Coulomb interaction is not separated from the transverse waves in all reference frames. On the other hand, when the Lorentz gauge is used in a straightforward manner, one obtains quantities corresponding to both longitudinal and timelike photons, and in the first formulations it was necessary to remove these by means of an additional subsidiary condition which is not at all manifestly covariant. Moreover, these objects are important to the formal description of quantities such as the Coulomb interaction. In 1950 Gupta[14] and Bleuler[15] developed a formalism which resolved this difficulty by retaining the longitudinal and timelike photons in the theory, but which explicitly demonstrates that their contributions to physical quantities exactly cancel one another, and therefore do not correspond to any observable effects (in the free field).

As is the case with nonrelativistic QED, the covariant theory is plagued with divergences. Now, however, the manifest covariance is precisely what permits a systematic removal of these divergences to all orders, and the remaining finite contributions to physical quantities are observable and constitute the major success of QED. The self-energy problem arises in exactly the same way as in Chapter XIV, and the mass-renormalization program must be carried out as indicated there. In addition, there are divergences arising in connection with vacuum polarization which require a further renormalization in terms of the charge. Actually, a charge renormalization even becomes necessary when treating the self-energy problem. Dyson[16] showed that this program can be carried out to all orders, and that the resulting theory is then completely free of these divergences.

The infrared problem, of course, still persists, and is treated exactly as outlined in the last chapter and the references cited there. One further divergence difficulty of interest concerns the possibility of photon self-energies, which are now energetically allowed (i.e., virtual pair production). These photon self-energy parts are indeed divergent, but generally result from failure to maintain gauge invariance in the calculations (i.e., render it manifest!). When gauge invariance is invoked everywhere,[17] it is found that the photon self-energies contribute no observable effects to the theory, and the mass of the photon is identically zero. The experimental situation regarding the photon mass has recently been reviewed by Kobzarev and Okun,[18] who conclude that the Compton wavelength of the photon must exceed 30,000 km.

As a final point in this brief summary of the formal covariant properties of QED, let us return to the phenomenon of high-energy pair production. We can define "pure" QED as the theory of interacting electrons, positrons, and, perhaps, muons (see below). Unfortunately, QED is not really "pure," because *any* antiparticle pairs coupled to the electromagnetic field can be produced either actually or virtually. Consequently, in higher order it is necessary to understand the other interactions among these particles, known as the strong and weak interactions. The coupling constants for these phenomena are large, and an ordinary perturbation expansion for studying them is not at all useful; one must resort to several ingenious techniques, none of which has been an unqualified success. It is quite fortunate that these interactions enter electromagnetically only in very high order, so that the expansion in the weak coupling constant α has led to great success. Nevertheless, quantum electrodynamics cannot be considered a completely satisfactory theory until the weak and strong interactions themselves are well understood, and, in this sense, it is only "pure" QED which, with all its conceptual shortcomings, represents a quantum field theory *par excellence.*

B. RELATIVISTIC FORMS OF SOME PREVIOUS RESULTS

In the last few chapters we obtained the nonrelativistic descriptions of several charged-particle systems, along with their radiative properties. The resulting equations were valid only for particle speeds $v \ll c$. Although we are in no position to derive the corresponding expressions for high energies, it does seem appropriate to tabulate some of the results here so as to provide the reader with a comparison. Generally, correction terms to the highest order to be found in the literature are not included, but references are provided where appropriate.

The calculation of the Thomson cross section in Section XII-A explicitly accounted for the nonrelativistic nature of the process by requiring that the frequencies of initial and scattered photons be equal. On the other hand, we saw in Problem 3-1 that correct relativistic kinematics predicts a shift in frequency for the scattered electron, known as the Compton effect. Klein and Nishina[19] first calculated the correct relativistic cross section for this process, and we shall exhibit it only for the case in which the photon polarizations are not observed. If primes refer to the scattered photon, the differential cross section is

$$\frac{d\sigma}{d\Omega} = \frac{r_0^2}{4}\left(\frac{\omega'}{\omega}\right)^2\left(\frac{\omega}{\omega'} + \frac{\omega'}{\omega} - \sin^2\theta\right), \tag{16-2}$$

where θ is the scattering angle in the rest frame of the initial electron, and

$$\omega'/\omega = [1 + (\hbar\omega/mc^2)(1 - \cos\theta)]^{-1}. \tag{16-3}$$

An angular integration yields for the total cross section

$$\sigma = 2\pi r_0^2\left\{\frac{1+x}{x^3}\left[\frac{2x(1+x)}{1+2x} - \ln(1+2x)\right] + \frac{\ln(1+2x)}{2x} - \frac{1+3x}{(1+2x)^2}\right\}, \tag{16-4}$$

where $x \equiv \hbar\omega/mc^2$. One notes that this last expression goes over to Eq. (6-22) in the extreme relativistic limit. Radiative corrections to Compton scattering were discussed briefly in Chapter XV, and were first treated in detail by Brown and Feynman.[20] Finally, the major differences in angular distributions for the relativistic and nonrelativistic cases are illustrated in Fig. 20 (Chapter VI).

In the *Bremsstrahlung* process one is interested experimentally in high-energy incident electrons, and therefore in the relativistic form of Eqs. (12-23) and (12-24). We shall employ essentially the same notation as in Section XII-B, assign primes to scattered quantities, and note that all angles are between particle momenta and the photon wave vector. It is useful to define $\mathbf{q} = \mathbf{p} - \mathbf{p}' - \mathbf{k}$, $y = \gamma mc$, and the usual β and γ of special relativity (Chapter I). The relativistic Bethe–Heitler formula[21] is then

$$d\sigma = \frac{\alpha Z^2 r_0^2}{(2\pi)^2} \frac{(mc)^2}{|\hbar\mathbf{q}|^4} \left(\frac{p'}{p}\right) \frac{dk}{k}\, d\Omega'\, d\Omega_k$$
$$\times \left[\frac{\beta'^2 \sin^2\theta'}{(1-\beta'\cos\theta')^2}(4y^2 - \hbar^2 q^2) + \frac{\beta^2\sin^2\theta}{(1-\beta\cos\theta)^2}(4y'^2 - \hbar^2 q^2) \right.$$
$$- 2\,\frac{\beta\beta'\sin\theta'\sin\theta\cos\phi}{(1-\beta'\cos\theta')(1-\beta\cos\theta)}(4yy' - \hbar^2 q^2 + 2\hbar^2 k^2)$$
$$\left. + 2\hbar^2 k^2 \frac{(\gamma'/\gamma)\beta'^2\sin^2\theta' + (\gamma/\gamma')\beta^2\sin^2\theta}{(1-\beta'\cos\theta')(1-\beta\cos\theta)} \right]. \tag{16-5}$$

The extreme relativistic limit is of great importance, and the angular integrations can be readily done in that case to give

$$d\sigma = 4\alpha Z^2 r_0^2 \frac{dk}{k} \left[1 + \left(\frac{\gamma'}{\gamma}\right)^2 - \frac{2\gamma'}{3\gamma}\right] \left[\ln \frac{2mc\gamma\gamma'}{\hbar k} - \frac{1}{2}\right]. \tag{16-6}$$

The infrared problem for *Bremsstrahlung* was treated in Chapter XV, and electron screening is discussed by Bethe and Ashkin.[22] Again, comparison of the high- and low-energy results can be seen in Fig. 39. A detailed discussion of the theory and its agreement with experiment has been given by Koch and Motz,[23] and more recently by Bogdankevich and Niolaev.[24]

For electron scattering from a Coulomb field the nonrelativistic cross section is given by the first term of Eq. (12-31), the Rutherford formula. At high energies there are additional factors due to both spin and relativistic corrections, and these were first obtained correctly by McKinley and Feshbach.[25] In the second Born approximation

$$\frac{d\sigma}{d\Omega} = \frac{Z^2 r_0^2}{4} \frac{1}{(\beta^2\gamma)^2 \sin^4(\theta/2)}$$
$$\times \{1 - \beta^2\sin^2(\theta/2) + \alpha Z\beta\pi \sin(\theta/2)[1 - \sin(\theta/2)]\}, \tag{16-7}$$

and the second term in the brackets is a spin correction. For large values of αZ this result is no longer valid and the calulation must be redone using Coulomb wave functions. These and other details, as well as the experimental situation, are discussed by Corson and Hanson,[26] and the leading-order radiative corrections are exhibited in Eqs. (15-37) and (15-38).

Particle–particle Coulomb scattering is somewhat more complicated relativistically, depending on whether or not an antiparticle is involved. For two electrons, or two positrons, the process is known as Møller scattering,[27] and in the laboratory system the cross section is

$$\frac{d\sigma}{d\Omega} = r_0^2\left(4\frac{\gamma+1}{\beta^2\gamma}\right)^2 \frac{\cos\theta}{[2+(\gamma-1)\sin^2\theta]^2} \times\left[\frac{4}{(1-x^2)^2} - \frac{3}{1-x^2} + \frac{(\gamma-1)^2}{4\gamma^2}\left(1+\frac{4}{1-x^2}\right)\right], \tag{16-8}$$

where the polarization is not observed, and x is defined below, Eq. (16-10). Radiative corrections have been calculated through $O(\alpha)$ by both Tsai[28] and Anders,[29] and experiments[30] indicate that these remain very small contributions up to 500 MeV.

Electron–positron scattering is known as Bhabha scattering,[31] and for unpolarized particles in the laboratory system in which the electron is initially at rest

$$\begin{aligned}\frac{d\sigma}{d\Omega} = r_0^2\left(2\frac{\gamma+1}{\beta^2\gamma}\right)^2 &\frac{\cos\theta}{[2+(\gamma-1)\sin^2\theta]^2} \\ \times\Bigg\{&\frac{4}{(1-x)}\left[1-\frac{\gamma^2-1}{2\gamma^2}(1-x)+\frac{1}{2}\left(\frac{\gamma-1}{2\gamma}\right)^2(1-x)^2\right] \\ &-\frac{2}{1-x}\left(\frac{\gamma-1}{\gamma+1}\right)\left[\frac{2\gamma+1}{\gamma^2}+\frac{\gamma^2-1}{\gamma^2}x+\left(\frac{\gamma-1}{2\gamma}\right)^2(1-x)^2\right] \\ &+\left(\frac{\gamma-1}{\gamma+1}\right)^2\left[\frac{1}{2}+\frac{1}{\gamma}+\frac{3}{2\gamma^2}-\left(\frac{\gamma-1}{2\gamma}\right)(1-x)^2\right]\Bigg\},\end{aligned} \tag{16-9}$$

with

$$x = \frac{2-(\gamma+3)\sin^2\theta}{2+(\gamma-1)\sin^2\theta}. \tag{16-10}$$

The Møller and Bhabha processes are not as dissimilar as the forms of the cross sections would suggest. In the former the exclusion principle simplifies matters for the identical particles, whereas in the latter the charges render the particles distinguishable. Moreover, in electron–positron scattering virtual pair annihilation and pair creation can mediate the process, and, while kinematically equivalent to the exchange effect in Møller scattering, the pair effect leads to a slightly more extensive expression. Experimental confirmation of Bhabha scattering has been provided by Ashkin *et al.*[32]

The total cross section for the high-energy, K-shell photoelectric effect was first calculated by Sauter[33]:

$$\sigma_p = \frac{3}{2} Z^5\alpha^4 \left(\frac{mc}{\hbar k}\right)^5 \sigma_T(\gamma^2 - 1)^{3/2} \left\{\frac{4}{3} + \frac{\gamma(\gamma - 2)}{\gamma + 1} \times \left[1 - \frac{1}{2\gamma(\gamma^2 - 1)^{1/2}} \ln \frac{\gamma + (\gamma^2 - 1)^{1/2}}{\gamma - (\gamma^2 - 1)^{1/2}}\right]\right\}, \tag{16-11}$$

where we have used the same notation as in Eq. (13-31), to which Eq. (16-11) reduces in the nonrelativistic limit. Again it is assumed that the incident photon energy is much greater than the ionization energy; otherwise, one must use the correction factor of Stobbe.[34] When this factor is employed, one finds good agreement between theory and emperiment[35] in Cu, Mo, Ag, Ta, and Au.

Finally, it seems in order to say a few more words about the present status of the Lamb shift, since it represents a classic test of the validity of QED. Although relativistically there is a shift for states of nonzero angular momentum, we shall focus only on S states. A covariant calculation then leads to the following form for the shift in energy levels:

$$\Delta E_n = \frac{\alpha mc^2}{n^3} [A(Z\alpha)^4 \ln(Z\alpha) + B(Z\alpha)^4 + C(Z\alpha)^5 + D\alpha(Z\alpha)^4 + E(Z\alpha)^6 \ln^2(Z\alpha) + F(Z\alpha)^6 \ln(Z\alpha) + O(Z^6\alpha^6)] . \tag{16-12}$$

The coefficients A and B are just the second-order radiative corrections, the first of which is due to the Bethe sum calculated in the last chapter, while B contains corrections of this order due to vacuum polarization, Eq. (16-1), and the electron magnetic moment (see the following section). These two terms contribute about 1051 MHz to the shift. The coefficient C is due to relativistic corrections to the above ($\sim$7.14 MHz), whereas E is due in part to relativistic corrections to nonrelativistic wave functions (~ -0.52 MHz). The further higher-order term F contributes about 0.27 MHz. All of these terms arise from the one-photon effect and include small contributions from the reduced mass. On the other hand, the coefficient D consists partially of two-photon effects, along with fourth-order vacuum polarization and magnetic moment corrections, amounting to a few tenths of a megahertz. When the very small additional effects such as transverse virtual photons emitted by the proton, and finite size of the nucleus, are accounted for, one obtains the most recent theoretical result of 1057.57 ± 0.08 MHz, using the new[36] solid state value

$$\alpha^{-1} = 137.0359 \pm 0.0004\,. \qquad (16\text{-}13)$$

This is to be compared with the latest measured value[37] of 1057.86 ± 0.10. Whether the small discrepancy is indicating a low-energy breakdown of QED is not known at this time. For further details, the reader is referred to the theoretical review by Yennie[38] and a summary of the experimental situation by Hand.[39]

C. SOME INHERENTLY RELATIVISTIC PROCESSES

In the covariant theory there arise many new processes which have no nonrelativistic analogs, and we shall consider here just a few of these so as to indicate the scope of QED. A very basic relativistic phenomenon is the collision of two photons which produces an electron–positron pair. For unpolarized particles the total cross section has been calculated by Breit and Wheeler,[40] and Karplus and Neuman.[41] We give the pair-production cross section in the notation of Jauch and Rohrlich[6]:

$$\sigma = \tfrac{1}{2} r_0^2 \pi (1 - \beta^2)\left[(3 - \beta^4)\ln\left(\frac{1+\beta}{1-\beta}\right) - 2\beta(2 - \beta^2)\right]. \qquad (16\text{-}14)$$

Unfortunately, pair-production intensities from free photons appear to be too small to be easily observable. On the other hand, the process may well have important cosmological implications.[42]

If the two photons do not annihilate, but merely scatter, then the cross section is of higher order (nonlinear effect). The calculations lead to simple closed forms only in the nonrelativistic limit, and were first done by Euler,[43] and later included in the work of Karplus and Neuman.[41] In the center-of-mass system of the two photons the angular distribution is given by

$$\frac{d\sigma}{d\Omega} = \frac{139}{8100}\left(\frac{\alpha}{2\pi}\right)^2 r_0^2 \left(\frac{\hbar\omega}{mc^2}\right)^6 (3 + \cos^2\theta)^2\,, \qquad (16\text{-}15)$$

and the total cross section is

$$\sigma = \frac{973}{10125}\,\frac{\alpha^2}{\pi^2}\, r_0^2 \left(\frac{\hbar\omega}{mc^2}\right)^6. \qquad (16\text{-}16)$$

In these expressions $\hbar\omega$ is the center-of-mass photon energy and m is the electron mass, which arises from intermediate states in the calculation. Thus, we have required $\hbar\omega \ll mc^2$. Again, it is difficult to produce the required intensities for experimental observation, but recent work with high-intensity lasers is quite promising.

The process inverse to the above, annihilation of free electron–positron pairs, is perhaps more interesting because there is a fair amount of experimental evidence available for comparison. Dirac[44] first calculated the two-photon cross section for electrons at rest. If the polarizations are unspecified, the total cross section in the rest frame of the electron is

$$\sigma = r_0^2\pi \frac{1}{\beta^2\gamma^2(\gamma + 1)} \times \{(\gamma^2 + 4\gamma + 1)\ln[\gamma + (\gamma^2 - 1)^{1/2}] - \beta(\gamma + 3)\}\,. \tag{16-17}$$

Agreement with experiment is very good and has been considered by Colgate and Gilbert[45] at incident positron energies of 50, 100, and 200 MeV, Malamud *et al.*[46] at 650 MeV, and Braccini *et al.*[47] at 800 MeV. Three-photon annihilation has been studied by Ore and Powell,[48] and, as might be expected, the total cross section is of order α smaller than that of Eq. (16-17).

We have seen from the preceding discussion that electron–positron pairs are most characteristic of electrodynamic processes which are intrinsically relativistic. Another important phenomenon involving this pair is the electron–positron bound state, called *positronium*, which is probably the most electrodynamic of all systems. Because of the difference in masses, the Bohr radius is about half of that in hydrogen and the ground state is approximately 6.8 eV. As might be expected, the bound state is short lived, and the lifetime depends critically on the angular momentum state. Since energy-momentum conservation prohibits one-photon annihilation, it is found that the singlet state decays by two-photon emission, while the triplet annihilates to three photons. The lifetime of the singlet ground state (parapositronium) is

$$\tau_2 = 1.25 \times 10^{-10} \quad \text{sec}\,, \tag{16-18}$$

which is about 1000 times shorter than that for the triplet (orthopositronium):

$$\tau_3 = 1.386 \times 10^{-7} \quad \text{sec}\,. \tag{16-19}$$

Rather than pursue the detailed structure of positronium here, we refer the reader to the review article of De Benedetti and Corben[49] for selection rules, references, and the experimental picture.

It seems important to mention in passing that there exists another bound electrodynamic system of interest, which consists of a positive muon and an electron, called *muonium*. The existence of this system was first demonstrated experimentally by Hughes *et al.*[50] An essential point to be made here is that such a bound state cannot be studied satisfactorily by an

S-matrix expansion, and we must solve the relativistic two-body problem via the Dirac equation; this is also true for positronium. Nevertheless, if the assumption of regular Dirac particles is made, then the separation of the two hyperfine energy levels of the ground state of muonium can be calculated[51] from QED. Cleland *et al.*[52] have measured this splitting to high precision and have obtained essentially exact agreement with the theoretical value. In turn, the evidence strongly supports the contention that the muon is a Dirac particle with conventional electromagnetic coupling. We might mention that the system $\mu^+\mu^-$ should really be called muonium, but the necessary intensities are very difficult to obtain and the system has yet to be observed.

In 1933 Delbrück[53] suggested that photons could be scattered by an external field. The process is actually a radiative correction to Compton scattering,[54] but is important in its own right because it is observable. The only exact calculation of the cross section has been restricted to forward scattering, and is due to Rohrlich and Gluckstern.[55] One finds that the amplitude for Delbrück scattering is complex and that the cross section in the forward direction (for unpolarized photons) is

$$\frac{d\sigma}{d\Omega} = a_1^2 + a_2^2 , \tag{16-20}$$

where the real part

$$a_1 = \begin{cases} \dfrac{73}{72}\dfrac{1}{32}(\alpha Z)^2 r_0 \left(\dfrac{\hbar\omega}{mc^2}\right), & \hbar\omega \ll mc^2 \\[2ex] \dfrac{7}{18}(\alpha Z)^2 r_0 \left(\dfrac{\hbar\omega}{mc^2}\right), & \hbar\omega \gg mc^2 , \end{cases} \tag{16-21}$$

the imaginary part

$$a_2 = \begin{cases} \dfrac{1}{24}(\alpha Z)^2 r_0 \left(\dfrac{\hbar\omega}{mc^2} - 2\right)^3, & (\hbar\omega - 2mc^2) \ll mc^2 \\[2ex] \dfrac{7}{9\pi}(\alpha Z)^2 r_0 \left(\dfrac{\hbar\omega}{mc^2}\right)\left(\ln\dfrac{2\hbar\omega}{mc^2} - \dfrac{109}{42}\right), & \hbar\omega \gg 2mc^2 , \end{cases} \tag{16-22}$$

and the electronic mass arises as in photon–photon scattering. The real part corresponds to virtual pairs, while a_2 derives from real intermediate pairs. Delbrück scattering was first observed by Wilson,[56] and further experiments[57] have confirmed the reality of the process. The low-energy forms of the amplitudes appear to be excellent approximations up to $\hbar\omega \sim 2mc^2$. However, one must observe that Rayleigh and nuclear Thomson scattering can occur coherently with this process, and these amplitudes must be included with the above when comparison with experiment is to be made.

Unfortunately, it is not within the scope of this brief summary to discuss in more detail the wealth of very interesting physical phenomena related to the above relativistic processes, but the reader can pursue these points further and yet still steer clear of the detailed derivations by referring to the readable compendium by Roy and Reed.[58] There is, however, one further relativistic phenomenon which we should like to discuss here, and which represents a classic success of QED.

Delicate microwave experiments[59] first suggested that the intrinsic magnetic moment of the electron must be slightly different than the value predicted by the Dirac equation, which yields the *Bohr magneton*: $\mu_B = e\hbar/2mc$. In 1948 Schwinger[60] first showed that part of the radiative corrections of QED indeed corresponds to an additional magnetic moment of magnitude $\alpha/2\pi$, and the current calculation gives the expression

$$\mu = \mu_B\left(1 + \frac{\alpha}{2\pi} - 0.328\,\frac{\alpha^2}{\pi^2}\right). \tag{16-23}$$

The fourth-order correction was first calculated by Karplus and Kroll,[61] and numerically corrected by later calculations.[62] Its importance lies with the fact that it was the first α^2 calculation in QED, and demonstrated the efficacy of the renormalization program. With the aid of Eq. (16-13) and the notation $\Delta\mu \equiv (\mu - \mu_B)/\mu_B$, one finds the theoretical value

$$(\Delta\mu)_{\rm th} = 0.001159614 \pm (3 \times 10^{-9}), \tag{16-24a}$$

whereas the most recent experimental result of Wilkinson and Crane[63] is

$$(\Delta\mu)_{\rm exp} = 0.001159622 \pm (27 \times 10^{-9}). \tag{16-24b}$$

Even more recently Rich[64] has carried out a new data reduction and error analysis of the Wilkinson–Crane experiment and suggests that a better value is

$$(\Delta\mu)'_{\rm exp} = 0.001159557 \pm (30 \times 10^{-9}). \tag{16-24c}$$

A new experiment is currently underway in an attempt to resolve this discrepancy.[64]

The proton and neutron are also Dirac particles, of course, and the relativistic wave equation predicts a magnetic moment for the proton which differs from μ_B by replacing the electron mass by the proton mass, giving the *nuclear magneton*, μ_N. Since the neutron is charge neutral, no magnetic moment is indicated. Nevertheless, there is an additional, anomalous magnetic moment for the proton,

$$\mu_p = \mu_N + (1.79275 \pm 0.00003)\mu_N, \tag{16-25}$$

and that of the neutron is entirely anomalous[65]:

$$\mu_n = -(1.91329 \pm 0.00004)\mu_N . \tag{16-26}$$

These quantities are very difficult to calculate theoretically, because they depend in an essential way on the structure of the particles, and on the strong and weak interactions. On the other hand, by developing a suitable theory for these moments, one is likely to learn a good deal about the electromagnetic structure of nucleons.

In the relativistic theory the particle–particle electromagnetic interaction is *not* the point Coulomb interaction of Fig. 26f, but is described instead as an exchange of photons between the two charged particles. To an excellent approximation, one-photon exchange is quite adequate for obtaining agreement with the data in most cases. Nevertheless, one cannot reconcile the anomalous magnetic moment of the proton, and the structure thereby implied, with the view of electron–proton scattering, say, as that of one-photon exchange between *point* particles; at very high energies the proton structure should become evident ($\lambda = h/p$). Thus, it is to be expected that the amplitude for such a process will deviate from the expected due to both the magnetic moment of the proton and the cloud of virtual mesons surrounding the particle which gives rise to the nuclear force. The coupling of the photon to the proton may then take place over an extended region (the much smaller mass of the electron allows us to continue to view it as a point particle). Therefore, the amplitudes should be corrected by incorporating *form factors* in them, and when this is done correctly,[66] one finds the Rosenbluth equation[67] for the electron–proton scattering cross section:

$$\frac{d\sigma}{d\Omega} = \frac{e^2}{64\pi^2 E_0^2} \frac{\cos^2(\theta/2)}{\sin^4(\theta/2)} \frac{1}{1 + (2E_0/Mc^2)\sin^2(\theta/2)}$$
$$\times \left\{ F_1^2 + \frac{q^2}{4M^2c^2}[2(F_1 + 2\lambda_C^{-1}F_2)^2 \tan^2(\theta/2) + (2\lambda_C^{-1}F_2)^2] \right\}, \tag{16-27}$$

where the square of the 4-momentum transferred is

$$c^2q^2 = \frac{[2E_0 \sin(\theta/2)]^2}{1 + (2E_0/Mc^2)\sin^2(\theta/2)}, \tag{16-28}$$

the electron mass has been set equal to zero, M is the proton mass, λ_C is the proton Compton wavelength divided by 2π, and E_0 denotes the incident energy and θ the scattering angle of the electron in the laboratory system. By definition, $F_1(0) = e$, the renormalized physically observable charge, and $F_2(0) = \mu_a$, the anomalous magnetic moment of the proton. Thus, for a point particle with charge e and total magnetic moment $(g + 1)\mu_B$, the

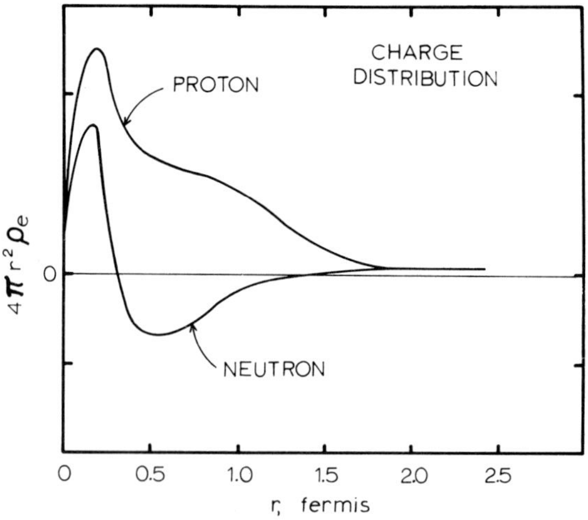

Fig. 51. Spatial distribution of charge for the proton and neutron based on electron–proton scattering determination of the form factors at high energies.

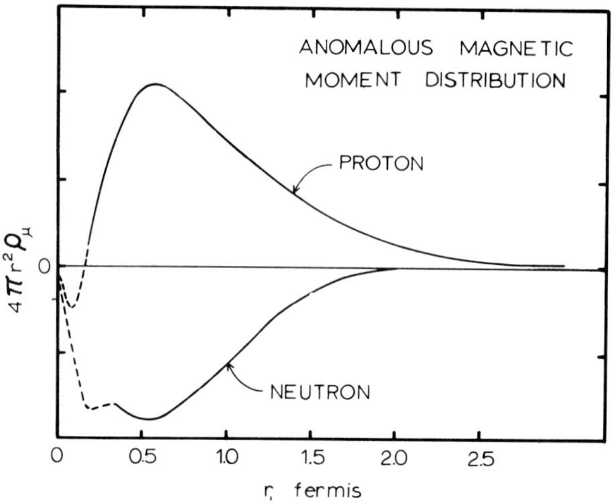

Fig. 52. Distribution of anomalous magnetic moments for the proton and neutron based on form-factor measurements.

charge form factor $F_1(q^2) = e$ and the magnetic form factor $F_2(q^2) = g\mu_B$, for *all* values of q^2. The particle is then said to have electromagnetic structure if and only if $F_1(q^2)$ or $F_2(q^2)$, or both, are not constant, and this is a precise definition of electromagnetic structure. Finally, we have used electron–proton scattering only as an example, and one can obviously define form factors for any particle.

The importance of the form factors is that they can be measured indirectly from scattering experiments on the one hand, and calculated from QED on the other. One therefore has a means for testing QED at high energies, or small distances. The Rosenbluth equation has been checked in detail by experiment in the nonrelativistic limit,[68] and the effects of multiple-photon exchange seem to be negligible, at least up to 1.3 GeV. Theoretical expressions for F_1 and F_2 can be calculated, and they are generally expressed as power series expansions in q^2 at low energies.[69] A great deal of experimental work toward determining form factors has been done by groups at both Stanford[70] and Cornell.[71] The one interesting point which we should like to bring out here is that from this data one can construct spatial distributions of charge and magnetic moment for the nucleons. The Cornell group has interpreted its measurements based on a core model of a nucleon, surrounded by a virtual meson cloud.[72] The pictures of the charge and moment distributions which arise are shown in Figs. 51 and 52. A remarkable feature of these curves is the charge distribution of the neutron, which is not at all zero, but only gives rise to a zero *total* charge.

D. CONCEPTUAL PROBLEMS AND THE VALIDITY OF QED

In recent years, as the large accelerators have gone into operation, several experiments designed primarily to test the validity of QED at high energies (small distances) have been performed. An important process of this type is the photoproduction of electron–positron pairs from carbon:

$$\gamma + \mathrm{C} \longrightarrow e^+ + e^- + \mathrm{C}\,. \tag{16-29}$$

The expected pair yield according to QED was calculated by Bjorken *et al.*[73] to within 3%, neglecting the Compton term only, and was first checked experimentally by Blumenthal *et al.*[74] This experiment indicated rather large deviations from the theoretical predictions and initiated somewhat of a crisis in the field. Nevertheless, the experiment has now been repeated twice and no deviations from QED are found within experimental error. Asbury *et al.*[75] find that first-order QED correctly predicts the e^+e^- yield for momentum transfers up to 400 MeV/c, and Alvensleben *et al.*[76] reach the same conclusions throughout a range of 150–900 MeV/c^2 for the invariant pair mass. In a similar experiment dePagter *et al.*[77] have measured the $\mu^+\mu^-$ production in the process analogous to (16-29) and find agreement with QED when both muons have energies in the range $2040 < E_\mu < 2400$ MeV. Finally, Liberman *et al.*[78] have measured the *Bremsstrahlung* of 9–13-GeV/c muons using a carbon-plate spark chamber as target, and find results consistent with QED for the region of invariant 4-momenta of the recoil muon–photon system between 200 and 650 MeV/c. Further tests of QED relate to the form factors discussed in the preceding section. Recent measurements of proton form factors by Albrecht *et al.*[79] have placed an upper bound of $F_2/\mu_p < 4 \times 10^{-3}$ on a hard core in the proton structure. Coward *et al.*[80] have studied electron–proton elastic scattering at very high-momentum transfers and found that good agreement with theory exists throughout the range $0.7 < q^2 < 25.0$ (GeV/c)^2. Moreover, the one-photon-exchange approximation appears to remain good at these energies, and the magnetic form factor continues to decrease as q^{-4}. These and other high-energy tests of QED are discussed in more detail in the reviews by Hand[39] and Panofsky.[81] All in all, quantum electrodynamics in its present form is probably valid up to center-of-mass energies of approximately 300 GeV (or down to distances of about 0.5 fermi), where the contributions from weak interactions become important.

On the one hand, we are tempted to be quite happy with our understanding of quantum electrodynamics and its excellent agreement with observation. On the other hand, the theory possesses conceptual flaws, such as the infinite renormalization constants, which we really do not understand at all. Although some recent attempts have been made to reformulate QED with finite renormalization,[82] further calculations have illustrated the futility of these attempts.[83] It must also be admitted that the vacuum is not completely understood, neither physically nor philosophically. Whether or not the vacuum fluctuations are intimately related to the (unobservable) zero-point energy remains an open question.

One of the bright moments in the development of QED occured when it was realized that, of necessity, the covariant theory had to be constructed *ab initio*, and that one could not merely generalize the classical formulation. The nonrelativistic limits of appropriate processes exist, to be sure, but the situation appears to be unidirectional. Moreover, the picture darkens when it is realized on what a fundamental level the confusion arises. For example, we have seen in Chapter XII that the cross section for Thomson scattering is due entirely to the $\mathbf{A}^2$ diagram of Fig. 26e, whereas the $\mathbf{p} \cdot \mathbf{A}$ diagrams of Figs. 27d and 27e exactly cancel one another in the nonrelativistic limit. However, in the covariant theory the $\mathbf{A}^2$ diagram does not contribute, and it is the $\mathbf{p} \cdot \mathbf{A}$ diagrams that give the Klein–Nishina formula, which in turn goes over into the Thomson formula in the nonrelativistic limit! It is not clear at this time why the physical pictures in the two domains should be so different, and perhaps it would become more transparent if one had a better fundamental understanding of something like Gell-Mann's principle of minimal electromagnetic coupling.[84] Nevertheless, we do indeed seem to be looking through a glass darkly.

Finally, we have several times observed that the success of QED lies primarily with the weakness of the electromagnetic coupling constant,

$\alpha = e^2/\hbar c$. While there is not a shred of evidence that the series expansions we have employed are convergent, they are very probably asymptotic. Yet this is an unpleasant situation, because in higher order the number of diagrams of order α^n becomes astronomical, and one can never hope to completely understand the theory in this form. One is tempted to agree with Feynman's conjecture that there is really no satisfactory quantum electrodynamics.[5] To be more precise, the present formulation of QED is entirely unable to account for the value of α. One observes that there are only three fundamental lengths in the theory: the classical electron radius, $r_0 = e^2/mc^2$; the electron Compton wavelength (divided by 2π), $\lambda_C = \hbar/mc$; and the first Bohr radius, $a_B = \hbar^2/me^2$. Consequently, there are only three possible dimensionless constants which can be formed from these three lengths. Surprisingly enough, they are essentially the same:

$$\lambda_C/a_B = \alpha\,, \qquad r_0/\lambda_C = \alpha\,, \qquad r_0/a_B = \alpha^2\,. \tag{16-30}$$

When we understand why this is so, maybe then we shall also *really* understand quantum electrodynamics.

REFERENCES

1. P. A. M. Dirac, *Proc. Roy. Soc.* (*London*) **A117,** 610 (1928).
2. S. Tomonaga, *Progr. Theoret. Phys.* (*Kyoto*) **1,** 27 (1946).
3. J. Schwinger, *Phys. Rev.* **73,** 416; **74,** 1439 (1948); **75,** 651; **76,** 790 (1949).
4. R. P. Feynman, *Phys. Rev.* **74,** 1430 (1948); **76,** 769 (1949).
5. R. P. Feynman, *Science* **153,** 699 (1966); J. Schwinger, *ibid.* **153,** 949 (1966); S. Tomonaga, *ibid.* **154,** 864 (1966).
6. J. M. Jauch and F. Rohrlich, "The Theory of Photons and Electrons." Addison-Wesley, Reading, Massachusetts, 1955.
7. W. E. Thirring, "Principles of Quantum Electrodynamics." Academic Press, New York, 1958.
8. A. I. Akhiezer and V. B. Berestetskii, "Quantum Electrodynamics." Wiley (Interscience), New York, 1965.
9. J. Schwinger, "Quantum Electrodynamics." Dover, New York, 1958.
10. R. P. Feynman, "Quantum Electrodynamics." Benjamin, New York, 1961.
11. S. S. Schweber, "An Introduction to Relativistic Quantum Field Theory." Harper, New York, 1961.
12. J. D. Bjorken and S. D. Drell, "Relativistic Quantum Mechanics." McGraw-Hill, New York, 1964.
13. E. A. Uehling, *Phys. Rev.* **48,** 55 (1935).
14. S. N. Gupta, *Proc. Phys. Soc.* (*London*) **A63,** 681 (1950).
15. K. Bleuler, *Helv. Phys. Acta* **23,** 567 (1950).
16. F. Dyson, *Phys. Rev.* **75,** 486, 1736 (1949).
17. See, e.g., J. Schwinger, *Phys. Rev.* **74,** 1439 (1948).
18. I. Yu. Kobzarev and L. B. Okun', *Soviet Phys. Usp.* (*English Transl.*) **11,** 338 (1968).
19. O. Klein and Y. Nishina, *Z. Physik* **52,** 853 (1929).
20. L. M. Brown and R. P. Feynman, *Phys. Rev.* **85,** 542 (1950).
21. H. A. Bethe and W. Heitler, *Proc. Roy. Soc.* (*London*) **A146,** 83 (1934); F. Sauter, *Ann. Physik* **20,** 404 (1934).

22. H. A. Bethe and J. Ashkin, "Experimental Nuclear Physics" (E. Segre, ed.). Wiley, New York, 1953.
23. H. W. Koch and J. W. Motz, *Rev. Mod. Phys.* **31,** 920 (1959).
24. O. V. Bogdankevich and F. A. Niolaev, "Methods in Bremsstrahlung Research." Academic Press, New York, 1966.
25. W. A. McKinley and H. Feshbach, *Phys. Rev.* **74,** 1959 (1948).
26. D. R. Corson and A. O. Hanson, *Ann. Rev. Nucl. Sci.* **3,** 67 (1953).
27. C. Møller, *Ann. Physik* **14,** 568 (1932).
28. Y. S. Tsai, *Phys. Rev.* **120,** 269 (1960).
29. T. B. Anders, *Nucl. Phys.* **59,** 127 (1964).
30. E. B. Dally, *Phys. Rev.* **123,** 1840 (1961).
31. H. J. Bhabha, *Proc. Roy. Soc.* (*London*) **A154,** 195 (1935).
32. A. Ashkin, L. A. Page and W. M. Woodward, *Phys. Rev.* **94,** 357 (1954).
33. F. Sauter, *Ann. Physik* **9,** 217; **11,** 454 (1931).
34. See, e.g., G. W. Grodstein, *Natl. Bur. Std.* (*U.S.*) *Circ. No. 583* (1957).
35. W. F. Titus, *Phys. Rev.* **115,** 351 (1959).
36. W. H. Parker, B. N. Taylor, and D. N. Langenberg, *Phys. Rev. Letters* **18,** 287 (1967).
37. R. T. Robiscoe, *Phys. Rev. Letters* **17,** 69 (1966).
38. D. R. Yennie, *Brandeis Summer Inst. Theoret. Phys.* **1,** 165 (1963).
39. L. Hand, *Proc. Intern. Conf. Particles and Fields, Rochester, New York, 1967,* p. 21. Wiley (Interscience), New York, 1967.
40. G. Breit and J. A. Wheeler, *Phys. Rev.* **46,** 1087 (1934).
41. R. Karplus and M. Neuman, *Phys. Rev.* **83,** 776 (1951).
42. R. J. Gould and G. P. Schreder, *Phys. Rev. Letters* **16,** 252 (1966); *Phys. Rev.* **155,** 1404 (1967).
43. H. Euler, *Ann. Physik* **26,** 398 (1936).
44. P. A. M. Dirac, *Proc. Cambridge Phil. Soc.* **26,** 361 (1930).
45. S. A. Colgate and F. C. Gilbert, *Phys. Rev.* **89,** 790 (1953).
46. E. Malamud, R. Weill, and J. G. McEwen, *Helv. Phys. Acta* **34,** 497 (1961).
47. P. L. Braccini, I. X. Ion, A. Stefanini, G. Torelli, and R. Torelli Tosi, *Nuovo Cimento* **29A,** 1215 (1963).
48. A. Ore and J. L. Powell, *Phys. Rev.* **75,** 1696 (1949).
49. S. De Benedetti and H. C. Corben, *Ann. Rev. Nucl. Sci.* **4,** 191 (1954).
50. V. W. Hughes, D. W. McColm, K. Zieck, and R. Prepost, *Phys. Rev. Letters* **5,** 63 (1960).
51. R. Karplus and A. Klein, *Phys. Rev.* **85,** 972 (1952); N. M. Kroll and F. Polock, *ibid.* **86,** 876 (1952); R. Arnowitt, *ibid.* **92,** 1002 (1953).
52. W. E. Cleland, J. M. Bailey, M. Eckhause, V. W. Hughes, R. M. Mobley, R. Prepost, and J. E. Rothberg, *Phys. Rev. Letters* **13,** 202 (1964).
53. M. Delbrück, *Z. Physik* **84,** 144 (1933).
54. See, e.g., J. M. Jauch and F. Rohrlich, "The Theory of Photons and Electrons," p. 380. Addison-Wesley, Reading, Massachusetts, 1955.
55. F. Rohrlich and R. L. Gluckstern, *Phys. Rev.* **86,** 1 (1952).
56. R. R. Wilson, *Phys. Rev.* **90,** 720 (1953).
57. A. M. Bernstein and A. K. Mann, *Phys. Rev.* **110,** 805 (1958); J. Moffat and M. W. Stringfellow, *Proc. Roy. Soc.* (*London*) **A254,** 242 (1960); R. Bosch, J. Lang, R. Müller, and W. Wölf, *Phys. Letters* **2,** 16 (1962); *Helv. Phys. Acta* **36,** 625 (1963).
58. R. R. Roy and R. D. Reed, "Interactions of Photons and Leptons with Matter." Academic Press, New York, 1968.
59. H. M. Foley and P. Kusch, *Phys. Rev.* **73,** 412 (1948); J. E. Nafe, E. B. Nelson, and

I. I. Rabi, *ibid.* **71,** 914 (1947); D. E. Nagle, R. S. Julian, and J. R. Zacharias, *ibid.* **72,** 971 (1947).
60. J. Schwinger, *Phys. Rev.* **73,** 416 (1948).
61. R. Karplus and N. M. Kroll, *Phys. Rev.* **77,** 536 (1950).
62. C. M. Sommerfield, *Phys. Rev.* **107,** 328 (1957); *Ann. Phys.* (*N.Y.*) **5,** 26 (1958); A. Peterman, *Helv. Phys. Acta* **30,** 407 (1957).
63. D. T. Wilkinson and H. R. Crane, *Phys. Rev.* **130,** 852 (1963).
64. A. Rich, *Phy. Rev. Letters* **20,** 967 (1968).
65. See, e.g., N. F. Ramsey, "Molecular Beams." Oxford Univ. Press (Clarendon), London and New York, 1956.
66. R. Hofstadter, *Ann. Rev. Nucl. Sci.* **7,** 231 (1957); S. D. Drell and F. Zachariasen, "Electromagnetic Structure of Nucleons." Oxford Univ. Press, London and New York, 1961.
67. M. N. Rosenbluth, *Phys. Rev.* **79,** 615 (1950).
68. R. Hofstadter, F. Bumiller and M. Yearian, *Rev. Mod. Phys.* **30,** 482 (1958).
69. D. R. Yennie, M. M. Levy, and D. G. Ravenhall, *Rev. Mod. Phys.* **29,** 144 (1957). See, also, R. Hofstadter, *Ann. Rev. Nucl. Sci.* **7,** 231 (1957); S. D. Drell and F. Zachariasen, "Electromagnetic Structure of Nucleons." Oxford Univ. Press, London and New York, 1961.
70. See, e.g., R. Hofstadter, C. DeVries, and R. Herman, *Phys. Rev. Letters* **6,** 290 (1961); R. Hofstadter and R. Herman, *ibid.* **6,** 293 (1961), and references in these two papers.
71. See, e.g., D. N. Olson, H. F. Schopper, and R. R. Wilson, *Phys. Rev. Letters* **6,** 286 (1961); R. M. Littauer, H. F. Schopper, and R. R. Wilson, *ibid.* **7,** 141 (1961), and references in these two papers.
72. R. M. Littauer, H. F. Schopper, and R. R. Wilson, *Phys. Rev. Letters* **7,** 144 (1961).
73. J. D. Bjorken, S. D. Drell, and S. Frautschi, *Phys. Rev.* **112,** 1409 (1958).
74. R. B. Blumenthal, D. C. Ehn, W. L. Faissler, P. M. Joseph, L. J. Lanzerotti, F. M. Pipkin, and D. G. Stairs, *Phys. Rev. Letters* **14,** 660 (1965); *Phys. Rev.* **144,** 1199 (1966).
75. J. G. Asbury, W. K. Bertram, U. Becker, P. Joos, M. Rohde, A. J. S. Smith, S. Friedlander, C. L. Jordan, and S. S. C. Ting, *Phys. Rev. Letters* **18,** 65 (1967).
76. H. Alvensleben, U. Becker, W. K. Bertram, M. Binkley, K. Cohen, C. L. Jordan, T. M. Knasel, R. Marshall, D. J. Quinn, M. Rohde, G. H. Sanders, and S. C. C. Ting, *Phys. Rev. Letters* **21,** 1501 (1968).
77. J. K. dePagter, J. I. Friedman, G. Glass, R. C. Chase, M. Gettner, E. von Goeler, R. Weinstein, and A. M. Boyarski, *Phys. Rev. Letters* **17,** 767 (1966).
78. A. D. Liberman, C. M. Hoffman, E. Engels, Jr., D. C. Imrie, P. G. Innocenti, R. Wilson, C. Zajde, W. A. Blanpied, D. G. Stairs, and D. Drickey, *Phys. Rev. Letters* **22,** 663 (1969).
79. W. Albrecht, H.-J. Behrend, H. Dorner, W. Glauger, and H. Hultschig, *Phys. Rev. Letters.* **18,** 1014 (1967).
80. D. H. Coward, H. DeStaebler, R. A. Early, J. Litt, A. Minten, L. W. Mo, W. K. H. Panofsky, R. E. Taylor, M. Breidenbach, J. I. Friedman, H. W. Kendall, P. N. Kirk, B. C. Barish, J. Mar, and J. Pine, *Phys. Rev. Letters* **20,** 292 (1968).
81. W. K. H. Panofsky, Electromagnetic interactions, Intern. Conf. Elementary Particles, Heidelberg, September, 1967.
82. K. Johnson, M. Baker, and R. Willey, *Phys. Rev.* **136,** B1111 (1964); *ibid.* **163,** 1699 (1967).
83. C. R. Hagen and M. A. Samuel, *Phys. Rev. Letters* **20,** 1405 (1968).
84. M. Gell-Mann, *Nuovo Cimento Suppl.* **4,** 848 (1956).

APPENDIX A THE COVARIANT DERIVATIVE

In order to obtain a more satisfactory intuitive feeling for the covariant derivative as defined in Chapter II, we wish to examine more carefully here the notion of change in a vector quantity when it is moved from point to point in an N-dimensional space. As discussed below Eq. (2-54), the lengths of vector components are not equal at two different points in a Riemannian space unless the g_{ij} are constants. In a three-dimensional Euclidean space a constant vector field can be defined by specifying the vector components to be constants at all points of the space, but the preceding observation indicates that this may not be possible in a Riemannian space. If one refers to the transformation law (2-9) for contravariant vectors, it is seen that if A^j is constant in one system, it is not necessarily so in another.

Consider first a general N-dimensional space in which a metric has not yet been introduced, and suppose the components of a contravariant vector A^j to be constant along a curve $x(p)$, with parameter p. Then, differentiating Eq. (2-9), we find

$$\frac{d\bar{A}^i}{dp} = \partial^2_{kl}\bar{x}^i \frac{dx^l}{dp} A^k . \tag{A-1}$$

The right-hand side contains the components of the vector in the original coordinate system, and it is more transparent if we rewrite Eq. (A-1) completely in terms of barred quantities. Thus,

$$\frac{d\bar{A}^i}{dp} = \bar{\Gamma}^i_{mj} \frac{d\bar{x}^m}{dp} \bar{A}^j , \tag{A-2}$$

where

$$\bar{\Gamma}^i_{mj} \equiv \partial^2_{kl}\bar{x}^i \cdot \bar{\partial}_m x^l \cdot \bar{\partial}_j x^k . \tag{A-3}$$

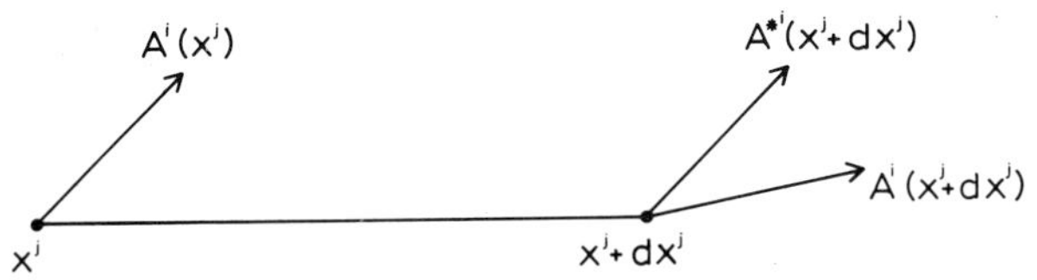

Fig. A-1. Parallel transplantation of a vector in an affine space.

and insert into this result the transplantation law (A-4):

$$(\partial_k g_{ij}) \frac{dx^k}{ds} A^i B^j + g_{ij} \Gamma^i_{mk} \frac{dx^m}{ds} A^k B^j + g_{ij} A^i \Gamma^j_{mk} \frac{dx^m}{ds} B^k = 0 .$$

Appropriate relabeling of this last expression yields

$$[\partial_k g_{ij} + g_{mj} \Gamma^m_{ki} + g_{im} \Gamma^m_{kj}] \frac{dx^k}{ds} A^i B^j = 0 ,$$

and, since the final three factors are arbitrary,

$$\partial_k g_{ij} + g_{mj} \Gamma^m_{ki} + g_{im} \Gamma^m_{kj} = 0 . \tag{A-14a}$$

If the indices (i, j, k) are permuted cyclically, we can obtain two similar equations:

$$\partial_j g_{ki} + g_{mi} \Gamma^m_{jk} + g_{jm} \Gamma^m_{ik} = 0 , \tag{A-14b}$$

$$\partial_i g_{jk} + g_{mk} \Gamma^m_{ij} + g_{km} \Gamma^m_{ji} = 0 . \tag{A-14c}$$

Equations (A-14) are now combined by adding (A-14b) and (A-14c), and subtracting (A-14a) from the sum. One finds that

$$\partial_i g_{jk} + \partial_j g_{ki} - \partial_k g_{ij} + 2\Gamma^m_{ji} g_{mk} = 0 ,$$

or

$$-\Gamma^m_{ji} = \tfrac{1}{2} g^{mk} (\partial_i g_{jk} + \partial_j g_{ki} - \partial_k g_{ij}) = \{m, ji\} , \tag{A-15}$$

the Christoffel symbol of the second kind.

The result (A-15) is now substituted into Eq. (A-4) to obtain the law of vector transplantation in a Riemannian space:

$$dA^i = -\{i, mn\} \, dx^m A^n , \tag{A-16}$$

which is now called the *law of parallel displacement*. It should be noted as a check that the transformation law for the connetions, (A-10), is identical to that for the Christoffel symbol of the second kind, (2-62), except for a sign.

With the preceding background, we are now in a position to consider the definition of derivative from an intuitively satisfying point of view. Again, we first consider the change in a vector field $A^i(x^j)$ as it is moved from one point to another in an affine space. If A^{*i} is the vector obtained by transplantation, we can compare the difference in the two vector fields when evaluated at $x^j + dx^j$ (see Fig. A-1):

$$A^i(x^j + dx^j) - A^{*i}(x^j + dx^j)\,. \tag{A-17}$$

To leading order in small quantities,

$$A^i(x^j + dx^j) = A^i(x^j) + \partial_k A^i \cdot dx^k + O(dx^k)^2\,. \tag{A-18}$$

For the transplanted vector

$$A^{*i}(x^j + dx^j) = A^i(x^j) + \Gamma^i_{kl} A^l\, dx^k\,. \tag{A-19}$$

Combining Eqs. (A-18) and (A-19), we have

$$\begin{aligned} &A^i(x^j + dx^j) - A^{*i}(x^j + dx^j) \\ &\quad = [\partial_k A^i - \Gamma^i_{kl} A^l]\, dx^k + O(dx^k)^2\,. \end{aligned} \tag{A-20}$$

The difference in the vector field at $x^j + dx^j$ and its transplanted value there has the appearance of a power-series expansion in dx^k, thereby satisfying our intuitive notion of derivative as $dx^j \to 0$ (or, $dx^k \to 0$). Thus, we define

$$\nabla_k A^i \equiv \partial_k A^i - \Gamma^i_{kl} A^l \tag{A-21}$$

as the *covariant derivative* of the vector field $A^i(x^j)$. From the quotient law, $\nabla_k A^i$ is clearly a second-rank, mixed tensor.

The relation (A-15) then permits us to write in a Riemannian space

$$\nabla_k A^i = \partial_k A^i + \{i, kl\} A^l\,, \tag{A-22}$$

representing the invariant change in A^i from one point to a neighboring point. In a similar manner, we find for a covariant vector field

$$\nabla_k B_i = \partial_k B_i - \{r, ki\} B_r\,. \tag{A-23}$$

In order to obtain the differentiation law for an arbitrary tensor field, recall that in a three-dimensional Euclidean space

$$\partial_k A_i B^j = A_i\, \partial_k B^j + B^j\, \partial_k A_i = A_i\, \nabla_k B^j + B^j\, \nabla_k A_i\,. \tag{A-24}$$

But the right-hand side of this equation is a tensor, so that the usual *product law* of differentiaton of vectors is a consistent definition in all coordinate systems.

Consider now a third-rank, mixed tensor $\mathsf{T}^{\alpha}_{\beta\gamma}$ and three arbitrary vectors A_α, B^β, and C^γ, and form the invariant $\mathsf{T}^{\alpha}_{\beta\gamma} A_\alpha B^\beta C^\gamma$. The gradient of this scalar,

$$\begin{aligned} w_l = \partial_l(\mathsf{T}^{\alpha}_{\beta\gamma} A_\alpha B^\beta C^\gamma) &= (\partial_l \mathsf{T}^{\alpha}_{\beta\gamma}) A_\alpha B^\beta C^\gamma + \mathsf{T}^{\alpha}_{\beta\gamma}(\partial_l A_\alpha) B^\beta C^\gamma \\ &\quad + \mathsf{T}^{\alpha}_{\beta\gamma} A_\alpha(\partial_l B^\beta) C^\gamma + \mathsf{T}^{\alpha}_{\beta\gamma} A_\alpha B^\beta(\partial_l C^\gamma)\,, \end{aligned} \tag{A-25}$$

is a covariant vector. Another covariant vector can be defined as

$$\begin{aligned} v_l &= T^{\alpha}_{\beta\gamma}\,\nabla_l(A_\alpha B^\beta C^\gamma) \\ &= T^{\alpha}_{\beta\gamma}(\nabla_l A_\alpha)B^\beta C^\gamma + T^{\alpha}_{\beta\gamma}A_\alpha(\nabla_l B^\beta)C^\gamma + T^{\alpha}_{\beta\gamma}A_\alpha B^\beta(\nabla_l C^\gamma)\,. \end{aligned} \tag{A-26}$$

Now form the vector difference of these two quantities:

$$\begin{aligned} w_l - v_l = {}&(\partial_l T^{\alpha}_{\beta\gamma})A_\alpha B^\beta C^\gamma + T^{\alpha}_{\beta\gamma}\{r, \alpha l\}B^\beta C^\gamma A_r \\ &- T^{\alpha}_{\beta\gamma}\{\beta, lr\}A_\alpha B^r C^\gamma - T^{\alpha}_{\beta\gamma}\{\gamma, lr\}A_\alpha B^\beta C^r\,, \end{aligned} \tag{A-27}$$

where we have used Eqs. (A-22) and (A-23). Relabeling dummy indices, we find that

$$\begin{aligned} w_l - v_l = {}&[\partial_l T^{\alpha}_{\beta\gamma} + \{\alpha, ls\}T^{s}_{\beta\gamma} \\ &- \{s, l\beta\}T^{\alpha}_{s\gamma} - \{s, l\gamma\}T^{\alpha}_{\beta s}]A_\alpha B^\beta C^\gamma\,. \end{aligned} \tag{A-28}$$

Since the three vectors A_α, B^β, and C^γ are arbitrary, and the left-hand side of Eq. (A-28) is a covariant vector, the quotient law implies that the quantity

$$\nabla_l T^{\alpha}_{\beta\gamma} \equiv \partial_l T^{\alpha}_{\beta\gamma} + \{\alpha, ls\}T^{s}_{\beta\gamma} - \{s, l\beta\}T^{\alpha}_{s\gamma} - \{s, l\gamma\}T^{\alpha}_{\beta s}\,, \tag{A-29}$$

is a fourth-rank mixed tensor, which the calculation has also shown to be the covariant derivative of $T^{\alpha}_{\beta\gamma}$. That is, if the tensor $T^{\alpha}_{\beta\gamma}$ is represented symbolically by T, and the product $A_\alpha B^\beta C^\gamma$ by A, then we have written

$$w_l = \partial_l(TA) = \nabla_l(TA) = T(\nabla_l A) + (\nabla_l T)A\,,$$

since TA is an invariant and the covariant and partial derivatives are identical for an invariant. Also,

$$v_l = T(\nabla_l A)\,,$$

so that

$$w_l - v_l = (\nabla_l T)B\,,$$

and the quantity in square brackets in Eq. (A-28) must indeed be the covariant derivative of $T^{\alpha}_{\beta\gamma}$.

The procedure used to derive Eq. (A-29) goes through for any tensor, so that one may conclude that the general expression (2-77) for the covariant derivative of an arbitrary tensor field is quite valid. Moreover, the definition of derivative in a Riemannian space can now be considered to be on an intuitively sound basis.

REFERENCES

1. See, e.g., R. Adler, M. Bazin, and M. Schiffer, "Introduction to General Relativity," p. 44. McGraw-Hill, New York, 1965.
2. H. Weyl, "Space, Time, Matter," Section 14. Dover, New York, 1952.

APPENDIX B GROUND STATE OF THE MANY-ELECTRON SYSTEM

In order to further establish the great generality and utility of the Fock space formulation of Chapter X, we here undertake a calculation of the ground-state energy of a system of many electrons. It is possible, in fact, to treat the system in the beginning quite generally as an electrically neutral collection of N electrons and N positive ions contained in a cubical volume V. In a somewhat more convenient notation, the many-body Hamiltonian of Chapter X can be written

$$\mathsf{H}_N = \sum_{i,j} \omega_{ij} a_i^\dagger a_j + \tfrac{1}{2} \sum_{ijkl} a_i^\dagger a_j^\dagger \mathsf{V}_{ijkl} a_l a_k = \mathsf{H}_0 + \mathsf{V}_2 \,, \tag{B-1}$$

where

$$\omega_{ij} = \int \phi_i^*(\mathbf{x})[-(\hbar^2/2m)\,\nabla^2]\phi_j(\mathbf{x})\, d^3x \,, \tag{B-2}$$

$$\mathsf{V}_{ijkl} = \int \phi_i^*(\mathbf{x})\phi_j^*(\mathbf{y})\mathsf{V}_2(\mathbf{x},\mathbf{y})\phi_k(\mathbf{x})\phi_l(\mathbf{y})\, d^3x\, d^3y \,. \tag{B-3}$$

The multicomponent nature of the system is then incorporated into these equations by allowing the sums over states in Eq. (B-1) to implicitly include a sum over particle types.

For the present problem the two-body interaction V_2 is the Coulomb interaction

$$\mathsf{V}_2(\mathbf{x},\mathbf{y}) = Z_\alpha Z_\beta e^2/|\,\mathbf{x} - \mathbf{y}\,| \,, \tag{B-4}$$

where Z_α is the charge number of an α-type particle (in this problem $Z_\alpha = \pm 1$). The expectation value of the Hamiltonian in an arbitrary state of the system can be written

$$\langle \mathsf{H}_N \rangle = \langle \mathsf{H}_0 \rangle + \langle \mathsf{V}_2 \rangle \,. \tag{B-5}$$

Let us consider first the free-particle term

$$\langle \mathsf{H}_0 \rangle = \sum_{i,j} \omega_{ij} \, \langle n \cdots n_i \cdots | a_i^\dagger a_j | n_1 \cdots n_i \cdots \rangle \,. \tag{B-6}$$

The states i and j may refer to either electrons or ions, but, in any case, the matrix element vanishes unless $i = j$. Thus, the product of the two operators is effectively the number operator, and, since this is diagonal, we can write

$$\langle \mathsf{H}_0 \rangle = \sum_i n_i \omega_i \,, \qquad \omega_i \equiv \omega_{ii} \,. \tag{B-7}$$

In the same manner

$$\langle \mathsf{V}_2 \rangle = \tfrac{1}{2} \sum_{ijkl} \langle n_1 \cdots n_i \cdots | a_i^\dagger a_j^\dagger a_l a_k | n_1 \cdots n_i \cdots \rangle \, \mathsf{V}_{ijkl} \,. \tag{B-8}$$

Exercise. Show that the matrix element in Eq. (B-8) vanishes unless $l = i$, $k = j$; *or* unless $k = i$, $l = j$, in the case of identical particles.

Then,

$$\langle \mathsf{V}_2 \rangle = \tfrac{1}{2} \sum_{i,j} [\mathsf{V}_{ijij} \, \langle n_1 \cdots n_i \cdots | a_i^\dagger a_j^\dagger a_j a_i | n_1 \cdots n_i \cdots \rangle$$
$$+ \, \delta_{\alpha\beta} \mathsf{V}_{ijji} \, \langle n_1 \cdots n_i \cdots | a_i^\dagger a_j^\dagger a_i a_j | n_1 \cdots n_i \cdots \rangle] \,. \tag{B-9}$$

One can now apply the commutation relations for fermions or bosons to the matrix elements. This bit of algebra is left to the reader, who will find that, finally,

$$\langle \mathsf{V}_2 \rangle = \tfrac{1}{2} \sum_{\alpha,\beta} \sum_{i,j} n_i n_j (\mathsf{V}_{ijij} + \varepsilon_\alpha \, \delta_{\alpha\beta} \mathsf{V}_{ijji}) \,, \tag{B-10}$$

where we have now indicated the sum over all different particle types explicitly. In this respect we shall arbitrarily assign i to the α label and j to the β label, and note that the second term in Eq. (B-10) vanishes unless $\alpha = \beta$.

We have been fairly explicit in writing out the multicomponent form of Eq. (B-10) because it is now possible to make an important observation concerning the term V_{ijij}. To do this, let us choose to work in the momentum representation and take for the complete set of single-particle functions the plane waves

$$\phi_{\mathbf{k}}(\mathbf{x}) = [\exp(i\mathbf{k} \cdot \mathbf{x})]/V^{1/2} \,. \tag{B-11}$$

One easily confirms from Eq. (B-3), then, that V_{ijij} is actually independent of i and j or, in this case, $\mathbf{k}$ and $\mathbf{k}'$.* The first term in Eq. (B-10) is therefore proportional to

$$e^2 \sum_{\alpha} Z_\alpha N_\alpha \sum_{\beta} Z_\beta N_\beta \,, \tag{B-12}$$

where N_α is the number of α-type particles in the system. But we have agreed to consider the system to have overall charge neutrality, so that the quantity of Eq. (B-12) is identically zero. Hence, in Eq. (B-10) only the exchange term contributes, and the multicomponent nature of the system is manifested only as a sum of similar terms for each type of particle.

Having evoked the condition of charge neutrality in detail, we can now specialize to a system of electrons and ignore the presence of the ions except for their role of providing a uniform background of positive charge to preserve charge neutrality. Thus, for this so-called *electron gas*, Eq. (B-10) reduces to

$$\langle \mathsf{V}_2 \rangle = -\tfrac{1}{2} \sum_{i,j} \mathsf{V}_{ijji} n_i n_j \,. \tag{B-13}$$

Equations (B-5), (B-7), and (B-13) now allow us to calculate the energy of the system.

We shall use perturbation theory to make the calculation, in which case the leading-order contribution is found by computing the expectation value of the Hamiltonian using *unperturbed* state vectors, and these can be chosen to be the free-particle waves of Eq. (B-11). We are interested particularly in the ground state of the system, and so one might ask if these wave functions are an adequate first approximation in this state. It will be assumed here that they are, and a Hartree–Fock calculation[1] proves that plane waves satisfy the equations for minimum energy of the many-electron system. The ground-state nature of the problem can be built into our calculation by invoking the Pauli principle, which prohibits more than one electron from occupying the same state. Thus, we assume that the momentum states are filled in a spherically symmetric manner up to a maximum value of k_{F}, called the *Fermi momentum*. That is, the total number of electrons in the system is (using the momentum representation)

$$N = \sum_{\mathbf{k}} n_{\mathbf{k}} \,, \qquad n_{\mathbf{k}} = \begin{cases} 1, & |\mathbf{k}| < k_{\mathrm{F}} \\ 0, & |\mathbf{k}| > k_{\mathrm{F}} \,. \end{cases} \tag{B-14}$$

The sum can be converted to an integral in the usual manner (infinite volume limit), but it must be remembered that sums over states implicitly

* To make this confirmation, it may help to refer to the evaluation (11-15).

contain sums over spin states, so that in this case we obtain an additional factor of two. Hence,

$$N = 2\frac{V}{(2\pi)^3}\int d^3k = 2\frac{V}{(2\pi)^3}(4\pi)\int_0^{k_F} k^2\,dk = \frac{V}{3\pi^2}k_F{}^3\,. \qquad \text{(B-15)}$$

Therefore, the density of particles in the system is

$$n = N/V = k_F{}^3/3\pi^2\,, \qquad \text{(B-16)}$$

which is usually taken as a more general definition of the Fermi momentum k_F.

From Eq. (B-13) the ground-state expectation value of the interaction is

$$\begin{aligned}\langle \mathsf{V}_2\rangle &= -\tfrac{1}{2}\sum_{\mathbf{k}_1\mathbf{k}_2} n_{\mathbf{k}_1}n_{\mathbf{k}_2}\mathsf{V}_{\mathbf{k}_1\mathbf{k}_2\mathbf{k}_2\mathbf{k}_1}\\ &= -V\frac{4\pi e^2}{(2\pi)^6}\int \frac{n_{\mathbf{k}_1}n_{\mathbf{k}_2}}{|\mathbf{k}_1-\mathbf{k}_2|^2}\,d^3k_1\,d^3k_2\,, \end{aligned} \qquad \text{(B-17)}$$

where the matrix element of the Coulomb interaction has been evaluated with the aid of Eq. (11-15).

With the limits implied by Eq. (B-14) the integral is evaluated as follows:

$$\int \frac{n_{\mathbf{k}_1}n_{\mathbf{k}_2}}{|\mathbf{k}_1-\mathbf{k}_2|^2}\,d^3k_1\,d^3k_2 = -8\pi^2\int_0^{k_F} k_2\,dk_2\int_0^{k_F} k_1\ln\left|\frac{k_1-k_2}{k_1+k_2}\right|dk_1\,. \qquad \text{(B-18)}$$

One now verifies the following two integrations:

$$\int_0^{k_F} k_1\ln\left|\frac{k_1-k_2}{k_1+k_2}\right|dk_1 = -k_Fk_2 + \frac{k_F{}^2-k_2{}^2}{2}\ln\left|\frac{k_F-k_2}{k_F+k_2}\right|\,; \qquad \text{(B-19a)}$$

$$\int_0^{k_F} k_2\left[-k_Fk_2 + \frac{k_F{}^2-k_2{}^2}{2}\ln\left|\frac{k_F-k_2}{k_F+k_2}\right|\right]dk_2 = -\frac{1}{2}k_F{}^4\,. \qquad \text{(B-19b)}$$

As a result of these calculations, we have, finally,

$$\langle \mathsf{V}_2\rangle = -[2Ve^2/(2\pi)^3]k_F{}^4\,. \qquad \text{(B-20)}$$

Proceeding in the same manner, one easily finds that

$$\langle \mathsf{H}_0\rangle = (\hbar^2V/10\pi^2m)k_F{}^5\,, \qquad \text{(B-21)}$$

so that from Eq. (B-5)

$$\langle \mathsf{H}_N\rangle = E_0 = [V/(2\pi)^3][(4\pi\hbar^2/5m)k_F{}^5 - 2e^2k_F{}^4]\,. \qquad \text{(B-22)}$$

Equation (B-22) gives the leading-order contributions to the expectation value of the Hamiltonian in the ground state of the system of electrons. The first term, of course, is merely the kinetic energy of the free electrons restricted by the Pauli principle, whereas the term proportional to e^2 is called the *exchange term* and is just the free-particle expectation value of the interaction in the ground state. This was first calculated by Wigner.[2] As we have seen, there is no direct term, due to the overall charge neutrality.

The quantity which is usually considered measurable, however, is the energy per particle. To obtain this number, we define the *Fermi energy* as

$$E_F = \hbar^2 k_F{}^2/2m\,. \tag{B-23}$$

Then, dividing both sides of Eq. (B-22) by the total number of particles N, and employing Eq. (B-16), one finds

$$E_0/N = \tfrac{3}{5}E_F - (3/4\pi)e^2 k_F\,. \tag{B-24}$$

Finally, it is conventional to introduce the only dimensionless parameter available for describing this system in the ground state, which is related to the number of particles contained in a volume of radius equal to the Bohr radius:

$$r_s = [(4\pi/3)na_B{}^3]^{-1/3}\,, \tag{B-25}$$

where $a_B = \hbar^2/me^2$. It is now appropriate to measure the energy per particle in units of rydbergs ($=e^2/2a_B$), so that evaluation of the constants yields

$$E_0/N \simeq (2.21)r_s^{-2} - (0.916)r_s^{-1}\,. \tag{B-26}$$

This final expression we have obtained for the ground-state energy per particle of the many-electron system is an excellent approximation to the behavior of electrons in metals in many cases. However, when one attempts to obtain more precision by means of including higher-order terms, the ordinary perturbation methods break down. It is at this point that the very sophisticated techniques of many-body theory developed in recent years become important, by means of which the perturbation series is rearranged and certain types of terms are summed to infinite order. Thus, Gell-Mann and Brueckner[3] first demonstrated how to correctly obtain the next two terms in the expression (B-26) by summing the so-called ring diagrams, which physically introduces the screening of the Coulomb interaction, effectively reducing it to a short-range interaction due to many-body effects. These next two terms are $\ln r_s$ and a constant.

One notes, however, that the result of this entire calculation appears to be an expansion in increasing positive powers of r_s and $\ln r_s$. Such an expansion can be expected to be valid only at high densities ($r_s \ll 1$), according to Eq. (B-25). In the average metal we find that for electrons $2 \lesssim r_s \lesssim 6$, and so it would seem that the entire method of calculation is really unsuited for the electronic ground-state energy in metals. An adequate procedure for overcoming these difficulties has yet to be developed.

REFERENCES

1. D. R. Hartree, *Proc. Cambridge Phil. Soc.* **24,** 89 (1928); V. Fock, *Z. Physik* **61,** 126 (1930).
2. E. P. Wigner, *Phys. Rev.* **46,** 1002 (1934).
3. M. Gell-Mann and K. A. Brueckner, *Phys. Rev.* **106,** 364 (1957).

APPENDIX C THE QUANTUM THEORY OF SCATTERING

To develop a formal theory of scattering, we begin by writing the Hamiltonian in terms of a perturbation:

$$\mathsf{H} = \mathsf{H}_0 + \mathsf{V}\,, \tag{C-1}$$

where H_0 is a time-independent operator describing the unperturbed system. The perturbation V may or may not depend on the time, but, for the case of scattering, it is taken to be time-independent. Even so, time-dependent perturbation theory can still be applied to the system. The operator H_0 is assumed to possess the eigenvalue equation

$$\mathsf{H}_0\psi_n = E_n\psi_n\,. \tag{C-2}$$

If the solutions to Eq. (C-2) are known, then the solution to the corresponding equation when H_0 is replaced by the Hamiltonian (C-1) can be written as the expansion

$$\psi(t) = \sum_n a_n(t)[\exp(-itE_n/\hbar)]\psi_n\,, \tag{C-3}$$

where $\psi(t)$ describes the system at some time after the perturbation has been applied. The sum is to be interpreted symbolically, and could be taken as an integral for continuum states. The Schroedinger equation of motion is

$$i\hbar\,\partial_t\psi(t) = \mathsf{H}\psi(t)\,, \tag{C-4}$$

and, if Eq. (C-3) is substituted into this expression, one obtains the equation determining the coefficients:

$$i\hbar\,\partial_t a_m(t) = \sum_n \langle m\,|\,\mathsf{V}\,|\,n\rangle\, a_n(t)\exp(i\omega_{mn}t)\,, \tag{C-5}$$

with

$$\omega_{mn} \equiv (E_m - E_n)/\hbar\,. \tag{C-6}$$

Equation (C-5) can be solved by an iteration procedure for an initial state $|\mathbf{a}\rangle$ and a final state $|\mathbf{b}\rangle$, the latter state being assumed relevant at $t = +\infty$. This implies the initial conditions

$$a_{\mathbf{a}}(-\infty) = 1\,, \qquad a_{\mathbf{b}}(-\infty) = 0\,, \qquad \mathbf{a} \neq \mathbf{b}\,, \tag{C-7}$$

and the approximate solution is

$$a_{\mathbf{b}}(t) = \delta_{\mathbf{ba}} - (i/\hbar)\langle \mathbf{b} \,|\, \mathsf{V} \,|\, \mathbf{a}\rangle \int_{-\infty}^{t} \exp(i\omega_{\mathbf{ba}}t')\, dt'\,, \tag{C-8}$$

since V is taken to be time independent.

This last expression is an important result because, according to Eq. (C-3), $a_{\mathbf{b}}(t)$ is the probability amplitude for a transition from the state $|\mathbf{a}\rangle$ into the state $|\mathbf{b}\rangle$. Unfortunately, it is also a meaningless result, since the integral diverges at the lower limit. The situation is not hopeless, however, because Eq. (C-8) can be modified by means of an appropriate limiting process. That is, we shall replace the latter expression with

$$a_{\mathbf{b}}(t) = \delta_{\mathbf{ba}} - (i/\hbar)\langle \mathbf{b} \,|\, \mathsf{R} \,|\, \mathbf{a}\rangle \int_{t_0}^{t} \exp(i\omega_{\mathbf{ba}}t' + \varepsilon t')\, dt'\,,$$
$$\varepsilon^{-1} \gg t \gg -\varepsilon^{-1}\,, \tag{C-9}$$

where we agree to first take the limit $t_0 \to -\infty$ *and then* the limit $\varepsilon \to 0$. Furthermore, we have introduced the unknown operator R in place of V in hopes that the former can be determined in a way which renders the final solution independent of a perturbation expansion. Of course, this forces us to devise a means for determining the *reaction-matrix* elements $\mathsf{R}_{\mathbf{ba}}$.

Nevertheless, we are now able to perform the integral in Eq. (C-9) by taking the limit $t_0 \to -\infty$, and we find for the transition probability when $\mathbf{b} \neq \mathbf{a}$

$$|\,a_{\mathbf{b}}(t)\,|^2 = \frac{|\,\mathsf{R}_{\mathbf{ba}}\,|^2 e^{2\varepsilon t}}{\hbar^2(\omega_{\mathbf{ba}}^2 + \varepsilon^2)}\,. \tag{C-10}$$

The transition rate into the state $\mathbf{b} \neq \mathbf{a}$ is then

$$w_{\mathbf{ba}} \equiv \frac{d}{dt}\,|\,a_{\mathbf{b}}(t)\,|^2 = \frac{2\varepsilon}{\omega_{\mathbf{ba}}^2 + \varepsilon^2}\,\frac{e^{2\varepsilon t}\,|\,\mathsf{R}_{\mathbf{ba}}\,|^2}{\hbar^2}\,, \tag{C-11}$$

and we must now take the limit $\varepsilon \to 0$. A well-known representation of the δ function is given by

$$\delta(x) = \frac{1}{\pi} \lim_{\varepsilon \to 0} \frac{\varepsilon}{x^2 + \varepsilon^2}\,,$$

and we obtain

$$w_{\mathbf{ba}} = (2\pi/\hbar)\,\delta(E_{\mathbf{b}} - E_{\mathbf{a}})\,|\,\mathsf{R}_{\mathbf{ba}}\,|^2\,, \tag{C-12}$$

which is just Eq. (11-39). One then derives Eq. (11-40) in a straightforward manner.

In order to complete the justification of Eq. (C-12), it remains to be shown that we can always calculate $\mathsf{R}_{\mathbf{ba}}$, at least in principle. As a first step, we substitute the solution (C-9) back into the differential equation (C-5):

$$\mathsf{R}_{\mathbf{ba}} = \mathsf{V}_{\mathbf{ba}} + \sum_n \frac{\mathsf{V}_{\mathbf{b}n}\mathsf{R}_{n\mathbf{a}}}{E_{\mathbf{a}} - E_n + i\hbar\varepsilon}\,, \tag{C-13}$$

and it is to be remembered that eventually $\varepsilon \to 0$. Now define a set of state vectors $\{\phi_i^{(+)}\}$ by means of the scalar product

$$\mathsf{R}_{\mathbf{ba}} \equiv \int \psi_{\mathbf{b}}^*(\mathbf{r})\mathsf{V}(\mathbf{r})\phi_{\mathbf{a}}^{(+)}(\mathbf{r})\,d^3r\,, \tag{C-14}$$

and substitute into Eq. (C-13). Since the result must hold for *all* **b**, one finds

$$\begin{aligned}\phi_{\mathbf{a}}^{(+)} &= \psi_{\mathbf{a}} + \sum_n \frac{1}{E_{\mathbf{a}} - \mathsf{H}_0 + i\hbar\varepsilon}\psi_n \int \psi_n^*(\mathbf{r})\mathsf{V}(\mathbf{r})\phi_{\mathbf{a}}^{(+)}(\mathbf{r})\,d^3r \\ &= \psi_{\mathbf{a}} + \frac{1}{E_{\mathbf{a}} - \mathsf{H}_0 + i\hbar\varepsilon}\mathsf{V}\phi_{\mathbf{a}}^{(+)}\,,\end{aligned} \tag{C-15}$$

the last line following from the completeness of the set $\{\psi_n\}$. Equation (C-15) is one of the *Lippmann–Schwinger equations*[1] and, coupled with Eq. (C-14), completes the proof that we can always determine $\mathsf{R}_{\mathbf{ba}}$, at least in principle. That is, one solves Eq. (C-15) for $\phi_{\mathbf{a}}^{(+)}$ and substitutes into Eq. (C-14) in order to calculate $\mathsf{R}_{\mathbf{ba}}$. Note that $\psi_{\mathbf{b}}$ is a solution of Eq. (C-2) and is presumably known.

The state vectors $\phi_{\mathbf{a}}^{(+)}$ are easily identified by operating on the Lippmann–Schwinger equation with the operator $(E_{\mathbf{a}} - \mathsf{H}_0 + i\hbar\varepsilon)$ and taking the limit $\varepsilon \to 0$. With the aid of Eq. (C-2) one finds

$$(E_{\mathbf{a}} - \mathsf{H}_0)\phi_{\mathbf{a}}^{(+)} = \mathsf{V}\phi_{\mathbf{a}}^{(+)}\,, \tag{C-16}$$

and $\phi_{\mathbf{a}}^{(+)}$ is just an eigenstate of the complete Hamiltonian with eigenvalue $E_{\mathbf{a}}$. Thus, $\phi_{\mathbf{a}}^{(+)}$ is an outgoing scattered wave function, and $E_{\mathbf{a}}$ is an eigenvalue of *both* H and H_0. It should be noted, however, that only the continuum of H coincides with the spectrum of H_0, since H may also have bound states, while H_0 does not (by assumption).

As mentioned above, Eq. (C-15) represents only one of the Lippmann–Schwinger equations. In a similar manner one can develop the following expressions:

$$\phi_{\mathrm{a}}^{(-)} = \phi_{\mathrm{a}} + \frac{1}{E_{\mathrm{a}} - \mathsf{H}_0 - i\hbar\varepsilon} \mathsf{V}\phi_{\mathrm{a}}^{(-)} , \tag{C-17}$$

$$\phi_{\mathrm{a}}^{(0)} = \phi_{\mathrm{a}} + \mathrm{P}\left(\frac{1}{E_{\mathrm{a}} - \mathsf{H}_0}\right) \mathsf{V}\phi_{\mathrm{a}}^{(0)} , \tag{C-18}$$

where we have introduced an identity for the Cauchy principle value

$$\mathrm{P}\left(\frac{1}{x}\right) \equiv \frac{1}{2} \lim_{\varepsilon \to 0+} \left(\frac{1}{x + i\varepsilon} + \frac{1}{x - i\varepsilon}\right). \tag{C-19}$$

The function $\phi_{\mathrm{a}}^{(-)}$ is an incoming scattered wave, and $\phi_{\mathrm{a}}^{(0)}$ is a standing-wave solution. The entire set of Lippmann–Schwinger equations can then be written as

$$\phi_{\mathrm{a}}^{(\mu)} = \phi_{\mathrm{a}} + \frac{1}{E_{\mathrm{a}} - \mathsf{H}_0 + i\hbar\mu\varepsilon} \mathsf{V}\phi_{\mathrm{a}}^{(\mu)} , \tag{C-20}$$

where it is understood that the principle value is to be taken when $\mu = 0$.

Exercise. With the aid of Eq. (C-20) show that

$$(\phi_{\mathrm{b}}^{(+)}, \phi_{\mathrm{a}}^{(+)}) = (\phi_{\mathrm{b}}^{(-)}, \phi_{\mathrm{a}}^{(-)}) = \delta_{\mathrm{ba}} . \tag{C-21}$$

Note that the sets $\{\phi_{\mathrm{a}}^{(+)}\}$ and $\{\phi_{\mathrm{b}}^{(-)}\}$ together do *not* form a complete set, because H may also possess a set of bound states, which are orthogonal to the scattering states.

Next, let us expand the outgoing states in terms of the incoming states, which can be done because any bound states will be orthogonal to the set $\{\phi_{\mathrm{a}}^{(+)}\}$:

$$\phi_{\mathrm{a}}^{(+)} = \sum_n \phi_n^{(-)} \mathsf{S}_{n\mathrm{a}} . \tag{C-22}$$

From Eq. (C-21) we obtain

$$\mathsf{S}_{\mathrm{ba}} = (\phi_{\mathrm{b}}^{(-)}, \phi_{\mathrm{a}}^{(+)}) , \tag{C-23}$$

which defines the matrix elements of the *scattering matrix*, S. The S matrix was first introduced by Wheeler[2] in connection with problems of nuclear structure, but in the present context it serves to unify the formal theory of scattering.

Now, the S matrix must be diagonal in the energy representation, which follows from Eq. (C-23) and the observation that any two eigenstates of H corresponding to distinct energy eigenvalues are orthogonal. Hence, we can write

$$\mathsf{S}_{\mathrm{ba}} = \delta_{\mathrm{ba}} + \delta(E_{\mathrm{b}} - E_{\mathrm{a}})\mathsf{T}_{\mathrm{ba}} , \tag{C-24}$$

where T_{ba} is assumed to be nonsingular at $E_b = E_a$. Using a well-known representation of the δ function, we can rewrite this last expression as

$$\mathsf{S}_{ba} = \delta_{ba} + \frac{1}{2\pi i}\left(\frac{1}{E_b - E_a - i\hbar\varepsilon} - \frac{1}{E_b - E_a + i\hbar\varepsilon}\right)\mathsf{T}_{ba}\,, \tag{C-25}$$

in the limit $\varepsilon \to 0$. Let us now substitute this expression, along with the Lippman–Schwinger equations, into the definition (C-22). After a fair amount of algebra and a recollection of Eq. (C-14), one finds the relation

$$\begin{aligned}\frac{1}{E_b - \mathsf{H}_0 + i\hbar\varepsilon}\sum_n \psi_n \mathsf{R}_{nb} &= \frac{1}{2\pi i}\left(\frac{1}{E_b - \mathsf{H}_0 - i\hbar\varepsilon} - \frac{1}{E_b - \mathsf{H}_0 + i\hbar\varepsilon}\right)\sum_n \psi_n \mathsf{T}_{nb} \\ &\quad + \sum_n \frac{1}{E_n - \mathsf{H}_0 - i\hbar\varepsilon}\mathsf{V}\phi_n^{(-)}\mathsf{S}_{nb}\,.\end{aligned} \tag{C-26}$$

Since the operators in this equation are linearly independent, we can compare coefficients and obtain the relation

$$\mathsf{T}_{nb} = -2\pi i \mathsf{R}_{nb}\,. \tag{C-27}$$

Therefore, from Eq. (C-24) we find the relation between the S matrix and the reaction matrix,

$$\mathsf{S}_{ba} = \delta_{ba} - 2\pi i\, \delta(E_b - E_a)\mathsf{R}_{ba}\,, \tag{C-28}$$

which is just Eq. (11-38).

This last result has an important consequence if we return to Eq. (C-9) and take the limits $t_0 \to -\infty$, $\varepsilon \to 0$, $t \to +\infty$, and recall the integral representation of the δ function. Then,

$$a_b(\infty) = \langle \mathbf{b} \,|\, \mathsf{S} \,|\, \mathbf{a} \rangle = \mathsf{S}_{ba}\,. \tag{C-29}$$

One also recalls the definition of the time-evolution opertor $\mathsf{U}(t, t_0)$ in Eq. (11-30) in the interaction picture. With the aid of the definition (11-19) and the expansion (11-42) it is now an easy matter to refer to Eq. (C-8) and observe that for $\mathbf{b} \neq \mathbf{a}$, and in the limit of very weak interactions,

$$a_b(\infty) = \big(\psi_b, \mathsf{U}(\infty, -\infty)\psi_a\big)\,. \tag{C-30}$$

Comparison of the last two equations then proves the relation (11-35):

$$\langle \mathbf{b} \,|\, \mathsf{S} \,|\, \mathbf{a} \rangle = \langle \mathbf{b} \,|\, \mathsf{U}(\infty, -\infty) \,|\, \mathbf{a} \rangle\,. \tag{C-31}$$

Finally, if we use plane-wave states and the momentum representation, we can calculate the differential scattering cross section from Eq.

(11-40), because the density of states is well known in this case. One can then identify $f_{\mathbf{k}}(\hat{\mathbf{k}}')$ in terms of R_{ba} and therefore derive Eqs. (11-36) and (11-37). We leave this calculation to the reader.

REFERENCES

1. B. A. Lippmann and J. Schwinger, *Phys. Rev.* **79**, 469 (1950).
2. J. A. Wheeler, *Phys. Rev.* **52**, 1107 (1937).

Index

Numbers in italics refer to the pages on which the complete references are listed.

A

B

PURE AND APPLIED PHYSICS

A Series of Monographs and Textbooks

Consulting Editors

H. S. W. Massey
University College, London, England

Keith A. Brueckner
University of California, San Diego
La Jolla, California

1. F. H. Field and J. L. Franklin, Electron Impact Phenomena and the Properties of Gaseous Ions.
2. H. Kopfermann, Nuclear Moments, English Version Prepared from the Second German Edition by E. E. Schneider.
3. Walter E. Thirring, Principles of Quantum Electrodynamics. Translated from the German by J. Bernstein. With Corrections and Additions by Walter E. Thirring.
4. U. Fano and G. Racah, Irreducible Tensorial Sets.
5. E. P. Wigner, Group Theory and Its Application to the Quantum Mechanics of Atomic Spectra. Expanded and Improved Edition. Translated from the German by J. J. Griffin.
6. J. Irving and N. Mullineux, Mathematics in Physics and Engineering.
7. Karl F. Herzfeld and Theodore A. Litovitz, Absorption and Dispersion of Ultrasonic Waves.
8. Leon Brillouin, Wave Propagation and Group Velocity.
9. Fay Ajzenberg-Selove (ed.), Nuclear Spectroscopy. Parts A and B.
10. D. R. Bates (ed.), Quantum Theory. In three volumes.
11. D. J. Thouless, The Quantum Mechanics of Many-Body Systems.
12. W. S. C. Williams, An Introduction to Elementary Particles.
13. D. R. Bates (ed.), Atomic and Molecular Processes.
14. Amos de-Shalit and Igal Talmi, Nuclear Shell Theory.
15. Walter H. Barkas. Nuclear Research Emulsions. Part I.
    Nuclear Research Emulsions. Part II. *In preparation*
16. Joseph Callaway, Energy Band Theory.
17. John M. Blatt, Theory of Superconductivity.
18. F. A. Kaempffer, Concepts in Quantum Mechanics.
19. R. E. Burgess (ed.), Fluctuation Phenomena in Solids.
20. J. M. Daniels, Oriented Nuclei: Polarized Targets and Beams.
21. R. H. Huddlestone and S. L. Leonard (eds.), Plasma Diagnostic Techniques.
22. Amnon Katz, Classical Mechanics, Quantum Mechanics, Field Theory.
23. Warren P. Mason, Crystal Physics in Interaction Processes.
24. F. A. Berezin, The Method of Second Quantization.
25. E. H. S. Burhop (ed.), High Energy Physics. In four volumes.

26. L. S. Rodberg and R. M. Thaler, Introduction to the Quantum Theory of Scattering.

27. R. P. Shutt (ed.), Bubble and Spark Chambers. In two volumes.

28. Geoffrey V. Marr, Photoionization Processes in Gases.

29. J. P. Davidson, Collective Models of the Nucleus.

30. Sydney Geltman, Topics in Atomic Collision Theory.

31. Eugene Feenberg, Theory of Quantum Fluids.

32. Robert T. Beyer and Stephen V. Letcher, Physical Ultrasonics.

33. S. Sugano, Y. Tanabe, and H. Kamimura, Multiplets of Transition-Metal Ions in Crystals.

34. Walter T. Grandy, Jr., Introduction to Electrodynamics and Radiation.

In preparation

J. Killingbeck and G. H. A. Cole, Physical Applications of Mathematical Techniques.

Herbert Uberall, Electron Scattering from Complex Nuclei. Parts A and B.